# Implantable Biomedical Microsystems
## Design Principles and Applications

**Swarup Bhunia**

**Steve J.A. Majerus**

**Mohamad Sawan**

AMSTERDAM • BOSTON • HEIDELBERG • LONDON • NEW YORK • OXFORD
PARIS • SAN DIEGO • SAN FRANCISCO • SINGAPORE • SYDNEY • TOKYO
William Andrew is an imprint of Elsevier

William Andrew is an imprint of Elsevier
The Boulevard, Langford Lane, Kidlington, Oxford, OX5 1GB, UK
225 Wyman Street, Waltham, MA 02451, USA

**Notices**
Knowledge and best practice in this field are constantly changing. As new research and experience broaden our understanding, changes in research methods, professional practices, or medical treatment may become necessary.

Practitioners and researchers must always rely on their own experience and knowledge in evaluating and using any information, methods, compounds, or experiments described herein. In using such information or methods they should be mindful of their own safety and the safety of others, including parties for whom they have a professional responsibility.

To the fullest extent of the law, neither the Publisher nor the authors, contributors, or editors, assume any liability for any injury and/or damage to persons or property as a matter of products liability, negligence or otherwise, or from any use or operation of any methods, products, instructions, or ideas contained in the material herein.

**British Library Cataloguing-in-Publication Data**
A catalogue record for this book is available from the British Library

**Library of Congress Cataloging-in-Publication Data**
A catalog record for this book is available from the Library of Congress

ISBN: 978-0-323-26208-8

For information on all William Andrew publications
visit our website at http://store.elsevier.com/

Typeset by SPi Global, India

Printed and bound in the United States of America

Working together
to grow libraries in
developing countries

www.elsevier.com • www.bookaid.org

# Contents

## PART I  DESIGN PRINCIPLES FOR BIOIMPLANTABLE SYSTEMS

# Contributors

**Abhishek Basak**
Department of Electrical Engineering and Computer Science, Case Western Reserve University, Cleveland, Ohio, USA

**Niloy Bhadra**
Department of Biomedical Engineering, Case Western Reserve University, Cleveland, Ohio, USA

**Swarup Bhunia**
Department of Electrical Engineering and Computer Science, Case Western Reserve University, Cleveland, Ohio, USA

**Margot S. Damaser**
Advanced Platform Technology Center of Excellence, Louis Stokes Cleveland Department of Veterans Affairs Medical Center; Glickman Urological and Kidney Institute, Cleveland Clinic; Department of Biomedical Engineering, Cleveland Clinic, Cleveland, Ohio, USA

**Manuel Delgado-Restituto**
Institute of Microelectronics of Sevilla, Sevilla, Spain
University of Seville, Sevilla, Spain

**Dominique M. Durand**
Urology Institute, University Hospitals Case Medical Center, Cleveland, Ohio, USA

**Yazan Dweiri**
Urology Institute, University Hospitals Case Medical Center, Cleveland, Ohio, USA

**Elizabeth K. Ferry**
Urology Institute, University Hospitals Case Medical Center, Cleveland, Ohio, USA

**Paul C. Fletter**
Advanced Platform Technology Center of Excellence, Louis Stokes Cleveland Department of Veterans Affairs Medical Center, Cleveland, Ohio, USA

**Bishoy Gad**
Orthopaedic and Rheumatology Institute, Cleveland Clinic, Cleveland, Ohio, USA

**Maysam Ghovanloo**
School of Electrical and Computer Engineering, Georgia Institute of Technology, Atlanta, GA, USA

**Allison Hess-Dunning**
Rehabilitation Research and Development, Louis Stokes Cleveland VA Medical Center, Cleveland, Ohio, USA

**Grant Hoffman**
Innovations Institute, Cleveland Clinic, Cleveland, Ohio, USA

**Niraj K. Jha**
Princeton University, Princeton, New Jersey, USA

**Younghyun Kim**
Purdue University, West Lafayette, Indiana, USA

**Wen Ko**
Case Western Reserve University, Cleveland, Ohio, USA

**Shem Lachhman**
MediMEMS, LLC, Shaker Heights, Ohio, USA

**Hyung-Min Lee**
School of Electrical and Computer Engineering, Georgia Institute of Technology, Atlanta, GA, USA

**Woosuk Lee**
Purdue University, West Lafayette, Indiana, USA

**Steve J.A. Majerus**
Advanced Platform Technology Center of Excellence, Louis Stokes Cleveland Department of Veterans Affairs Medical Center; Department of Electrical Engineering and Computer Science, Case Western Reserve University, Cleveland, Ohio, USA

**Iryna Makovey**
Glickman Urological and Kidney Institute, Cleveland Clinic, Cleveland, Ohio, USA

**Arnaldo Mendez**
Polystim Neurotechnologies Lab, Department of Electrical Engineering, Polytechnique, Montreal, Quebec, Canada

**Anand Raghunathan**
Purdue University, West Lafayette, Indiana, USA

**Vijay Raghunathan**
Purdue University, West Lafayette, Indiana, USA

**Alberto Rodríguez-Pérez**
Institute of Microelectronics of Sevilla, Sevilla, Spain
University of Seville, Sevilla, Spain

**Mohamad Sawan**
Polystim Neurotechnologies Lab, Department of Electrical Engineering, Polytechnique, Montreal, Quebec, Canada

**Randy Scherer**
DEKRA Certification Incorporate, Chalfont, Pennsylvania, USA

**Tina Vrabec**
Department of Biomedical Engineering, Case Western Reserve University, Cleveland, Ohio, USA

**Peng Wang**
Case Western Reserve University, Cleveland, Ohio, USA

**Brian Wodlinger**
Electrical and Computer Engineering Department, University of Utah, Salt Lake City, Utah, USA

**Darrin J. Young**
Electrical and Computer Engineering Department, University of Utah, Salt Lake City, Utah, USA

**Hui Zhu**
Advanced Platform Technology Center of Excellence, Louis Stokes Cleveland Department of Veterans Affairs Medical Center; Urology Service, Louis Stokes Cleveland Department of Veterans Affairs Medical Center; Glickman Urological and Kidney Institute, Cleveland Clinic, Cleveland, Ohio, USA

**Christian A. Zorman**
Department of Electrical Engineering and Computer Science, Case Western Reserve University, Cleveland, Ohio, USA

# Preface

This book is intended as a reference book on bioimplantable systems for researchers, clinical practitioners, and students at both undergraduate level and graduate level. It can also serve as a textbook for graduate students for relevant courses. It is suggested that this book is used as a comprehensive source of reference materials related to design, deployment, and understanding of diverse implantable systems.

This 1st edition makes fundamental and unique contributions in the area of implantable systems design by integrating materials on (1) systems design issues, design space exploration, and optimization of major building blocks and (2) several case studies on significant applications of these systems. The book covers the design principles of bioimplantable systems including efficient implementation of sensing and stimulation electrodes; ultralow-power signal processing electronics; design techniques to achieve security, privacy, and reliability of implants; system integration, packaging, and power delivery; and regulatory and chronic implantation issues associated with the deployment of an implant in clinical practice. Comprehensive coverage of different design principles and methods is complemented with five important case studies, which represent the spectrum of applications of biomedical implants for clinical diagnosis and treatment, ranging from peripheral neural interfaces and intracortical recording to bladder pressure monitoring, implantable imaging unit, and nerve conduction block. These case studies demonstrate how different design principles can be employed in the effective design of practical implants that interface with different internal body parts to serve myriad clinical purposes.

The chapters are contributed by prominent researchers and practitioners in relevant fields. We believe this book will greatly help its readers to understand and analyze existing implants of diverse natures. Furthermore, it would enable development of complex, intelligent, and effective implants for many emerging applications and serve as a valuable resource in translational and clinical research related to the development of promising biomedical implants.

**Swarup Bhunia**
**Steve Majerus**
**Mohamad Sawan**

# Design Principles for Bioimplantable Systems

# Introduction

1

**Swarup Bhunia\*, Steve J.A. Majerus\*,†, Mohamad Sawan‡**

*\*Department of Electrical Engineering and Computer Science, Case Western Reserve University, Cleveland, Ohio, USA*

*†Advanced Platform Technology Center of Excellence, Louis Stokes Cleveland Department of Veterans Affairs Medical Center; Cleveland, Ohio, USA*

*‡Polystim Neurotechnologies Lab, Department of Electrical Engineering, Polytechnique, Montreal, Quebec, Canada*

## CHAPTER CONTENTS

With great advances in electronics and electrode technologies, it has become possible to realize implantable biomedical microsystems that interface with the internal body parts to monitor and manipulate their activities. One of the major success stories in the field of implantable systems is the cardiac pacemaker, in which over one million pacemakers were installed or replaced worldwide in 2009 [1]. Today, miniaturized

Bhunia et al. Implantable Biomedical Microsystems. http://dx.doi.org/10.1016/B978-0-323-26208-8.00001-7

wireless implantable systems are changing the face of biomedical research and clinical practices through the development of intelligent pacemakers, cochlear implants, neuroprostheses, brain–computer interfaces, deep organ pressure sensors, and precise drug delivery units. New and exciting applications of implantable systems enabled by the technology advances are emerging, such as implantable contraceptives, which can be implanted under a woman's skin to release a small dose of levonorgestrel, a hormone, every day over a period of 16 years and can be remotely controlled [2], or implantable miniaturized imaging devices that can help effective diagnosis and/or monitoring of a disease, including cancer [3].

These systems are making broad scientific and translational impact and are saving or enhancing the lives of millions through clinical diagnosis and therapeutics for complex diseases. On the other hand, they are providing invaluable tools to increase our scientific understanding of different body parts including the brain and central nervous system. For example, implantable neural interfaces are being used for neural recording and stimulation as in functional electrical stimulation (FES) to assist patients in grasping, standing, or urination. Deep brain stimulation has been shown to be an effective treatment for Parkinson's disease, and its long-reaching benefits are being realized in treatment methods for epilepsy, psychological disorders, and even drug addiction, among other debilitating diseases. Concurrently to offering patient care, these systems are providing researchers with an enhanced toolbox to probe the underlying mechanics of complex physiological systems.

Although the field of implantable electronic medical devices is fairly old—simple radio transmitters were implanted with early commercial transistors in 1959 [4]—decades of microelectronic technology innovations have recently permitted the development of fully autonomous implants suitable for chronic application, as shown in Figure 1.1. The complexity of implanted devices has chronologically tracked the trend of ever-shrinking electronics. Early implants from 1960 to 1975 were mainly analogue telemeters [5] and simple pacemakers [6], while devices developed in the 1980s integrated precise, programmable digital logic to incorporate additional important functionalities. This permitted the development of semiautonomous stimulation devices, such as the cochlear implant [7] and peripheral FES systems [8].

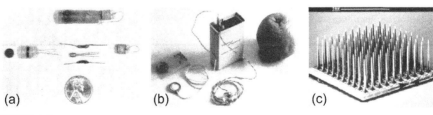

(a)  (b)  (c)

**FIGURE 1.1**

Implantable systems have increased in complexity with advances in electronics and electrode technologies, from (a) an implantable transmitter in 1959 [4], to (b) a cochlear implant in 1982 [7], to (c) a 100-channel microelectrode neural recording array in 1999 [11].

However, a lack of miniature, onboard sensing on these implants limited their application to open-loop devices requiring external control to achieve different treatment modalities. With the advent of microelectromechanical systems and micromachining techniques in the 1990s [9], implantable microsystems could finally integrate wireless telemetry, low-power digital control, and analogue sensing of the biological environment, and this confluence of technologies has led to an explosion of research and development efforts in the field of bioimplantable systems.

Today, researchers are designing diverse implantable devices for different bodily systems, as indicated in Figure 1.2. While some implantable devices fulfill a security role similar to biometric identification (e.g., PositiveID©, VeriMED, or Dangerous Things© xM1 implantable RFID tag) and use the host body mostly as a containment/

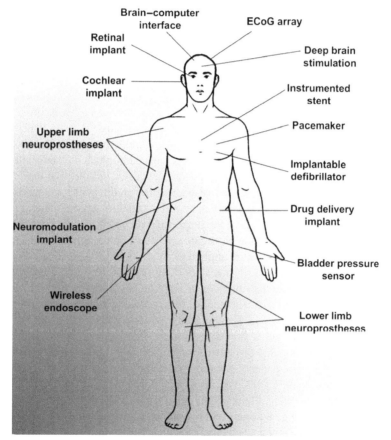

**FIGURE 1.2**

Bioimplantable systems have been designed across the breadth of the medical field. This figure highlights some common applications, but is by no means a complete depiction of the myriad electronic implants developed by researchers. Many of these implants are already being used by medical practitioners as effective options of disease treatment.

transport vessel, the focus of this book is on implantable systems that can realize translational success in clinical disciplines. The functionality of these systems is rooted in the interaction between the human body parts and electrodes, sensors, and actuators such that a bridge between physiology and electronics is created. Currently, such systems are used in a broad range of clinical applications but most commonly for disease treatment and diagnosis. In the future, however, as implant devices decrease in size and cost and integrate higher level of computation intelligence, the potential for ubiquitous physiological sensing will greatly increase, permitting implantable systems to operate at the point of care. Ultimately, implantable systems are expected to be so small, powerful, affordable, easy to implant, and suitable for chronic operation that preemptive implantation for health monitoring into otherwise healthy patients will enable preventative care and early diagnosis of the onset of serious diseases (e.g., cancer and cardiovascular problems) [10].

While implantable devices provide tremendous opportunities in clinical success and scientific research, they call for the development of new technology platforms for implementing these complex systems under extremely tight design constraints in terms of power, size, biocompatibility, and robust operation for long-term chronic deployment. The technology platform needs to consider several common design principles and address many application-specific requirements as well. Any implantable device must be designed with deference to physiological traits that limit the device size, placement method and location, sensing modality, and RF power density. Furthermore, these engineering considerations must be implemented beneath an umbrella of regulatory issues (imposed by governmental institutions like the FDA and FCC). An implantable device can be highly effective on the bench and animal trials but must have a path toward regulatory approval in order to achieve clinical success. Finally, the long-term usefulness of the device and its potential side effects must be well understood, through both the physiological basis of the implant and the stability of the biocompatible encapsulation applied to it.

Designers of bioimplantable systems are often required to possess a set of multidisciplinary skills to realize the ambitious goals of highly complex and powerful next-generation implantable devices. The book aims to provide comprehensive coverage on the fundamental design principles for implantable systems and several major application case studies, which demonstrate how these systems can be designed and optimized for an application considering the respective design objectives. It includes detailed descriptions of design issues, components, parameters, and the optimization of each component in an implantable system. These components include recording devices, analogue front ends for signal conditioning, digital signal processing (DSP) electronics for real-time *in situ* signal analysis (e.g., for noise removal and pattern recognition), drug delivery or stimulation units, and wireless telemetry devices. It covers an important set of system design examples including central and peripheral neural interfaces, bladder pressure monitoring, implantable imaging device for internal organs, and electrical conduction block for peripheral nerves to achieve pain management. While these case studies only scratch the surface of the field of bioimplantable systems, they offer valuable practical insight into the design approach surrounding different potent applications.

The book is organized into two integrated parts with a total of 13 chapters including the summary chapter. The content of each part as well as each chapter in the part is intended to provide necessary background to readers new in the field and detailed discussions of state-of-the-art techniques for experts in the topic. The first half of the book is dedicated to the fundamental principles and approaches for designing an implantable system. The second half focuses on several implantable system design examples. This part covers design requirements, application-specific implementation challenges, and appropriate design choices for diverse application areas including neural interfaces, nerve conduction block, and bladder pressure monitoring. We will now provide a brief overview of each part of this book and the corresponding chapters.

## PART I: DESIGN PRINCIPLES FOR BIOIMPLANTABLE SYSTEMS

The first part of the book covers design of all major components in an implantable microsystem including electrodes for sensing/stimulation, analogue electronics for signal conditional and telemetry, and DSP hardware. It also covers challenges and solutions associated with the packaging and integration of the components. Finally, it presents chronic implantation and regulatory issues related to deployment of implants and security/reliability issues in these devices and corresponding solutions.

## CHAPTER 2: ELECTRICAL INTERFACES FOR RECORDING, STIMULATION, AND SENSING

Miniaturized implantable systems provide an important interface with the internal body parts including the central nervous system for interpreting and controlling their activities. To achieve a biointerface that establishes a link between the electronic system and organs, electrodes are used for sensing (i.e., recording physiological parameters) and stimulation for controlling activity. Chapter 2 focuses on the design of electrodes for bioimplantable systems with specific emphasis on electrodes for neural interfaces. It covers the primary design considerations and types of different neural electrodes, their properties, and the implementation processes.

## CHAPTER 3: ANALOGUE FRONT-END AND TELEMETRY SYSTEMS

The analogue front-end system is responsible for amplifying, filtering, signal conditioning, and digitizing *in vivo* physiological signals. Chapter 3 focuses on the key analogue components in an implant and describes their design principles. Analogue electronics with low power dissipation and high performance are critical for bioimplantable systems. The chapter describes design methodologies to effectively address these important requirements.

## CHAPTER 4: **SIGNAL PROCESSING HARDWARE**

The next chapter presents design challenges and methodologies for signal processing in implantable systems. With growing need of higher computational intelligence, these systems tend to integrate complex signal processing engines for real-time online processing of sensed body signals. The signal processing hardware performs two major tasks: compression of sensed data and recognition of meaningful patterns from recorded data. The chapter outlines the primary tasks for the signal processing hardware and then focuses on the algorithms as well as design/implementation of the signal processing hardware.

## CHAPTER 5: **ENERGY MANAGEMENT INTEGRATED CIRCUITS FOR WIRELESS POWER TRANSMISSION**

Wireless power transmission is one of the few viable techniques to power up implantable medical devices (IMDs) across the skin without any direct electrical contact between the energy source and the IMD. This technology not only has been adopted across a range of devices in the consumer and industrial fields but also has been applied to approved medical devices such as cochlear implants and brain–computer interfaces. This chapter summarizes the mechanisms for wireless power transmission while providing details on the optimization of the power transmission to improve overall system efficiency and to increase robustness to interfering factors. Key energy management circuits are presented for receiving AC input through the inductive link and passing it through power-efficient converters before producing a regulated supply voltage for the IMD. Power storage components, including rechargeable batteries and supercapacitors, are also discussed in terms of their characteristics, charging mechanisms, and circuit applicability.

## CHAPTER 6: **SYSTEM INTEGRATION AND PACKAGING**

The human body is a harsh environment for implanted electronics, and implants must be properly encapsulated to protect devices from aqueous bodily fluids and to protect the body from potentially harmful materials within the implant. This chapter first reviews the historically proven hermetic-box packaging technologies for devices such as pacemakers and the design considerations required. Next, a summary of nonhermetic packaging techniques—applied through various layers of polymeric films—is provided along with general considerations for material selection, application, and system design for successful encapsulation. Finally, a case study on the design, testing, and application of a nonhermetic micropackaging technology is presented, with measured results and estimated *in vivo* lifetimes based on accelerated lifetime tests.

## CHAPTER 7: CLINICAL AND REGULATORY CONSIDERATIONS OF IMPLANTABLE MEDICAL DEVICES

Development of an IMD starts with an aim of addressing a clinical need and improving patient outcomes while minimizing risks. Clinical considerations for device development focus on unique patient population needs and limitations, the implant environment, local tissue response to the device, and need for durability or biodegradability. Additionally, the design must account for the techniques of implantation or explanation while minimizing risk of infections and addressing the management plan for possible complications. Finally, each device is a subject to rigorous review by the Food and Drug Administration. This chapter will review basic clinical considerations of an implantable medical system and will discuss regulatory principles and steps of the device approval process.

## CHAPTER 8: RELIABILITY AND SECURITY OF IMPLANTABLE AND WEARABLE MEDICAL DEVICES

This chapter covers the reliability and security challenges in bioimplantable systems and presents solutions to address them. With increasing functional complexity and connectivity of modern implants, long-term reliability and security of these devices have emerged as paramount concerns. This chapter discusses the vulnerability of these systems to comprise patient safety and privacy due to security attacks on the hardware/software components in implants. Finally, the chapter surveys potential solutions and discusses their merits and demerits.

## PART II: APPLICATIONS OF BIOIMPLANTABLE SYSTEMS

The second part of the book covers some major case studies of bioimplantable systems spanning diverse application areas. It includes two instances of neural implants: peripheral nerve sensor and intracortical neural recording unit, along with case studies covering a bladder pressure monitoring unit, nerve conduction block for acute pain management, and implantable imaging unit for early detection of disease onset.

## CHAPTER 9: ELECTRICAL BIOSENSORS: PERIPHERAL NERVE SENSORS

While neural signals originate in the central nervous system with the peripheral nervous system mostly used for signal transmission, recording from peripheral nerves has several advantages. Some of the advantages are that lower motor signals do not require decoding, the peripheral nerves carry information from biological sensors, and peripheral electrodes have historically proven to be safe and effective, to name a few. This chapter focuses on the construction and use of two types of peripheral electrodes: cuff-style electrodes, which wrap around peripheral nerve fibers without penetrating them, and penetrating/sieve electrodes, which are more invasive but provide stronger signal levels. Finally, an overview of the amplification considerations for using peripheral electrodes is provided to guide users in selecting the optimum amplifier parameters for maximum signal gain.

## CHAPTER 10: ELECTRODES FOR ELECTRICAL CONDUCTION BLOCK OF PERIPHERAL NERVE

Faulty neuronal firing caused by CNS lesions can lead to adverse effects in organs of the peripheral nervous system, such as chronic pain, motor function loss, abnormal limb posture, and spasticity. The pathologies of these symptoms can be severe, ranging from stroke to traumatic brain injury, and are commonly treated with pharmacological agents or irreversible surgical procedures. Due to the side effects of current treatments, there is a growing field of nerve block research that promises reliable, rapid, nontoxic, and reversible treatment of these neurological conditions. This chapter highlights two of the most promising technologies for conduction nerve block: kilohertz frequency alternating current (KHFAC) and short-duration direct current (DC) block. The present state of research with these block techniques is summarized, with details covering the selection of peripheral electrode type, physiological activation hypotheses, safety considerations, and the differences between KHFAC and DC nerve block.

## CHAPTER 11: IMPLANTABLE BLADDER PRESSURE SENSOR FOR CHRONIC APPLICATION

This chapter describes the design, fabrication, and testing of a wireless bladder pressure-sensing system for chronic, point-of-care applications such as urodynamics or closed-loop neuromodulation. The system consists of a miniature implantable device and an external RF receiver and wireless battery charger. The implant is small enough to be cystoscopically implanted within the bladder wall, where it is securely held and shielded from the urine stream. The implant consists of a custom application-specific integrated circuit, pressure transducer, rechargeable battery, and wireless telemetry and recharging antennas. Acute *in vivo* evaluation of the pressure-sensing system in anesthetized canine and feline models demonstrated that the system can accurately capture lumen pressure from a submucosal implant location. Further application of the system in an ambulatory canine model has confirmed feasibility, but ambulatory issues such as implant migration and superimposition of motion artifacts on received data must be considered.

## CHAPTER 12: NEURAL RECORDING INTERFACES FOR INTRACORTICAL IMPLANTS

This chapter presents another important application, namely, an intracortical neural recording interface. The implant consumes very low power and is inductively powered. It has 64 channels of neural signal recording and processing of the recorded data. The recording system implements an autocalibration mechanism for each channel to enable configuring the transfer characteristics of the recording site. The implant has the capability to transmit either uncompressed data or feature vectors extracted by an embedded digital signal processor.

## CHAPTER 13: IMPLANTABLE IMAGING SYSTEM FOR AUTOMATED MONITORING OF INTERNAL ORGANS

The next chapter focuses on implantable imaging systems, which can capture diagnostic images of internal body parts. These systems are primarily targeted to early detection of an anomaly in an organ. The development of miniaturized ultralow-power imaging systems is enabled by advancement in image sensors and integrated electronics that are needed for controlling the image sensors and *in situ* processing of sense data. The chapter describes different imaging modalities and applications for implantable imagers. It then focuses on the design and analysis of an ultrasound-based implantable imaging system for early detection of a malignant growth in various organs.

## REFERENCES

[1] Mond H, Proclemer A. The 11th world survey of cardiac pacing and implantable cardioverter-defibrillators calendar year 2009—A World Society of Arrhythmia's project. Pacing Clin Electrophysiol 2011;34(8):1013–27.

[2] Lee D. 'Remote control' contraceptive chip available 'by 2018'. BBC News, http://www.bbc.com/news/technology-28193720; 2014 [accessed July 2014].

[3] Basak A, Bhunia S. Implantable ultrasonic imaging assembly for automated monitoring of internal organs. IEEE Trans Biomed Circuits Syst 2014; [in press].

[4] Mackay R. Radio telemetering from within the human body. IRE Trans Med Electron 1959;6(2):100–5.

[5] Ko W, Neuman M. Implant biotelemetry and microelectronics. Science 1965;156(3773):351–60.

[6] McCombs R, Herrod C, Mackay R. An electronic cardiac defibrillator and pacemaker. Rev Sci Instrum 1954;25(4):378–9.

[7] White RL. Review of current status of cochlear prostheses. IEEE Trans Biomed Eng 1982;29(4):233–8.

[8] Thorpe G, Peckham P, Crago P. A computer-controlled multichannel stimulation system for laboratory use in functional neuromuscular stimulation. IEEE Trans Biomed Eng 1985;32(6):363–70.

[9] Wise K. Integrated microelectromechanical systems: a perspective on MEMS in the 90s. In: Proceedings of the IEEE micro electro mechanical systems, 1991 (MEMS '91). An investigation of micro structures, sensors, actuators, machines and robots; 1991.

[10] Strickland E. Medtronic wants to implant sensors in everyone. IEEE Spectrum, http://spectrum.ieee.org/tech-talk/biomedical/devices/medtronic-wants-to-implant-sensors-in-everyone/?utm_sourcc–tcchalcrt&utm_mcdium–cmail&utm_campaign=061214; 2014 [accessed June 2014].

[11] Rousche P, Normann R. Chronic intracortical microstimulation (ICMS) of cat sensory cortex using the Utah intracortical electrode array. IEEE Trans Rehabil Eng 1999;7(1):56–68.

# Electrical interfaces for recording, stimulation, and sensing

2

**Allison Hess-Dunning***, **Christian A. Zorman**†

*Rehabilitation Research and Development, Louis Stokes Cleveland VA Medical Center,
Cleveland, Ohio, USA
†Department of Electrical Engineering and Computer Science, Case Western Reserve University,
Cleveland, Ohio, USA

## CHAPTER CONTENTS

## 2.1 INTRODUCTION

Brain–machine interfaces (BMIs) or brain–computer interfaces have garnered much enthusiasm given their potential to revolutionize the quality of life for those with motor impairment and a range of neurological disorders. With these systems [1,2], brain activity can be detected and then transduced into functional signals for the control of

Bhunia et al. Implantable Biomedical Microsystems. http://dx.doi.org/10.1016/B978-0-323-26208-8.00002-9

computer cursors [3–5] and robotic arms [6,7]. BMIs also provide treatment for neurological diseases and disorders, including epilepsy [8], depression [9], and Parkinson's disease [10–12]. Similar strategies have been used to provide sensory input (i.e., auditory, visual, and tactile) in situations where these senses have otherwise been impaired. For example, electrical stimulation of the auditory nerve through cochlear implants, the most widely implemented type of neural interface, has enabled sound perception in individuals with hearing impairment [13–16]. Efforts are currently underway to develop neural interfaces and associated circuitry for retinal stimulation to enable visual perception in individuals with visual impairment [17–20]. There have been recent reports that tactile sensation can be restored in upper extremity amputees by connecting peripheral nerve electrode devices to sensors on an artificial hand [21]. The important subcomponents of these systems are the neural interface electrodes, signal processing circuitry, decoding software, command software, and the object under control. This chapter focuses on the neural interface itself, or the biotic/abiotic interface.

Neural interfaces provide the essential link between the electronic systems designed for electrical stimulation, electrical recording, or neurochemical detection to the particular region of the nervous system to which the system is designed. There are a number of different ways to classify neural interfaces—central nervous system or peripheral nervous system, penetrating or nonpenetrating, brain interface or spinal cord interface, and brain surface electrodes or intracortical probe. Neural interfaces have been in development for several decades [22–24]. The evolution of the neural electrode has leveraged heavily from technologies developed initially for nonbiomedical applications, resulting in a wide variety of technologies for interfacing to the central and peripheral nervous systems, with varying degrees of invasiveness, spatial resolution, and temporal resolution.

Conventional neural recording devices use electrically conducting, chemically stable, and biologically benign materials such as platinum, tungsten, or iridium oxide for electrodes and associated circuitry to sense the electrical neural activity (i.e., action potentials or field potentials). Action potentials can be detected with intracortical microelectrodes [25], while field potentials can be detected with either intracortical electrodes or brain surface electrodes [26]. Recording these signals allows for basic investigations of neuron behavior and communication [27] and clinical applications in which the neural signals can be decoded to determine intent [28,29]. This intent can then be used as a control signal for BMIs to restore independence to those with impaired motor function [2,30].

A second functionality associated with neural interfaces is neural stimulation. Neural stimulation involves injecting current pulses into the neural tissue or nerves to externally control brain activity [31], sensory perception [13,15,20–22,32], or motor function [22,33,34]. Neural stimulation devices are used for treatment of seizures [35] and Parkinson's disease [12]. Likewise, such devices are used to restore sensory and motor function using functional electrical stimulation systems.

Neural interfaces for neurochemical detection serve to measure the release and removal of neurotransmitters, such as serotonin and dopamine, in the brain. It is known that these biomolecules are an essential component in neural communication; however, devices designed to exploit this modality are not yet mature to the point of clinical implementation. However, *in vivo* monitoring of these neurotransmitters is gaining both popularity and support with the development of new materials and enabling electronics [36,37]. Sensing is typically performed using some form of cyclic voltammetry; therefore, the best electrode materials for neurochemical detection to date are carbon-based, such as carbon fiber microelectrodes [38,39] and semimetallic diamond [40], due to their chemical stability in biological media, inherent biocompatibility, low baseline current, wide water window, and high analyte sensitivity.

## 2.2 ELECTRODE DESIGN CONSIDERATIONS

There are a number of design challenges to advancing neural interfaces to the point of routine clinical implementation and long-term chronic use, which ideally would exceed 10 years. Many of the envisioned clinical applications for neural potential recording, stimulation, and neurochemical sensing require lifelong use of the device. Replacement of faulty devices is not a consideration to be taken lightly, as implantation of all neural interfaces currently requires invasive procedures, especially intracortical probes implanted directly into the brain. In many cases, the implantation of neural interfaces in a clinical setting will be considered an elective procedure. Though these technologies have the potential to greatly enhance the quality of life of many individuals, it is important to minimize the implantation risk.

One issue facing the routine clinical implementation of electrodes for neural recording is the very small extracellular potentials ($\sim 50\,\mu V$) that must be detected, requiring an electrode with both high sensitivity and selectivity. As the size of a typical neuron soma in the human brain ranges from 4 to $100\,\mu m$ in diameter, highly selective neural recordings require electrodes approximately $40$–$400\,\mu m^2$ in size [41]. However, as the electrode size decreases, the electrochemical impedance and the electrode noise level also increase, which can be detrimental to the sensitivity of the electrode. Further increasing the electrode–tissue impedance may also reduce the recording quality, potentially rendering the electrode completely ineffective. Tissue reactions to the damage caused by the insertion and presence of the neural electrode can increase the electrode–tissue impedance [42].

Electrode impedance is also a critical issue for electrical stimulation. Neural stimulation involves injection of charge across the organic/inorganic interface formed at the surface of the electrode. At this interface, charge must be transferred from electron charge carriers in the electrode to ionic charge carriers in the biological fluid. The process is essentially electrochemical in nature, and therefore, the electrode potential that drives current injection must be kept below the threshold for electrolysis of water (electrochemical water window) to prevent undesirable and irreversible formation of $H_2$ and $O_2$ gases. For a particular desired injection current, high electrode impedance necessitates high electrode voltages, potentially risking electrolysis. This issue becomes particularly challenging in device architectures that utilize small-diameter electrodes since impedance is proportional to surface area.

From a structural perspective, a wide range of neural interfaces have been developed based on the method of manufacture and intended use. The field of neural interfacing was built upon assembled structures made from metal foils and/or micro-diameter wires, and such designs are still the most widely used today [43,44]. From a practical perspective, the component size and assembly techniques ultimately limit the use of these devices in clinical settings to situations that tolerate relatively large form factors and moderate electrode densities. Recently, however, microfabrication techniques developed originally for silicon-integrated circuits and microelectromechanical systems (MEMS) but adapted for medical implants make high-density, microscale, electrode arrays possible [45]. The first MEMS-based neural interfaces were made from silicon; however, micromachinable and biocompatible polymers have emerged as structural materials for applications that require mechanical flexibility. Polymers have an inherently higher rate of moisture absorption, which may be particularly detrimental to the long-term performance and viability of thin-film microelectrode arrays. The absorption of a conductive electrolytic fluid will increase capacitive coupling between adjacent leads, causing electrical cross talk between electrodes, and may also corrode metallic interconnects. Therefore, polymeric passivation and encapsulation layers require highly moisture-resistant polymers, and these polymer insulation layers should not degrade upon long-term exposure to biological systems.

Electrical interfacing to the brain is traditionally accomplished by placing electrical conductors in contact with the tissue. The least invasive technology is the conventional electroencephalogram (EEG), in which electrode arrays made from films or foils are placed atop the scalp. The most invasive approach uses penetrating devices (i.e., microwires and shanks) that are designed for single-unit recording but require penetration and insertion through the brain's pia mater. Electrocorticogram (ECoG) electrode arrays lie between EEG and penetrating cortical electrodes in terms of invasiveness. ECoG electrode arrays are made of films or foils designed to lie on top of the tissue and are typically placed not only beneath the scalp but also on the surface of the cortex. Clinically, ECoG electrode arrays are often used to monitor brain function in applications such as detection of focal epileptic activity prior to resection surgery.

Electrical interfacing to the peripheral nervous system likewise ranges in invasiveness. Approaches for stimulation in the periphery range from surface electrodes, to percutaneous electrodes for direct muscle stimulation, to cuff and intraneural electrodes for peripheral nerve stimulation. Peripheral nerve stimulation is a more efficient technique than muscle-based stimulation since one peripheral nerve array can stimulate many muscles to produce a variety of motions, while muscle-based stimulation requires that electrodes are placed in or on each muscle to be activated. Peripheral nerve stimulation takes advantage of peripheral nerves remaining intact, despite disruption between the central nervous system and the peripheral nervous system. The peripheral nerves can be stimulated with electrical pulses to produce useful muscle contractions or excite sensory nerve fibers.

Peripheral nerve electrodes are designed as either intraneural electrodes or extraneural electrodes. Extraneural electrodes, often called "cuff electrodes," typically wrap around the circumference of the nerve. These nonpenetrating electrodes are used to record and/or stimulate groups of nerve fascicles in a manner that is minimally invasive to the nerve. Intraneural electrodes are used to electrically stimulate and record from small numbers of nerve fibers. These penetrating electrodes typically have the form of one or two needles or wires [46,47], but multielectrode arrays (MEAs) have also been used in intraneural peripheral nerve stimulation [48–51]. The selectivity demonstrated by electrodes of this type is high compared with extraneural electrodes, but since these are penetrating electrodes, the potential for short- and long-term nerve damage is of greater concern than with nonpenetrating electrodes.

## 2.3 ELECTRODE DESIGNS

### 2.3.1 MICROWIRE PROBES

The first intracortical electrodes consisted of microwires with diameters as small as 25 μm [52,53] that could be either used independently or assembled into microwire arrays [54,55]. As mentioned previously, these devices are assembled from discrete components. Each microwire is encased in a polymer coating (i.e., polyimide) that forms an electrically insulating and chemically protective sheath around the conductor. Each microwire forms a single electrode contact, which is typically located at one end of the wire by selective removal of the sheath in that region. Williams *et al.* described a microwire-based MEA that was developed to evaluate long-term neural recording performance of such structures [54]. The $3 \times 11$ array was constructed of 35 μm diameter tungsten wire encased in 7 μm thick polyimide. Contacts were made by removing ~2 mm of polyimide from one end of each wire by thermal decomposition. The opposite end of each wire was soldered to a standard slot connector that provided connection to associated recording electronics. A thin layer of epoxy was used to secure the wires and encapsulate the slot connector.

The microwire-based approach benefits from the following: (1) wires of appropriate diameter and material are commercially available, (2) the mechanical durability of the microwires enables reliable device structures as compared with thin-film metallic electrodes/interconnects, (3) the relatively simple assembly process enables a wide range of possible configurations that can be customized for the application, (4) interfacing with external wiring leverages mature techniques developed for conventional electronics and thus is straightforward and reliable, and (5) formation of nonpenetrating surface electrodes can readily be fabricated by spot-welding metal foils to the microwires and encapsulating the resulting structures in silicone. The main disadvantage of the microwire architecture relates to the fact that the wire diameter and assembly techniques ultimately limit the density of contacts and form factor of the device.

### 2.3.2 SILICON-BASED DEVICES

While microwire arrays have proven to be an effective means of neural interfacing and, in fact, are still in use today, there are significant advantages in moving toward microfabricated neural interfaces, regardless of the application. Microfabrication, like the integrated circuit technologies from which it was developed, enables the repeatable, batch fabrication of devices with high feature density and small size.

In addition, microfabrication is generous in design flexibility as small, microscale probes, and much larger, centimeter-scale devices can be fabricated using the same methods. Thus, applying microfabrication techniques to the manufacture of neural interfaces is a logical extension of the technology.

Microfabrication techniques, including photolithography, thin-film deposition, and wet and dry etching, support batch processing, allowing for many devices to be produced at the same time. The fabrication parameters, such as photoresist thickness, exposure time, metal thickness, and etch, can be closely controlled to reduce the variability from device to device. Further, because microfabrication techniques and tools do not rely on manual manipulation to position each electrode, they enable the fabrication of dense arrays, which is valuable for selectivity in stimulation and recording in some neural applications. Probes built on silicon substrates have the advantage of being inherently IC-compatible. As such, on-chip circuitry for signal processing can be placed directly on the silicon probe [45] or placed on the probe using common IC packaging techniques such as flip-chip bonding.

Many of the microfabricated neural interfaces are designed as cortical interfaces—implanted directly into the brain for recording and stimulation. The most popular microfabricated neural interfaces are the planar Michigan intracortical probe (Michigan probe) and the three-dimensional Utah microelectrode array (Utah array). Both use silicon substrates and silicon bulk micromachining techniques to form the probes. Microfabricated silicon-based neural electrodes have been developed by other groups, but these are often variations of the Michigan probe and Utah array designs, with variations in shape, fabrication sequence, or materials [56,57].

The Michigan probe is based on a conventional planar design, leveraging heavily the advantages of conventional monolithic fabrication [45]. These devices incorporate large numbers (i.e., eight) of individually addressable electrodes along the shank of a single needlelike spike. The shank is fabricated utilizing conventional Si bulk micromachining techniques combined with monolithic fabrication of conductors and insulators, thus enabling a much higher degree of design flexibility than can be achieved in microwire arrays. Freestanding devices can incorporate multiple shanks, and three-dimensional arrays can be realized by incorporating several multiple shank devices onto a scaffold-like package [58,59]. Fabrication from standard Si wafers using IC-compatible technology has enabled the development of both passive probes, in which all preamplifiers and other signal processing circuitry are off-chip [45], and active probes, which incorporate some signal processing circuitry directly on-chip [60].

Passive Michigan probes are fabricated on a standard p-type (100) Si wafer. The probe areas are heavily doped using a deep boron diffusion through an oxide mask. After stripping the oxide, a stress-balanced multilayer structure of thermal oxide, silicon nitride, and silicon dioxide is deposited to serve as an insulation layer between electrodes and the conductive Si substrate. The conductive layers, which are composed of either tantalum or poly-Si, are deposited and patterned, followed by the deposition of a second oxide–nitride–oxide dielectric stack. The top dielectrics are patterned using reactive-ion etching (RIE) to expose the recording sites and connector pads. The electrode sites and contact pads are then coated with thin-film

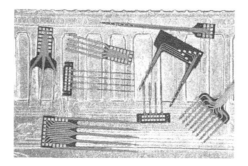

**FIGURE 2.1**

An assortment of Michigan probes [61].

Au and patterned by lift-off. A plasma etch is then used to remove the field dielectric layers, prior to the release of the probes by first thinning the Si wafer using an isotropic chemical etch and then defining the remaining probe geometry in ethylenediamine pyrocatechol, which is selective to lower levels of boron-doped Si [45].

A variety of Michigan probe architectures are shown in Figure 2.1 [61]. A typical probe for chronic implants consists of one to four shanks connected to a flexible silicon ribbon cable. The Michigan probe has been further developed to include microfluidic channels embedded into the substrates for drug delivery [41]. A microelectrode array developed by Hoogerwerf and Wise [58] uses an array of the planar Michigan probes to allow for true 3-D recording. This array is composed of a microfabricated silicon platform, two spacer bars, and the planar probes. Each probe is inserted through the slots in the silicon platform to produce an array of spikes.

An example of the Utah array is shown in Figure 2.2. This device is a 3-D structure composed of a $10 \times 10$ array of 1.5 mm long electrically isolated Si needles that protrude from a single Si platform [62–64]. The Utah arrays are fabricated from a 3 in. 1.7 mm thick n-type silicon wafer. Thermomigration is used to create paths of $p^+$-type silicon traversing the thickness of the silicon wafer. These paths will become

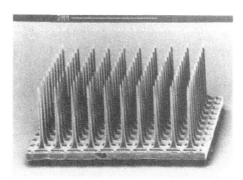

**FIGURE 2.2**

Scanning electron micrograph of Utah Intracortical Electrode Array [62].

the needles in the array and are electrically isolated by the back-to-back p–n junctions that result from the $p^+$ paths in the n-type wafer. A dicing saw is used to cut 1.5 mm into the wafer along the n-type grid to form a $10 \times 10$ array of $p^+$-type rectangular columns. Next, a two-step etch process is then used to etch the columns into the needle shapes. First, the columns are etched isotropically in an $HF/H_2NO_3$ solution. In the second step, the tips of the columns were sharpened and polished in the same solutions. Gold and platinum films are deposited on the tips of the needles through a metal foil mask. Thus, about 1.0 mm of each needle is coated with gold and platinum. The array is insulated with polyimide by placing the array onto a surface with the electrodes pointing upward and then flooding the array with a polyimide solution. This leaves a thick layer of polyimide on the base and a thinner layer along each electrode. A metal foil mask technique is again used to mask all but 0.5 mm at the tip of the electrodes such that the polyimide at the tip of each electrode can be removed by an oxygen plasma etch. Slight variations of this process have been developed, such as the investigation of amorphous SiC [65] or parylene [66] as the passivation layer, but the general structure of the Utah array has not changed appreciably over time.

Silicon-based devices like the Michigan probe or Utah array have the advantages of being inherently biocompatible, since silicon is recognized as a biocompatible material. Further, these devices have the ability to incorporate on-chip signal processing capabilities using integrated circuit technology developed for silicon substrates. However, the neural applications of these devices are limited partly because of the stiffness of silicon is high with a Young's modulus of approximately 190 GPa, as compared with brain, nerve, and other tissues that have Young's moduli on the order of 1–100 kPa. For chronic implantation, it can be expected that the neural tissue, whether it be the cortical tissue or nerve tissue, will experience some motion in the body. A stiff implanted device may not move with the tissue, resulting in a relative motion between the electrodes and the tissue. This relative motion may lead to poor recordings or stimulation as the electrode positioning with respect to the nerve or neurons may change. Furthermore, this relative motion may result in tissue damage. Moreover, a rigid substrate such as a silicon chip cannot conform to the nonplanar geometry of the brain or peripheral nerves. As a result, flexible substrates have been investigated for both cortical and peripheral nerve interfaces.

Implementation of micromachined intracortical electrodes has been limited primarily to animal studies, but there are some studies involving the use of Utah arrays implanted into humans [7]. Michigan probes, in a variety of designs, are available for purchase through NeuroNexus, but these are primarily for basic science research in animal studies.

### 2.3.3 POLYMER-BASED DEVICES

#### 2.3.3.1 General considerations
Microfabrication techniques have been used for the creation of Si-based MEAs for neural interfacing, including peripheral nerve electrodes and intracortical electrodes [32,63,67,68], as these techniques enable batch processing, design flexibility, and

small, dense features that cannot easily be achieved by hand fabrication. Initially, silicon substrates were used for microfabricated neural interfaces [33,45,56,59,62,63,69] in light of its established fabrication processes, lack of biotoxicity, stiffness sufficient for penetrating the cortical tissue without buckling, and ability to integrate signal processing circuitry directly on-chip [25]. However, mechanically rigid intracortical probes induce large strains on the neural tissue during micromotion of the brain [70,71], which can lead to tissue damage. This tissue damage is hypothesized to be the cause of the development of an interfering cellular sheath that has been observed to form around neural probes, limiting their viability to only a few months [72–74]. Strain induced on the neural tissue may be reduced with the use of mechanically flexible polymer substrates for intracortical probes [18,75]. It is thought that polymer-based substrates may be preferable to silicon in terms of functional reliability exceeding 10 years.

Flexible biomedical microdevices for chronic implantation require a biocompatible, compliant substrate that exhibits a high tensile strength and a high moisture resistance. The high tensile strength is required to ensure that the device has adequate mechanical robustness to survive the potentially harsh implantation procedure and the forces that may be applied during chronic implantation. An insulation layer, which may or may not be the same as the substrate material, must also be biocompatible, flexible, and highly resistant to moisture absorption. Moisture absorption of the substrate and insulation layers must be minimized to prevent cross talk between electrodes and electrode corrosion. Polymer materials have a lower Young's modulus than Si-based materials by at least 2 orders of magnitude, yielding mechanical flexibility in microscale devices. Mechanically flexible materials are especially advantageous in bioimplantable systems applications, as they are better able to conform to the irregular topographies and track the movement of biological structures. Biomedical microdevices have employed polyimide [76–78], parylene [79], SU-8 [80], and polydimethylsiloxane (PDMS) [81]. A summary of the electrical and mechanical properties of polymers used in microfabrication is shown in Table 2.1.

**Table 2.1** Select Properties of Polymers Used in MEMS-Based Neural Interfaces

|  | Dielectric Constant | Water Absorption % | Young's Modulus (GPa) | Tensile Strength (MPa) |
|---|---|---|---|---|
| Benzocyclobutene [152] | 2.7 | 0.2 | 2 | 85 |
| Parylene C [153] | 3.10 | 0.06 | 2.75 | 69 |
| PDMS [154] | 3 | n/a | $7.5 \times 10^{-4}$ | <12.5 |
| Polyimide [155] | 2.9 | <0.1 | 8.5 | 350 |
| Polynorbornene [156] | 2.42 | 0.07 | 0.8 | 18 |
| SU-8 [157] | 5.07 | n/a | 4.0 | 51 |
| Liquid crystal polymer [158] | 2.9 | 0.04 | 2.4 | 120 |

Benzocyclobutene (BCB), polyimide, polynorbornene (PNB), and SU-8 are spin-castable and photodefinable. BCB has been investigated for use in MEMS packaging [82] and neural interfaces [83]. Polyimide is one of the more popular materials for microfabricated, flexible structures in general, including flexible neural interfaces [84], because it is available in a large number of formulations, providing a wide variety of material properties for the same base polymer. It is available as a spin-castable and photodefinable solution and as a foil (Kapton®). Polyimides are known for their temperature stability, with some formulations able to withstand temperatures as high as 815 °C for short periods or 350 °C for extended time [85]. PNB was developed for microelectronic packaging applications [86], with limited use in freestanding micro-devices. PNB has been used in conjunction with liquid crystal polymer (LCP) as a capping layer in microfabricated peripheral nerve interface designs [68]. SU-8 was originally developed for use as a photoresist but has gained the most popularity as a soft-lithography molding material for PDMS [80,87]. Thus, its typical use is as an enabling material for fabrication of non-SU-8 MEMS devices, but it has been used as the structural material in some bioMEMS applications requiring a material with a higher Young's modulus, including penetrating electrodes [88].

### 2.3.3.2 Polyimide

Polyimide is one of the more popular polymers used as substrate and insulation layers in flexible microfabricated nerve electrodes. Rodriguez et al. employed polyimide as the substrate and insulating materials in a tubular, self-sizing peripheral nerve cuff that is capable of both recording and stimulation of the peripheral nerves [89]. The same group developed a sieve electrode of the same materials developed to interface with a regenerating peripheral nerve [90]. Boppart et al. described a flexible, planar microelectrode array to record potentials in brain slices [91]. The innovation in this device was the etching of perforations through the polyimide to allow the artificial cerebrospinal fluid to better circulate to the recording surface of the tissue and increase the viability of the brain slices used in the recordings. Microfabricated, flexible electrode arrays on polyimide with perforations were also developed by González and Rodríguez [92], although these electrodes were designed for use in nerve and muscle tissue recordings. The popularity of polyimide as a substrate and passivation material in microfabricated, flexible electrode arrays is further exemplified by Rousche et al. [93], who implemented a Michigan probe-like design in polyimide. A flexible cortical electrode array was developed by Takeuchi et al. [94]. The probes were fabricated on the planar substrate but then bent under the influence of a magnetic field to create a three-dimensional array.

By leveraging the advantages of micromachining and miniaturization, microfabricated ECoG (micro-ECoG) arrays can be mass-produced, made compatible with existing commercially available off-board amplifiers and recording equipment, and scaled up for applications requiring a large number of channels. Microfabricated 252-channel arrays with percutaneous leads have been realized and have exhibited no degradation in the quality of recorded ECoG signals up to 4.5 months after implantation [95] (Figure 2.3). The techniques used in the fabrication of polyimide-based

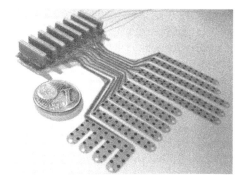

**FIGURE 2.3**

A 252-channel passive ECoG electrode array with connector hardware for external interfacing [95].

nerve electrodes are similar to the techniques used in surfacing micromachining of silicon-based MEMS devices. Therefore, special tooling or radically new techniques are not required for the fabrication of this type of device. However, as can be seen from Table 2.1, polyimide has one of the highest rates of moisture absorption in comparison with the other polymers used in microfabrication. At this rate of moisture absorption, it is likely that the device will remain viable for only a couple years, at best. To address this limitation, amorphous SiC thin films have been successfully used in polyimide-based devices to provide long-term moisture barrier protection [96].

### 2.3.3.3 Polydimethylsiloxane

Arguably, the most commonly used polymer in neural interfacing is polydimethylsiloxane, or PDMS. PDMS is a silicon-based material and is attractive for neural interfaces because of its mechanical flexibility, biocompatibility, and chemical inertness. The Young's modulus of PDMS is approximately 3 orders of magnitude lower than other materials used in MEMS applications and can be tuned by varying the base-to-curing agent ratio during preparation steps. PDMS is not photodefinable, instead often being patterned using micromolding. However, PDMS can be spin-coated and patterned using a technique similar to metal lift-off or with a laser. PDMS exhibits a very low tensile strength compared with other biocompatible polymers. Additionally, metal adhesion to PDMS is poor. Therefore, PDMS-based electrode arrays require complicated processes to promote metal adhesion or will suffer from metal delamination over time [97]. Further, PDMS shrinks when cured, which can cause difficulties in aligning subsequent layers to the PDMS base structure [97].

The flat interface nerve electrode (FINE) is a peripheral nerve cuff electrode [34]. This particular nerve cuff is unique in that it takes advantage of the oblong cross section of the nerve to maximize the interfacial area between the electrode array and the nerve. While other cuff electrodes, such as the spiral cuff [22,57], utilize a circular cross section, the FINE has a rectangular cross section. This rectangular cross section brings the electrodes into closer proximity to the nerve fascicles than is possible in the electrodes with circular cross sections. This increases the stimulation selectivity of the nerve fascicles for a given current pulse magnitude, as compared with cuff electrodes with circular cross sections. This design may also reduce the amount of current required for stimulation compared with the required current for cuff electrodes, reducing the amount of power needed for stimulation. The FINE devices are made by first molding a silicone elastomer into the correct shape. Eight or twelve platinum foil contacts ($0.5 \times 0.5 \times 0.050$ mm) are embedded in the silicone and spot-welded to Teflon-coated stainless steel wire. A 0.4 mm diameter window is opened through the silicone to the contacts.

While hand-fabricated nerve electrode arrays like the FINE have proven to be successful in selective stimulation of peripheral nerves, these devices do have significant limitations. First, fabrication of each device requires significant skill and a large amount of time to produce. Hand-fabricated devices inherently suffer from device-to-device variability with respect to electrode size, window size, and electrode placement, as well as variations in aspects such as electrode wrinkling and wire bending from handling and variations in the electrode-wire weld. The density of contacts is limited by the size of electrodes that can be handled practically and the precision with which they can be positioned. These limitations have resulted in a push toward the development of microfabricated electrode arrays for neural stimulation and recording.

ECoG electrode arrays are commonly used in the treatment of epilepsy to locate seizure centers in patients that do not respond favorably to drug therapy. ECoG electrode arrays are typically placed on the surface of the cortex to monitor brain function in applications such as the detection of focal epileptic activity prior to resection surgery. Commercially available ECoG electrode arrays are constructed from silicone elastomer and platinum–iridium foil contacts using conventional fabrication methods, and thus, their overall size and functionality are limited by achievable contact sizes and spacing on the order of millimeters and thicknesses of the same scale. For example, standard clinical ECoG grids by Ad-Tech Medical Instrument Corporation contain disk electrodes that are 3 mm in diameter with a 10 mm center-to-center spacing. An experimental "micro-ECoG" grid was recently manufactured by Ad-Tech Medical Instrument Corporation for a study involving the recording of human motor cortical activity that showed successful decoding of individual finger movements [98]. By pushing the limits of their manufacturing technology, 1.5 mm diameter disk electrodes with 4 mm center-to-center spacing were realized in the experimental array.

Although challenging, PDMS has been used as a substrate material in microfabricated electrodes, such as the structure shown in Figure 2.4 [99]. The device is fabricated on a glass slide that is coated with a thin Au/Ti antiadhesion layer that is used to release the device at the end of the fabrication process. The device incorporates a printed circuit board (PCB) to serve as the connection point between the thin-film interconnects and wiring to external electronics (bottom of Figure 2.4). To planarize the device, three layers of PDMS comprise the released device, with the bottom layer being the same thickness as the PCB. Oxygen plasma treatments were

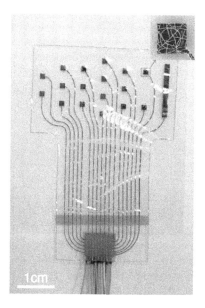

**FIGURE 2.4**

A 16-channel PDMS-based MEA [99].

used to improve PDMS adhesion between layers. Gold electrodes/interconnects were created by electron-beam evaporation combined with SU-8-based lift-off. Windows to expose the Au electrodes were created by molding the top PDMS layer using patterned photoresist. The periphery of the device was defined by cutting with a razor and released by simply peeling off from the glass substrate.

### 2.3.3.4 Parylene
Poly(p-xylylene), also known as parylene, is vapor-deposited, provides a conformal coating, and is available in several varieties, with parylene C being particularly popular due to its biocompatibility [100]. Parylene exhibits a high resistance to moisture absorption compared to polyimide, but a lower tensile strength. Parylene is not spin-castable and not photodefinable. Instead, parylene is vapor-deposited and can be patterned using RIE in an oxygen plasma. There are concerns, however, regarding the adhesion properties of parylene. Further, because parylene is vapor-deposited, it can be difficult to produce high-quality, thick layers [101]. Additionally, there are reports of parylene coatings cracking after long-term implantation [102]. A parylene-based spinal array is shown in Figure 2.5 [18].

### 2.3.3.5 Liquid crystal polymer
Based on the properties detailed in Table 2.1, LCP is arguably the best choice as a polymer for use in neural interfacing due to its relatively high tensile strength and low moisture absorption, yet LCP is not a material that has been widely used in the

2 mm

**FIGURE 2.5**

Flexible parylene-based electrode array for spinal cord simulation and recording [18].

development of neural interfaces. This may be attributed to LCP being difficult to process using standard microfabrication tools and techniques. In contrast to polymers that can be vapor-deposited or spin-casted, LCP is available as a cross-linked polymer in pressed sheets. Because one of the principal uses of LCP is as a flexible PCB substrate, it can be acquired from commercial vendors replete with copper cladding. A multilayered LCP device with embedded thin-film metal electrodes is very difficult to produce in comparison with the same device design made in polyimide. LCP layers are stacked using lamination, which requires high pressure and a temperature near the melting point of LCP (295 °C). Patterned copper structures made from the cladding layers (typically 18 μm in thickness) remain viable through a lamination process, but this may not be true for thin-film, biocompatible metals with thickness on the order of 0.25 μm. As such, the development and fabrication of a multilayered LCP structure with embedded thin-film electrodes are very challenging tasks.

Of all the candidate polymers, LCP has the most desirable properties, especially with respect to tensile strength and moisture absorption. A multilayered LCP device with embedded thin-film metal electrodes must be fabricated by lamination. Lamination requires high pressures and temperatures that can cause catastrophic damage to thin-film metal traces. As a result, all-LCP neural electrode devices consist of a single LCP layer with unprotected electrode structures [103] or are laminated at the level of individual devices [159]. To form a capping layer on LCP substrates, polymer thin films that can be deposited and patterned using conventional methods. Figure 2.6 is an example of an LCP-based micro-ECoG array that uses a laser-patterned 50 μm thick LCP sheet as the substrate, lift-off-patterned 100 nm thick sputter-deposited Pt as electrodes/interconnects/contact pads, and a photolithographically patterned ~5 μm thick parylene film as a capping layer. Testing in phosphate-buffered saline at 1 kHz and 10 mV, the measured impedance of each channel was approximately 4 kΩ.

### 2.3.3.6 Polymer nanocomposites

Compared to their rigid silicon counterparts, polymer-based microdevices have a wide range of mechanical, chemical, and electrical properties. Flexible polymeric

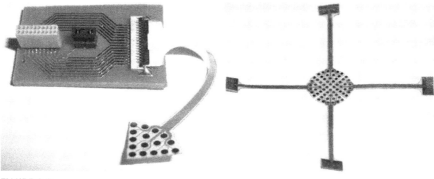

**FIGURE 2.6**

Microfabricated 64-channel ECoG electrode array using LCP substrate material. (Left) 16-channel array connected to a PCB; (right) high-density 64-channel array.

materials are especially attractive for implantable biomedical microdevices, which must operate reliably in the harsh biological environment without harming the surrounding biological tissue. The ability to tune the mechanical, electrical, and surface properties via the addition of various functional groups to polymer chains is a unique advantage of polymeric materials, as modifying polymer properties enables the formulation of a material well suited for a given application.

In addition to modification of the polymer itself, the material properties, including mechanical, electrical, and surface properties, can be further tuned by integrating nanoparticles into the polymer matrix. Carbon nanotubes have been used to enhance the conductivity and stiffness of polymers, including polyimide [104], polycarbonate [105], nylon [106], poly(vinyl alcohol) [106], and PDMS [105]. Cellulose nanocrystals have been used to reinforce latex [107] and polypropylene [108]. Integrating polytetrafluoroethylene nanoparticles into polymers, such as SU-8, rendered the material superhydrophobic [109]. While these material properties can be tuned by changing the concentration of the nanofiller within the polymer matrix, once a concentration is chosen and the material is formed, the material properties are fixed and cannot be reversibly altered.

New functionality may be incorporated into MEMS-based devices if the material properties can be tuned *in situ*. However, none of the materials described above have this capability. A thermally switchable Au nanoparticle/hydrogel composite was reported, with an order-of-magnitude increase in electrical conductivity with a small increase in temperature (22–25 °C) [110]. Some hydrogels are able to display a tunable, reversible storage or shear modulus in response to temperature [111] or by swelling [112], but the maximum stiffness offered by these materials is limited to the megapascal range, which is too compliant for applications requiring penetrating electrodes.

A bio-inspired, chemoresponsive polymer nanocomposite with switchable stiffness has provided new capabilities for MEMS-based neural interfaces. The nanocomposite is composed of a low elastic modulus poly(vinyl acetate) (PVAc) matrix embedded with high-modulus cellulose nanocrystalline (CNC) rods [113–116]. The overall stiffness of the nanocomposite (PVAc–CNC) is modulated by interactions between the CNCs; in the dehydrated state, the CNCs form a network that gives the nanocomposite rigidity, and in the wet state, interactions between CNCs are displaced by water and the material undergoes a reduction in elastic modulus from ~4 GPa to ~12 MPa. PVAc–CNC provides enabling properties for intracortical neural probes because it is initially sufficiently rigid to penetrate through the pia without buckling, but becomes tissue-like in mechanical properties after absorbing fluids in the brain after deployment [117], thus reducing damage caused by chronic strain on the tissue. A functional cortical probe with mechanical properties dominated by PVAc–CNC was fabricated using a combination of thin-film deposition, photolithographic patterning, and laser micromachining [118–120,160]. A freestanding, 50 μm thick PVAc–CNC film was adhered to a silicon wafer and then coated with a 1 μm thick vapor-deposited parylene-C film that served to insulate metallic traces and protect the PVAc–CNC from chemical exposure during subsequent process steps. A 50 nm thick Ti adhesion layer and a 250 nm thick Au conducting layer were then sputter-deposited and then wet-etched using iodine-based Au etchant and buffered oxide etchant through a photoresist etch mask. A second 1 μm thick parylene capping layer was then deposited and patterned using an oxygen plasma etch to open windows to the electrode sites and the connector contact pads. The overall shape of the probe was then patterned using a micromachining laser. Devices were released by gently peeling from the silicon wafer with the aid of a razor blade. A released, two-contact device is shown in Figure 2.7.

### 2.3.3.7 Issues

Though polymer-based microfabricated neural interfaces have provoked a reduced biological response compared to Si-based probes [121,122], these devices still face many challenges associated with long-term implantation, including a much higher moisture absorption rate than inorganic materials [123,124] and potential chemical- and thermal-process incompatibilities [125]. Further, typical electrodes patterned onto polymer substrates are thin-film metals (100–300 nm thick) deposited by physical vapor deposition, which are prone to corrosion, degradation, or delamination over time in the harsh physiological environment [126].

Intracortical neural interface designs are still very much evolving, especially with respect to microfabricated planar structures. Some of these developments include

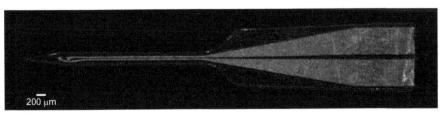

**FIGURE 2.7**

Two contact cortical probes fabricated from a stimuli-responsive PVAc nanocomposite.

the use of "fuzzy" conductive polymer electrodes to reduce electrode impedance by greatly increasing the effective surface area without increasing the footprint of the electrode and thus not decreasing its selectivity toward certain neurons [127,128]. Different probe architectures have been considered, such as the use of a parylene structure with a lattice-like platform with subcellular feature sizes to reduce encapsulation [129]. A number of different ideas are being implemented to make neural interfaces for neural recording camouflage into its surroundings, without provoking an immune or foreign body response [70,130].

Boron-doped diamond (BDD) microelectrodes have improved mechanical stability and chemical stability in the harsh physiological environment [131], compared with thin-film metal electrodes. Additionally, BDD microelectrodes have a number of properties that make them appealing for use in electrochemical sensing applications, including neurochemical detection of dopamine and serotonin [132,133]. However, microcrystalline diamond can be grown only at temperatures exceeding 1000 °C, thus prohibiting direct growth on a compliant polymer substrate. To overcome this process incompatibility of two very dissimilar materials, an inverted fabrication process was developed to produce mechanically flexible diamond-on-polymer microelectrode arrays for neural interfacing applications [134–136]. BDD microelectrodes were selectively grown by hot-filament chemical vapor deposition on a 1 μm thick thermal $SiO_2$ film on a silicon wafer. Next, a 1 μm thick PNB film was spin-cast and patterned to open access windows to the BDD. Next, a trilayer metallic film (20 nm thick Cr, 250 nm thick Cu, and 20 nm thick Cr) was sputter-deposited and patterned photolithographically. A 50 μm thick photodefinable PNB structural layer was then deposited and patterned to define the overall device geometry. Devices were released by dissolving the underlying $SiO_2$ film in HF. Examples of released devices are shown in Figure 2.8.

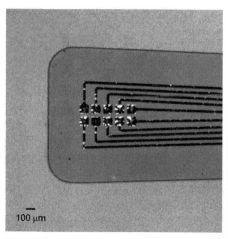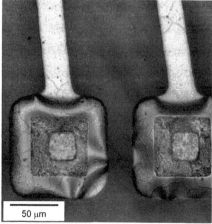

**FIGURE 2.8**

Diamond-on-polymer neural electrode array: (left) global view; (right) close-up showing diamond contacts and metal interconnects.

## 2.4 EMERGING DESIGN TRENDS

### 2.4.1 NEW MATERIALS AND DESIGNS FOR ENHANCED BIO-INTEGRATION

Emerging neural sensing devices employ new materials and geometries developed for bio-integration. The objective of these strategies is to minimize tissue encapsulation and promote neuron growth close to the electrodes. A sinusoidal parylene probe was designed to allow for brain deformations by lengthening and shortening and demonstrated stable recordings for up to 2 years of implantation [137]. A fishbone-structured compliant polyimide probe was reinforced with a biodegradable silk coating to stiffen the probe to facilitate insertion [138]. Similarly, a highly compliant parylene-based probe was coated with a tyrosine-derived polymer that can be quickly resorbed into the body after positioning [130]. The stiffening effect of the biodegradable coatings, followed by an increase in compliance following resorbing, is mimicked by materials that soften after implantation in the tissue [113,116,118,139,140]. These materials are initially stiff and do not require coatings or shuttles to facilitate penetration through the pia yet become very compliant after exposure to the physiological environment. All of these strategies are driving toward implantable neural electrodes and sensors that maintain functionality for clinically relevant time frames (>10 years).

### 2.4.2 WAVEGUIDES FOR IMPLANTED OPTOGENETIC STIMULATION SYSTEMS

New tools and modalities are in development, producing novel ways to interact with biological systems. Optogenetics, in particular, is an emerging method for stimulation that uses light of a particular wavelength to control neurons genetically modified to exhibit sensitivities to the particular wavelength. Due to the light absorption of neural tissues, light has to be delivered to the target with a penetrating waveguide or generated at the source. Optical fibers ~200 μm in diameter have been implanted several millimeters into the neural tissue to deliver light to the target neurons [141]. More recently, there have been efforts toward monolithic integration of waveguides in microelectrode arrays [142–145] with a material set that includes SU-8 [144–146], silicon oxynitride [147], and silicon carbide [148]. Material choice concerns include biocompatibility, stability of optical properties in physiological environment, sufficient transmission of optogenetics wavelength and minimal propagation loss, and high refractive index. Some designs and waveguide core materials have made use of the environment to serve as waveguide cladding, while others have integrated a cladding as part of the monolithic waveguide. In general, these integrated waveguides are smaller than the optical fibers, and more than a single waveguide can be integrated on a single device, providing enhanced spatial control.

For additional details about important aspects regarding neural electrode array technologies, the reader is encouraged to refer to the following detailed review papers [149–151].

# REFERENCES

[1] Schwartz AB. Cortical neural prosthetics. Annu Rev Neurosci 2004;27(1):487–507.

[2] Nicolelis MAL. Brain–machine interfaces to restore motor function and probe neural circuits. Nat Rev Neurosci 2003;4(5):417–22.

[3] Taylor DM, Tillery SIH, Schwartz AB. Direct cortical control of 3D neuroprosthetic devices. Science 2002;296(5574):1829–32.

[4] Musallam S, Corneil BD, Greger B, Scherberger H, Andersen RA. Cognitive control signals for neural prosthetics. Science 2004;305(5681):258–62.

[5] Santhanam G, Ryu S, Yu B, Afshar A, Shenoy K. A high-performance brain–computer interface. Nature 2006;442(7099):195–8.

[6] Velliste M, Perel S, Spalding M, Whitford A, Schwartz A. Cortical control of a prosthetic arm for self-feeding. Nature 2008;453(7198):1098–101.

[7] Hochberg LR, Serruya MD, Friehs GM, Mukand JA, Saleh M, Caplan AH, et al. Neuronal ensemble control of prosthetic devices by a human with tetraplegia. Nature 2006;442(7099):164–71.

[8] Theodore W, Fisher R. Brain stimulation for epilepsy. Lancet Neurol 2004; 3(2):111–8.

[9] Mayberg H, Lozano A, Voon V, McNeely H, Seminowicz D, Hamani C, et al. Deep brain stimulation for treatment-resistant depression. Neuron 2005;45(5):651–60.

[10] Jahanshahi M, Ardouin C, Brown R, Rothwell J, Obeso J, Albanese A, et al. The impact of deep brain stimulation on executive function in Parkinson's disease. Brain 2000;123(6):1142.

[11] Rodriguez-Oroz M, Obeso J, Lang A, Houeto J, Pollak P, Rehncrona S, et al. Bilateral deep brain stimulation in Parkinson's disease: a multicentre study with 4 years follow-up. Brain 2005;128(10):2240.

[12] Benabid A. Deep brain stimulation for Parkinson's disease. Curr Opin Neurobiol 2003;13(6):696–706.

[13] Wilson BS, Finley CC, Lawson DT, Wolford RD, Eddington DK, Rabinowitz WM. Better speech recognition with cochlear implants. Nature 1991;352(6332):236–8.

[14] Shepherd RK, Hatsushika S, Clark GM. Electrical stimulation of the auditory nerve: the effect of electrode position on neural excitation. Hear Res 1993;66(1):108–20.

[15] Zeng F-G, Rebscher S, Harrison WV, Sun X, Feng H. Cochlear implants. In: Greenbaum E, Zhou D, editors. Implantable neural prostheses 1. New York: Springer; 2009. p. 85–116.

[16] Rauschecker JP, Shannon RV. Sending sound to the brain. Science 2002;295(5557): 1025–9.

[17] Zrenner E. Will retinal implants restore vision? Science 2002;295(5557):1022.

[18] Rodger DC, Fong AJ, Li W, Ameri H, Ahuja AK, Gutierrez C, et al. Flexible parylene-based multielectrode array technology for high-density neural stimulation and recording. Sens Actuators B Chem 2008;132(2):449–60.

[19] Weiland JD, Humayun MS. Visual prosthesis. Proc IEEE 2008;96(7):1076–84.

[20] Yanai D, Weiland JD, Mahadevappa M, Greenberg RJ, Fine I, Humayun MS. Visual performance using a retinal prosthesis in three subjects with retinitis pigmentosa. Am J Ophthalmol 2007;143(5):820–7 e822.

[21] Raspopovic S, Capogrosso M, Petrini FM, Bonizzato M, Rigosa J, Di Pino G, et al. Restoring natural sensory feedback in real-time bidirectional hand prostheses. Sci Transl Med 2014;6(222):222ra19.

[22] Naples GG, Mortimer JT, Scheiner A, Sweeney JD. A spiral nerve cuff electrode for peripheral nerve stimulation. IEEE Trans Biomed Eng 1988;35(11):905–16.

[23] Starr A, Wise KD, Csongradi J. An evaluation of photoengraved microelectrodes for extracellular single-unit recording. IEEE Trans Biomed Eng 1973;bme-20(4):291.

[24] Marg E, Adams JE. Indwelling multiple micro-electrodes in the brain. Electroencephalogr Clin Neurophysiol 1967;23(3):277–80.

[25] Qing B, Wise KD. Single-unit neural recording with active microelectrode arrays. IEEE Trans Biomed Eng 2001;48(8):911–20.

[26] Ball T, Nawrot M, Pistohl T, Aertsen A, Schulze-Bonhage A, Mehring C. Towards an implantable brain–machine interface based on epicortical field potentials. Biomed Tech 2004;49:756–9.

[27] Seifritz E, Esposito F, Hennel F, Mustovic H, Neuhoff JG, Bilecen D, et al. Spatiotemporal pattern of neural processing in the human auditory cortex. Science 2002;297(5587):1706–8.

[28] Donoghue JP. Connecting cortex to machines: recent advances in brain interfaces. Nat Neurosci 2002;5:1085–8.

[29] Maynard EM, Nordhausen CT, Normann RA. The Utah Intracortical Electrode Array: a recording structure for potential brain–computer interfaces. Electroencephalogr Clin Neurophysiol 1997;102(3):228–39.

[30] Kotchetkov IS, Hwang BY, Appelboom G, Kellner CP, Connolly Jr. ES. Brain–computer interfaces: military, neurosurgical, and ethical perspective. Neurosurg Focus 2010;28(5):25–230.

[31] Perlmutter JS, Mink JW. Deep brain stimulation. Neuroscience 2006;29(1):229.

[32] Terasawa Y, Tashiro H, Uehara A, Saitoh T, Ozawa M, Tokuda T, et al. The development of a multichannel electrode array for retinal prostheses. J Artif Organs 2006;9(4):263–6.

[33] Branner A, Stein RB, Normann RA. Selective stimulation of cat sciatic nerve using an array of varying-length microelectrodes. J Neurophysiol 2001;85(4):1585–94.

[34] Tyler DJ, Durand DM. Functionally selective peripheral nerve stimulation with a flat interface nerve electrode. IEEE Trans Neural Syst Rehabil Eng 2002;10(4):294–303.

[35] Morrell M. Brain stimulation for epilepsy: can scheduled or responsive neurostimulation stop seizures? Curr Opin Neurol 2006;19(2):164.

[36] Roham M, Covey DP, Daberkow DP, Ramsson ES, Howard CD, Heidenreich BA, et al. A wireless IC for time-share chemical and electrical neural recording. IEEE J Solid-State Circuits 2009;44(12):3645–58.

[37] Chan H-Y, Aslam DM, Wiler JA, Casey B. A novel diamond microprobe for neuro-chemical and -electrical recording in neural prosthesis. J Microelectromech Syst 2009;18(3):511–21.

[38] Capella P, Ghasemzadeh B, Mitchell K, Adams RN. Nafion coated carbon fiber electrodes for neurochemical studies in brain tissue. Electroanalysis 1990;2(3):175–82.

[39] Pantano P, Morton TH, Kuhr WG. Enzyme-modified carbon-fiber microelectrodes with millisecond response times. J Am Chem Soc 1991;113(5):1832–3.

[40] Park J, Show Y, Quaiserova V, Galligan JJ, Fink GD, Swain GM. Diamond microelectrodes for use in biological environments. J Electroanal Chem 2005;583(1):56–68.

[41] Wise KD, Anderson DJ, Hetke JF, Kipke DR, Najafi K. Wireless implantable microsystems: high-density electronic interfaces to the nervous system. Proc IEEE 2005;92(1):76–97.

[42] Williams JC, Hippensteel JA, Dilgen J, Shain W, Kipke DR. Complex impedance spectroscopy for monitoring tissue responses to inserted neural implants. J Neural Eng 2007;4:410.

[43] Misra A, Burke J, Ramayya A, Jacobs J, Sperling M, Moxon K, et al. Methods for implantation of micro-wire bundles and optimization of single/multi-unit recordings from human mesial temporal lobe. J Neural Eng 2014;11(2):026013.

[44] Holinski BJ, Everaert DG, Mushahwar VK, Stein RB. Real-time control of walking using recordings from dorsal root ganglia. J Neural Eng 2013;10(5):056008.

[45] Najafi K, Wise KD, Mochizuki Y. A high-yield IC-compatible multichannel recording array. IEEE Trans Electron Devices 1985;32(7):1206.

[46] Nannini N, Horch K. Muscle recruitment with intrafascicular electrodes. IEEE Trans Biomed Eng 1991;38(8):769–76.

[47] Yoshida K, Horch K. Selective stimulation of peripheral nerve fibers using dual intrafascicular electrodes. IEEE Trans Biomed Eng 1993;40(5):492–4.

[48] Boretius T, Badia J, Pascual-Font A, Schuettler M, Navarro X, Yoshida K, et al. A transverse intrafascicular multichannel electrode (TIME) to interface with the peripheral nerve. Biosens Bioelectron 2010;26(1):62–9.

[49] Kundu A, Harreby KR, Yoshida K, Boretius T, Stieglitz T, Jensen W. Stimulation selectivity of the thin-film longitudinal intrafascicular electrode; (tfLIFE) and the transverse intrafascicular multi-channel electrode (TIME) in the large nerve animal model. IEEE Trans Neural Syst Rehabil Eng 2014;22(2):400–10.

[50] Lawrence SM, Dhillon G, Jensen W, Yoshida K, Horch KW. Acute peripheral nerve recording characteristics of polymer-based longitudinal intrafascicular electrodes. IEEE Trans Neural Syst Rehabil Eng 2004;12(3):345–8.

[51] Branner A, Normann RA. A multielectrode array for intrafascicular recording and stimulation in sciatic nerve of cats. Brain Res Bull 2000;51(4):293–306.

[52] Strumwasser F. Long-term recording from single neurons in brain of unrestrained mammals. Science 1958;127(3296):469–70.

[53] Palmer C. A microwire technique for recording single neurons in unrestrained animals. Brain Res Bull 1978;3(3):285–9.

[54] Williams J, Rennaker R, Kipke D. Long-term neural recording characteristics of wire microelectrode arrays implanted in cerebral cortex. Brain Res Protoc 1999;4(3):303.

[55] Kruger J, Caruana F, Dalla Volta R, Rizzolatti G. Seven years of recording from monkey cortex with a chronically implanted multiple microelectrode. Front Neuroeng 2010;4:12.

[56] Norlin P, Kindlundh M, Mouroux A, Yoshida K, Hofmann UG. A 32-site neural recording probe fabricated by DRIE of SOI substrates. J Micromech Microeng 2002;12(4):414.

[57] Kewley DT, Hills MD, Borkholder DA, Opris IE, Maluf NI, Storment CW, et al. Plasma-etched neural probes. Sens Actuators A Phys 1997;58(1):27–35.

[58] Hoogerwerf AC, Wise KD. A three-dimensional microelectrode array for chronic neural recording. IEEE Trans Biomed Eng 1994;41(12):1136–46.

[59] Bai Q, Wise KD, Anderson DJ. A high-yield microassembly structure for three-dimensional microelectrode arrays. IEEE Trans Biomed Eng 2000;47(3):281–9.

[60] Najafi K, Wise KD. An implantable multielectrode array with on-chip signal processing. IEEE J Solid-State Circuits 2002;21(6):1035–44.

[61] Wise KD. Silicon microsystems for neural science and neural prostheses. IEEE Eng Med Biol Mag 2005;24(5):22–9.

[62] Nordhausen CT, Maynard EM, Normann RA. Single unit recording capabilities of a 100 microelectrode array. Brain Res 1996;726(1–2):129–40.

[63] Campbell P, Jones K, Huber R, Horch K, Normann R. A silicon-based, three-dimensional neural interface: manufacturing processes for an intracortical electrode array. IEEE Trans Biomed Eng 1991;38(8):758–68.

[64] Rousche PJ, Normann RA. Chronic recording capability of the Utah Intracortical Electrode Array in cat sensory cortex. J Neurosci Methods 1998;82(1):1–15.

[65] Hsu JM, Tathireddy P, Rieth L, Normann AR, Solzbacher F. Characterization of a-SiCx: H thin films as an encapsulation material for integrated silicon based neural interface devices. Thin Solid Films 2007;516(1):34–41.

[66] Hsu JM, Rieth L, Normann RA, Tathireddy P, Solzbacher F. Encapsulation of an integrated neural interface device with Parylene C. IEEE Trans Biomed Eng 2009;56(1):23–9.

[67] Rodger DC, Weiland JD, Humayun MS, Tai Y-C. Scalable high lead-count parylene package for retinal prostheses. Sens Actuators B Chem 2006;117(1):107–14.

[68] Hess AE, Dunning J, Tyler D, Zorman CA. Development of a microfabricated flat interface nerve electrode based on liquid crystal polymer and polynorbornene multilayered structures. In: Proceedings—3rd international IEEE EMBS conference on neural engineering; 2007. p. 32–5.

[69] Wise KD, Angell JB. A low-capacitance multielectrode probe for use in extracellular neurophysiology. IEEE Trans Biomed Eng 1975;BME-22(3):212–9.

[70] Kim D, Viventi J, Amsden J, Xiao J, Vigeland L, Kim Y, et al. Dissolvable films of silk fibroin for ultrathin conformal bio-integrated electronics. Nat Mater 2010;9:511–7.

[71] Subbaroyan J, Martin DC, Kipke DR. A finite-element model of the mechanical effects of implantable microelectrodes in the cerebral cortex. J Neural Eng 2005;2(4):103.

[72] McConnell GC, Rees HD, Levey AI, Gutekunst CA, Gross RE, Bellamkonda RV. Implanted neural electrodes cause chronic, local inflammation that is correlated with local neurodegeneration. J Neural Eng 2009;6:056003.

[73] Kipke DR, Shain W, Buzsaki G, Fetz E, Henderson JM, Hetke JF, et al. Advanced neurotechnologies for chronic neural interfaces: new horizons and clinical opportunities. J Neurosci 2008;28(46):11830–8.

[74] Polikov VS, Tresco PA, Reichert WM. Response of brain tissue to chronically implanted neural electrodes. J Neurosci Methods 2005;148(1):1–18.

[75] Takeuchi S, Ziegler D, Yoshida Y, Mabuchi K, Suzuki T. Parylene flexible neural probes integrated with microfluidic channels. Lab Chip 2005;5:519–23.

[76] Kisban S, Herwik S, Seidl K, Rubehn B, Paul O, Ruther P, et al. Microprobe array with low impedance electrodes and highly flexible polyimide cables for acute neural recording. In: 29th annual international conference of the IEEE engineering in medicine and biology society, Lyon, France; 2007. p. 175–8.

[77] Metz S, Holzer R, Renaud P. Polyimide-based microfluidic devices. Lab Chip 2001;1(1):29–34.

[78] Engel J, Chen J, Liu C. Development of polyimide flexible tactile sensor skin. J Micromech Microeng 2003;13:359.

[79] Chen P-J, Rodger DC, Agrawal R, Saati S, Meng E, Varma R, et al. Implantable micromechanical parylene-based pressure sensors for unpowered intraocular pressure sensing. J Microelectromech Syst 2008;17(6):1342–51.

[80] Lorenz H, Despont M, Fahrni N, LaBianca N, Renaud P, Vettiger P. SU-8: a low-cost negative resist for MEMS. J Micromech Microeng 1997;7(3):121–4.

[81] Takao H, Miyamura K, Ebi H, Ashiki M, Sawada K, Ishida M. A MEMS microvalve with PDMS diaphragm and two-chamber configuration of thermo-pneumatic actuator for integrated blood test system on silicon. Sens Actuators A Phys 2005;119(2):468–75.

[82] Jourdain A, De Moor P, Pamidighantam S, Tilmans H. Investigation of the hermeticity of BCB-sealed cavities for housing (RF-) MEMS devices. In: Proceeding of the fifteenth IEEE international conference on micro electro mechanical systems; 2002. p. 677–80.

[83] Lee KK, He J, Singh A, Kim B. Benzocyclobutene (BCB) based intracortical neural implant. In: Proceedings of the international conference on MEMS, NANO, and smart systems; 2003. p. 418–22.

[84] Stieglitz T. Flexible biomedical microdevices with double-sided electrode arrangements for neural applications. Sens Actuators A Phys 2001;90(3):203.

[85] Hou T, Wilkinson S, Johnston N, Pater R, Schneiderk T. Processing and properties of IM7/LARC™-RP46 polyimide composites. High Perform Polym 1996;8(4):491.

[86] Shick RA, Jayaraman SK, Goodall BL, Rhodes LF, McDougall WC, Kohl P, et al. Avatrel(TM) dielectric polymers for electronic packaging. Adv Microelectron 1998;25(5):13–4.

[87] Zhang J, Tan K, Hong G, Yang L, Gong H. Polymerization optimization of SU-8 photoresist and its applications in microfluidic systems and MEMS. J Micromech Microeng 2001;11:20.

[88] Fernández LJ, Altuna A, Tijero M, Gabriel G, Villa R, Rodríguez MJ, et al. Study of functional viability of SU-8-based microneedles for neural applications. J Micromech Microeng 2009;19:025007.

[89] Rodriguez F, Ceballos D, Valero A, Valderrama E, Stieglitz T, Navarro X. Polyimide cuff electrodes for peripheral nerve stimulation. J Neurosci Methods 2000;98(2):105–18.

[90] Navarro X, Calvet S, Rodriguez F, Stieglitz T, Blau C, Buti M, et al. Stimulation and recording from regenerated peripheral nerves through polyimide sieve electrodes. J Peripher Nerv Syst JPNS 1997;3(2):91–101.

[91] Boppart SA, Wheeler BC, Wallace CS. A flexible perforated microelectrode array for extended neural recordings. IEEE Trans Biomed Eng 1992;39(1):37–42.

[92] González C, Rodríguez M. A flexible perforated microelectrode array probe for action potential recording in nerve and muscle tissues. J Neurosci Methods 1997;72(2):189–95.

[93] Rousche PJ, Pellinen DS, Pivin Jr. DP, Williams JC, Vetter RJ, Kipke DR. Flexible polyimide-based intracortical electrode arrays with bioactive capability. IEEE Trans Biomed Eng 2001;48(3):361–71.

[94] Takeuchi S, Suzuki T, Mabuchi K, Fujita H. 3D flexible multichannel neural probe array. J Micromech Microeng 2004;14(1):104.

[95] Rubehn B, Bosman C, Oostenveld R, Fries P, Stieglitz T. A MEMS-based flexible multichannel ECoG-electrode array. J Neural Eng 2009;6:036003.

[96] Kelly SK, Shire DB, Jinghua C, Gingerich MD, Cogan SF, Drohan WA, et al. Developments on the Boston 256-channel retinal implant. In: 2013 IEEE international conference on multimedia and expo workshops (ICMEW); 2013. p. 1–6.

[97] Paranjape M, Garra J, Brida S, Schneider T, White R, Currie J. A PDMS dermal patch for non-intrusive transdermal glucose sensing. Sens Actuators A Phys 2003;104(3):195–204.

[98] Wang W, Degenhart AD, Collinger JL, Vinjamuri R, Sundre GP, Adelson PD, et al. Human motor cortical activity recorded with micro-ECoG electrodes during individual finger movements. In: 31st annual international conference of the IEEE engineering in medicine and biology society, EMBS '09; 2009. p. 586–9.

[99] Guo L, Guvanasen GS, Liu X, Tuthill C, Nichols TR, DeWeerth SP. A PDMS-based integrated stretchable microelectrode array (is MEA) for neural and muscular surface interfacing. IEEE Trans Biomed Circuits Syst 2013;7(1):1–10.

[100] Song JS, Lee S, Jung SH, Cha GC, Mun MS. Improved biocompatibility of parylene C films prepared by chemical vapor deposition and the subsequent plasma treatment. J Appl Polym Sci 2009;112(6):3677–85.

[101] Lee H, Cho J. Development of conformal PDMS and Parylene coatings for microelectronics and MEMS packaging. In: Proceedings of IMECE, Orlando; 2005.

[102] Schmidt EM, Mcintosh JS, Bak MJ. Long-term implants of Parylene-C coated microelectrodes. Med Biol Eng Comput 1988;26:96–101.

[103] Lee CJ, Oh S, Song JK, Kim SJ. Neural signal recording using microelectrode arrays fabricated on liquid crystal polymer material. Mater Sci Eng C 2004;24:265–8.

[104] Park C, Ounaies Z, Watson KA, Crooks RE, Smith J. Dispersion of single wall carbon nanotubes by in situ polymerization under sonication. Chem Phys Lett 2002;364(3–4):303–8.

[105] Moniruzzaman M, Winey KI. Polymer nanocomposites containing carbon nanotubes. Macromolecules 2006;39(16):5194–205.

[106] Coleman JN, Khan U, Gun'ko YK. Mechanical reinforcement of polymers using carbon nanotubes. Adv Mater 2006;18(6):689–706.

[107] Favier V, Canova G, Cavaille J, Chanzy H, Dufresne A, Gauthier C. Nanocomposite materials from latex and cellulose whiskers. Polym Adv Technol 1995;6(5):351–5.

[108] Ljungberg N, Bonini C, Bortolussi F, Boisson C, Heux L, Cavaille J. New nanocomposite materials reinforced with cellulose whiskers in atactic polypropylene: effect of surface and dispersion characteristics. Biomacromolecules 2005;6(5):2732–9.

[109] Hong L, Pan T. Photopatternable superhydrophobic nanocomposites for microfabrication. J Microelectromech Syst 2010;19(2):246–53.

[110] Zhao X, Ding X, Deng Z, Zheng Z, Peng Y, Long X. Thermoswitchable electronic properties of a gold nanoparticle/hydrogel composite. Macromol Rapid Commun 2005;26(22):1784–7.

[111] Ozbas B, Rajagopal K, Haines-Butterick L, Schneider JP, Pochan DL. Reversible stiffening transition in β-hairpin hydrogels induced by ion complexation. J Phys Chem B 2007;111(50):13901–8.

[112] Santulli C, Patel S, Jeronimidis G, Davis F, Mitchell G. Development of smart variable stiffness actuators using polymer hydrogels. Smart Mater Struct 2005;14:434.

[113] Capadona JR, Shanmuganathan K, Tyler DJ, Rowan SJ, Weder C. Stimuli-responsive polymer nanocomposites inspired by the sea cucumber dermis. Science 2008;319:1370–4.

[114] Capadona JR, Tyler DJ, Zorman CA, Rowan SJ, Weder C. Mechanically adaptive nanocomposites for neural interfacing. MRS Bull 2012;37(06):581–9.

[115] Shanmuganathan K, Capadona JR, Rowan SJ, Weder C. Stimuli-responsive mechanically adaptive polymer nanocomposites. ACS Appl Mater Interfaces 2009;2(1):165–74.

[116] Shanmuganathan K, Capadona JR, Rowan SJ, Weder C. Bio-inspired mechanically-adaptive nanocomposites derived from cotton cellulose whiskers. J Mater Chem 2010;20(1):180.

[117] Harris JP, Hess AE, Rowan SJ, Weder C, Zorman CA, Tyler DJ, et al. In vivo deployment of mechanically adaptive nanocomposites for intracortical microelectrodes. J Neural Eng 2011;8(4):046010.

[118] Hess A, Capadona J, Shanmuganathan K, Hsu L, Rowan S, Weder C, et al. Development of a stimuli-responsive polymer nanocomposite toward biologically optimized, MEMS-based neural probes. J Micromech Microeng 2011;21:054009.

[119] Hess AE, Shanmuganathan K, Capadona JR, Hsu L, Rowan S, Weder C, et al. Mechanical behavior of microstructures from a chemo-responsive polymer nanocomposite based on cotton cellulose nanofibers. In: Technical digest—24th IEEE international conference on microelectromechanical systems, Cancun Mexico, January 23–27, 2011; 2011. p. 453–6.

[120] Hess AE, Zorman CA. Fabrication and characterization of MEMS-based structures from a bio-inspired, chemo-responsive polymer nanocomposite. In: Proceedings—2010 MRS fall meeting, Boston, MA, November 29–December 3, 2010; 2010 Abstract #S4.7. Also cited as MRS Online Proceedings Library, 2011, vol. 1299, pp. mrsf10-1299-s04-07.

[121] Mercanzini A, Cheung K, Buhl D, Boers M, Maillard A, Colin P, et al. Demonstration of cortical recording and reduced inflammatory response using flexible polymer neural probes. In: IEEE 20th international conference on microelectromechanical systems; 2008. p. 573–6.

[122] Mercanzini A, Cheung K, Buhl DL, Boers M, Maillard A, Colin P, et al. Demonstration of cortical recording using novel flexible polymer neural probes. Sens Actuators A Phys 2008;143(1):90–6.

[123] Blum NA, Carkhuff BG, Charles HK, Edwards RL, Meyer RA. Multisite microprobes for neural recordings. IEEE Trans Biomed Eng 1991;38(1):68.

[124] Lee K, Massia S, He J. Biocompatible benzocyclobutene-based intracortical neural implant with surface modification. J Micromech Microeng 2005;15:2149.

[125] Rogers JA, Someya T, Huang Y. Materials and mechanics for stretchable electronics. Science 2010;327(5973):1603–7.

[126] Chow AY, Pardue MT, Chow VY, Peyman GA, Liang C, Perlman JI, et al. Implantation of silicon chip microphotodiode arrays into the cat subretinal space. IEEE Trans Neural Syst Rehabil Eng 2002;9(1):86–95.

[127] Cui X, Lee VA, Raphael Y, Wiler JA, Hetke JF, Anderson DJ, et al. Surface modification of neural recording electrodes with conducting polymer/biomolecule blends. J Biomed Mater Res 2001;56(2):261–72.

[128] Cui X, Wiler J, Dzman M, Altschuler RA, Martin DC. In vivo studies of polypyrrole/peptide coated neural probes. Biomaterials 2003;24(5):777.

[129] Seymour JP, Kipke DR. Neural probe design for reduced tissue encapsulation in CNS. Biomaterials 2007;28(25):3594–607.

[130] Lewitus D, Smith KL, Shain W, Kohn J. Ultrafast resorbing polymers for use as carriers for cortical neural probes. Acta Biomater 2011;7(6):2483–91.

[131] Halpern JM, Xie S, Sutton GP, Higashikubo BT, Chestek CA, Lu H, et al. Diamond electrodes for neurodynamic studies in *Aplysia californica*. Diamond Relat Mater 2006;15(2–3):183–7.

[132] Martin HB, Argoitia A, Angus JC, Landau U. Voltammetry studies of single-crystal and polycrystalline diamond electrodes. J Electrochem Soc 1999;146(8):2959–64.

[133] Martin HB, Argoitia A, Landau U, Anderson AB, Angus JC. Hydrogen and oxygen evolution on boron-doped diamond electrodes. J Electrochem Soc 1996;143(6):L133–6.

[134] Hess A, Sabens DM, Martin HB, Zorman CA. Polycrystalline diamond-on-polymer microelectrode arrays for mechanically-flexible neural interfacing. In: Proceedings of the Hilton Head workshop 2010: a solid-state sensors, actuators and microsystems workshop, Hilton Head Island, SC, June 6–10, 2010; 2010.

[135] Hess AE, Sabens DM, Martin HB, Zorman CA. Polycrystalline diamond-on-polymer electrode arrays fabricated using a polymer-based transfer process. Electrochem Solid-State Lett 2010;13(11):J129–31.

[136] Hess AE, Sabens DM, Martin HB, Zorman CA. Diamond-on-polymer microelectrode arrays fabricated using a chemical release transfer process. J Microelectromech Syst 2011;20(4):867–75.

[137] Sohal HS, Jackson A, Jackson R, Clowry GJ, Vassilevski K, O'Neill A, et al. The sinusoidal probe: a new approach to improve electrode longevity. Front Neuroeng 2014;7.

[138] Wu F, Im M, Yoon E. A flexible fish-bone-shaped neural probe strengthened by biodegradable silk coating for enhanced biocompatibility. In: 2011 16th international solid-state sensors, actuators and microsystems conference; 2011. p. 966–9.

[139] Ware T, Simon D, Arreaga-Salas DE, Reeder J, Rennaker R, Keefer EW, et al. Fabrication of responsive, softening neural interfaces. Adv Funct Mater 2012;22(16):3470–9.

[140] Ware T, Simon D, Liu C, Musa T, Vasudevan S, Sloan A, et al. Thiol-ene/acrylate substrates for softening intracortical electrodes. J Biomed Mater Res Part B Appl Biomater 2014;102(1):1–11.

[141] Kravitz AV, Freeze BS, Parker PR, Kay K, Thwin MT, Deisseroth K, et al. Regulation of parkinsonian motor behaviours by optogenetic control of basal ganglia circuitry. Nature 2010;466(7306):622–6.

[142] Wu F, Stark E, Im M, Cho I-J, Yoon E-S, Buzsáki G, et al. An implantable neural probe with monolithically integrated dielectric waveguide and recording electrodes for optogenetics applications. J Neural Eng 2013;10(5):056012.

[143] Kanno S, Lee S, Harashima T, Kuki T, Kino H, Mushiake H, et al. Multiple optical stimulation to neuron using Si opto-neural probe with multiple optical waveguides and metal-cover for optogenetics. Conf Proc IEEE Eng Med Biol Soc 2013;2013:253–6.

[144] Rubehn B, Wolff SB, Tovote P, Lüthi A, Stieglitz T. A polymer-based neural microimplant for optogenetic applications: design and first in vivo study. Lab Chip 2013;13(4):579–88.

[145] Choi M, Choi JW, Kim S, Nizamoglu S, Hahn SK, Yun SH. Light-guiding hydrogels for cell-based sensing and optogenetic synthesis in vivo. Nat Photonics 2013;7(12):987–94.

[146] Cho IJ, Baac HW, Yoon E. A 16-site neural probe integrated with a waveguide for optical stimulation. In: 2010 IEEE 23rd international conference on micro electro mechanical systems (MEMS); 2010. p. 995–8.

[147] Zorzos AN, Boyden ES, Fonstad CG. Multiwaveguide implantable probe for light delivery to sets of distributed brain targets. Opt Lett 2010;35(24):4133–5.

[148] Register J, Muller A, King J, Weeber E, Frewin CL, Saddow SE. Silicon carbide waveguides for optogenetic neural stimulation. MRS Proc 2012;1433: mrss12-1433-h04-20.

[149] Fattahi P, Yang G, Kim G, Abidian MR. A review of organic and inorganic biomaterials for neural interfaces. Adv Mater 2014;26(12):1793.

[150] Cheung KC. Implantable microscale neural interfaces. Biomed Microdevices 2007;9:923–38.

[151] Cogan SF. Neural stimulation and recording electrodes. Annu Rev Biomed Eng 2008;10:275–309.

[152] Corning D. Cyclotene Advanced Electronics Resins; 2011.

[153] SCS Parylene Properties; 2010.

[154] Corning D. Sylgard 184 Silicone Elastomer; 2008.

[155] HD MicroSystems. PI-2600 series—low stress applications; 2009.

[156] Promerus. Avatrel 2585P Properties.

[157] MicroChem. SU-8 Permanent Photoresists; 2011.

[158] Rogers Corporation R/Flex 3600.

[159] Wang K, Liu C-C, Durand DM. Flexible nerve stimulation electrode with iridium oxide sputtered on liquid crystal polymer. IEEE Trans Biomed Eng 2009;56(1):6–14.

[160] Hess-Dunning A, Tyler D, Harris JP, Capadona JR, Weder C, Rowan SJ, Zorman CA. Microscale characterization of a mechanically adaptive polymer nanocomposite with cotton-derived cellulose nanocrystals for implantable BioMEMS. IEEE J Microelectromech Syst 2014;23(4):774–84.

# Analog front-end and telemetry systems

**Darrin J. Young**

*Electrical and Computer Engineering Department, University of Utah, Salt Lake City, Utah, USA*

## CHAPTER CONTENTS

## 3.1 INTRODUCTION

Integrated electronics with low power dissipation and high performance are critical for bio-implantable systems. Figure 3.1 provides a conceptual illustration of key electronic building blocks used for typical bioimplantable systems applications. A bio-implantable system usually performs two major functions: recording and stimulation. For recording, physiological parameters such as electroencephalography (EEG), extracellular neural signal (ENG), electrocardiogram (ECG), Electromyography (EMG), blood pressure (BP), body temperature, pH, glucose level, etc. are monitored by the corresponding sensors, which typically are implemented as a pair of sensing electrodes. The sensor-induced voltage is then amplified by a front-end amplifier or an amplifier array followed by an analog-to-digital converter (ADC). The digitized physiological signals are encoded and processed by a telemetry control unit prior to wireless transmission to an external receiver through a data telemetry circuit. For stimulation, a radio-frequency (RF) receiver first receives an external command signal, which typically specifies stimulation characteristics, sites, sequence, and timing. The command signal then will be decoded followed by a stimulation control unit, which controls the operation of an output stimulation circuit and its interface with an array of simulation electrodes, for example, $E_1$ to $E_N$, as shown in Figure 3.1.

Besides receiving command signals from an external device, an implantable system can also receive RF power along with the command. A rectifier circuit can be

Bhunia et al. Implantable Biomedical Microsystems. http://dx.doi.org/10.1016/B978-0-323-26208-8.00003-0

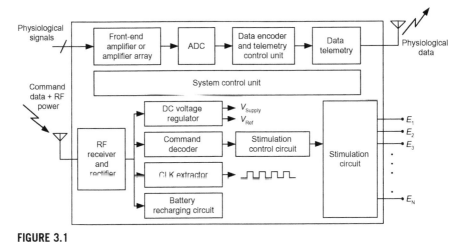

**FIGURE 3.1**

Conceptual illustration of electronic building blocks for bioimplantable systems applications.

implemented as a part of the RF receiver followed by a DC voltage regulator to convert the incoming RF power to a stable DC voltage, thus energizing the implantable system in a battery-less manner. Depending on applications, this approach can also be employed to recharge an implantable battery. Furthermore, the incoming RF signal exhibits a stable frequency; hence, a digital clock can be obtained from a clock extractor circuit to provide synchronization for the implant. The design principles of the aforementioned building blocks can be found in a variety of publications. This chapter provides a general discussion of the analog front-end system and telemetry system with a focus on bioimplantable systems applications.

## 3.2  ANALOG FRONT-END SYSTEM

As illustrated in Figure 3.1, the analog front-end system is responsible for amplifying, filtering, signal conditioning, and digitizing *in vivo* physiological signals. Physiological signals can exhibit an amplitude ranging from tens of microvolts ($\mu$V) to tens of millivolts (mV) with a bandwidth covering from near DC to a few kilohertz (kHz). For example, neural activity can produce spikes with amplitude as low as 10–20 $\mu$V and frequency bands of interest as high as 5 kHz, local field potentials (LFPs) can show amplitude on the order of a few millivolts with a bandwidth from 10 to 200 Hz, and EMG signals can be as large as tens of millivolts in amplitude. Therefore, it is challenging to design one analog front-end system that is compatible with a wide range of implant recording applications due to constraints on system power dissipation. In general, an analog front-end system is designed for a specific application with well-defined input signal characteristics, dynamic range, and required data rate while minimizing power dissipation. For simulation, the front-end system is designed to deliver biphasic electrical stimulation pulses across targeted nerves or muscles. Design attentions are devoted to ensure a well-controlled stimulation current level and proper operation of stimulation driver circuits. In certain applications, the impedance between a selected pair of stimulation electrodes is monitored during stimulation to indicate the electrodes' condition and to minimize stimulator power dissipation.

## 3.3 FRONT-END AMPLIFIER DESIGN

Physiological signals have most of their energy concentrated in low-frequency band. Therefore, they are susceptible to $1/f$ noise generated by the front-end amplifier. One common technique based on chopper-stabilized has been widely employed to suppress $1/f$ noise contribution at the front-end system output. Figure 3.2 depicts a general architecture of a chopper-stabilized design, where the incoming signal is first modulated by a high-frequency carrier signal, $f_c$, followed by an amplification stage. The output of the amplifier will be filtered by a band-pass filter (BPF) and demodulated to baseband.

A low-pass filter (LPF) is then applied to obtain the low-frequency signal of interest and reject the modulated amplifier's $1/f$ noise to ensure an adequate signal-to-noise ratio at the system output. This technique is effective to minimize the $1/f$ noise contribution, however, requiring an increased design complexity and power dissipation associated with the large bandwidth requirement for the amplifier.

Another low-noise front-end architecture has been adopted in bioimplantable systems applications. Figure 3.3 presents a general design topology, where an input physiological signal, $V_{in}$, detected by a pair of sensing electrodes is amplified through a capacitive feedback circuit consisting of an operational transconductance amplifier (OTA) and capacitive feedback elements $C_1$ and $C_2$. The $C_p$ shown in the figure represents the parasitic capacitance at the amplifier input terminal, which is mainly composed of OTA's input capacitance and parasitic capacitances associated with $C_1$ and $C_2$.

The input signal is capacitively coupled to $C_1$. Thus, any input DC offset, for example, offset signal from the sensing electrode–tissue interface, is removed. Since

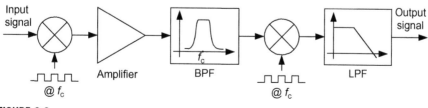

**FIGURE 3.2**

Chopper-stabilized front-end design architecture.

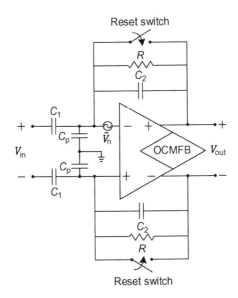

**FIGURE 3.3**

Capacitive feedback amplifier topology.

$C_1$ is in series with the electrode–tissue impedance, it should be made much smaller than the electrode–tissue impedance in order to minimize signal attenuation.

OTA's output common-mode voltage is typically set by an output common-mode feedback (OCMFB) circuit. The resistors, $R$, placed in parallel with $C_2$ are used to set the input common-mode voltage of the OTA. Consequently, the resistors and $C_2$ introduce a low-frequency cutoff, $f_L = \dfrac{1}{2\pi R C_2}$, for the amplifier. The resistors are typically implemented by using MOS–bipolar structures, which provide an area-efficient means of generating a large small-signal resistance. For example, resistance values larger than $10^{12}\,\Omega$ can be obtained with a resulting low-frequency cutoff on the order of 10 mHz, which is adequate for detecting LFPs. The large time constant associated with the low-frequency cutoff can, however, cause the amplifier to recover slowly from a large transient response. This undesirable effect can be mitigated by employing a reset switch in the feedback path to achieve a fast settling. Figure 3.4 presents the frequency response of the capacitive feedback amplifier, where the closed-loop midband gain, $A_v$, is determined by the ratio of $C_1$ and $C_2$, and the high-frequency roll-off is determined by $f_H = \dfrac{1}{A_v f_u}$, where $f_u$ is the open-loop unity-gain frequency of the OTA. In certain applications, $C_2$ may be implemented by a digitally controlled capacitor array to realize an adjustable midband gain to accommodate input signal amplitude variations over time.

The amplifier noise contribution can be modeled by an input-referred voltage noise generator, $V_n$, as shown in Figure 3.3. The closed-loop amplifier total output

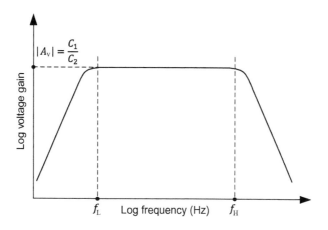

**FIGURE 3.4**

Capacitive feedback amplifier gain versus frequency.

voltage noise $\tilde{V}_{\text{out-n}}$ thus can be determined as $\tilde{V}_{\text{out-n}} = \tilde{V}_{\text{n}} \dfrac{1}{f} \sqrt{f_{\text{N}}}$, where $f$ is the

feedback factor of the amplifier set by $f = \dfrac{C_2}{C_1 + C_2 + C_{\text{P}}}$ and $f_{\text{n}}$ is the amplifier noise

bandwidth determined as $f_{\text{n}} = \dfrac{\pi}{2} f_{\text{u}} f$. This noise calculation is based on an assumption that the amplifier's noise is dominated by its thermal noise sources. Once the amplifier total output voltage noise is determined, an equivalent input voltage noise, $\tilde{V}_{\text{n-in}}$, can be readily derived to estimate the minimum detectable signal (MDS) at the system input as expressed by $V_{\text{MDS}} = \tilde{V}_{\text{n-in}} = \tilde{V}_{\text{n}} \dfrac{1}{f} \sqrt{f_{\text{N}}} \dfrac{1}{A_{\text{v}}}$.

An OTA can be designed in a single-stage or a multistage configuration depending on application requirements. Figure 3.5 presents a single-stage fully differential OTA design topology, which offers design simplicity with low-noise and low-power characteristics. P-channel MOSFETs, $M_1$ and $M_2$, are used to implement the input differential pair to achieve a reduced $1/f$ noise contribution. This is critical for sensing physiological signals with small amplitudes. The input transistors can be biased in subthreshold region to obtain an adequate small-signal transconductance, $g_{\text{m1}}$ or $g_{\text{m2}}$, without consuming a large DC bias current, hence low power dissipation. The low DC bias current will result in a reduced amplifier slew rate, which however can be accommodated in most biomedical applications due to the relatively low-speed requirements. The output loading capacitor, $C_{\text{L}}$, models OTA's output parasitic capacitance, feedback network capacitance, and input capacitance from subsequent stage. OTA's differential unity-gain frequency, $f_{\text{u}}$, can be determined as $f_{\text{u}} = \dfrac{g_{\text{m1}}}{2\pi C_{\text{L}}}$.

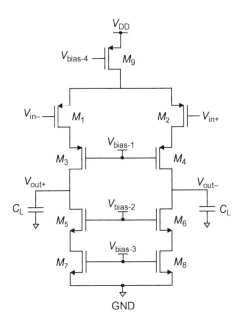

**FIGURE 3.5**

Single-stage OTA design topology.

It should be noted that an accurate determination of $C_L$ is important for achieving an accurate unity-gain frequency without the need of tuning OTA's small-signal transconductance. The amplifier input-referred thermal voltage noise power spectral density, $\tilde{V}^2_n$, can be estimated as $\tilde{V}^2_n = \frac{16}{3} k_B T \frac{1}{g_{m1}} \left(1 + \frac{g_{m7}}{g_{m1}}\right)$, where $k_B$ is Boltzmann's constant, $T$ is the absolute temperature in kelvin, and $g_{m7}$ is the small-signal transconductance of $M_7$. In a typical design where $g_{m1}$ is larger than $g_{m7}$, the amplifier input-referred thermal voltage noise power spectral density can be simplified as

$$\tilde{V}^2_n = \frac{16}{3} k_B T \frac{1}{g_{m1}} .$$

The single-stage architecture suffers from a limited open-loop gain and output signal swing range. The limited open-loop gain becomes a much pronounced drawback for achieving certain amplifier settling accuracy requirement in a modern CMOS process. To address these challenges, multistage design topologies are employed. Figure 3.6 shows a two-stage fully differential OTA design, where the first stage is similar to the topology presented in Figure 3.5. The first stage differential outputs drive the second stage in a complimentary manner to create a class AB operation. For example, the first stage differential outputs control the gates of $M_{10}$ and $M_{14}$ (or gates of $M_9$ and $M_{13}$) to increase the small-signal output current from the second stage, thus enhancing the overall transconductance without dissipating additional power. The enhanced transconductance from the second stage can improve the stability

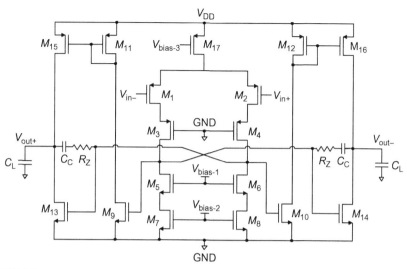

**FIGURE 3.6**

Two-stage OTA design topology.

of the overall circuit while driving a large output capacitive load. It can be shown that the amplifier input-referred thermal voltage noise power spectral density can be estimated as $\tilde{V}^2{}_n = \dfrac{16}{3} k_B T \dfrac{1}{g_{m1}}$, and the differential unity-gain frequency, $f_u$, can be determined as $f_u = \dfrac{g_{m1}}{2\pi C_c}$, where $C_c$ is the internal compensation capacitor shown in Figure 3.6, which is a well-controlled design parameter. The general design considerations described for the single-stage OTA design is also applicable for the multistage topology.

A number of physiological signals, for example, body temperature and BP, require the corresponding baseline information or DC signal to be detected and amplified. The interface circuit configuration presented in Figure 3.3 will not be adequate for such applications. Therefore, corresponding analog front-end designs are required for detecting such signals. In this chapter, BP sensing is used as an example for the discussion. BP can be detected by using piezoresistive pressure sensors employing microelectromechanical systems (MEMS) typically implemented in a Wheatstone bridge configuration. Figure 3.7 presents a simple front-end interface design, where a piezoresistive pressure sensor, powered by a DC supply voltage, delivers an output differential voltage signal representing the detected BP waveform including the DC baseline information. This signal then becomes the input signal, $V_{in}$, to a resistive feedback amplifier, which exhibits a gain factor of $-\dfrac{R_2}{R_1}$. The resistive feedback amplifier allows the BP waveform together with its baseline level to be amplified.

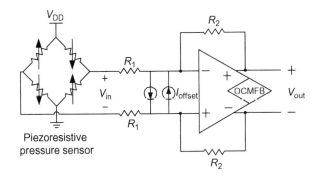

**FIGURE 3.7**

Piezoresistive pressure sensor interface electronics for BP detection.

The amplifier output common-mode voltage is designed to match the sensor output common-mode level to eliminate power dissipation through the resistive feedback network. In addition, input offset currents are applied between the amplifier input terminals to null out any residual offset voltage due to the sensor elements mismatch and amplifier inherent mismatch. To minimize $1/f$ noise contribution, P-channel MOSFETs can be employed to implement the amplifier input stage as illustrated in Figures 3.5 and 3.6.

Chopper-stabilized technique can also be applied to effectively suppress the amplifier's $1/f$ noise and DC offset effects. Figure 3.8 shows a chopper-stabilized design, where the piezoresistive pressure sensor is modulated by a high-frequency carrier signal operating at $f_c$. The modulated signal is then amplified by a gain stage followed by a demodulator and a LPF to retrieve the BP signal at $V_{out}$ while substantially suppressing $1/f$ noise and DC offset at the system output.

Besides employing piezoresistive pressure sensors to detect BP, MEMS capacitive pressure sensors have been developed and applied for BP detection. The sensor capacitance changes as a function of the BP exerted on the sensor diaphragm. Capacitive sensors offer a number of advantages such as zero DC power dissipation, high sensitivity, low temperature dependence, and low turn-on drift. However, they

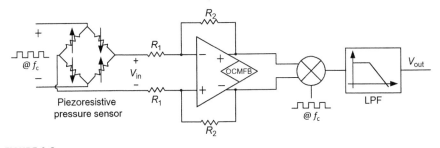

**FIGURE 3.8**

Chopper-stabilized interface design for piezoresistive pressure sensor.

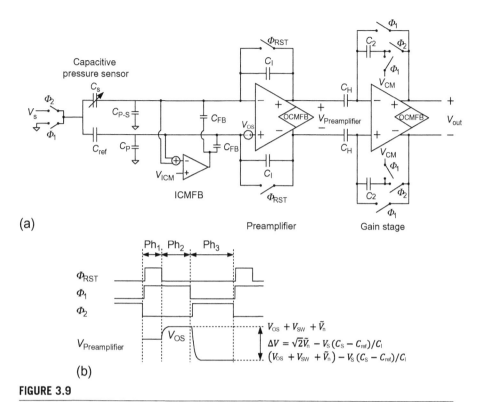

(a)    Preamplifier    Gain stage

(b)

**FIGURE 3.9**

Capacitive pressure sensor front-end interface circuit with operation timing diagram.
Capacitive pressure sensor interface circuit (a) design topology and (b) operation timing
diagram.

also exhibit certain drawbacks. For example, most capacitive pressure sensors are
single-ended devices lacking differential characteristics and are prone to parasitic
capacitance associated with long interconnect wires. Figure 3.9 presents a capacitive
pressure sensor interface topology with its operation timing diagram, where a capacitive pressure sensor, modeled as $C_s$, is interfaced with a preamplifier.

A stimulation voltage pulse with an amplitude of $V_s$ is applied during $\Phi_2$ phase to
convert the capacitance difference between $C_s$ and a reference capacitor, $C_{ref}$, to an
output voltage, $V_{Preamplifier}$. $V_{Preamplifier}$ can be determined as $V_{Preamplifier} = -\dfrac{C_s - C_{ref}}{C_I} V_s$,
where $C_I$ is an integration capacitor connected between the preamplifier input and
preamplifier output. The reference capacitor, $C_{ref}$, can be implemented by a digitally
controlled capacitor array to achieve a close match to the nominal capacitance value
of $C_s$, thus reducing offset voltage, hence the dynamic range requirement at the output of the preamplifier. The applied simulation voltage pulse will shift the amplifier
input common-mode voltage. Therefore, an input common-mode feedback (ICMFB)

circuit is necessary to ensure the input common-mode voltage to be maintained at the designed value of $V_{ICM}$. This is critical for eliminating any amplifier output offset voltage due to the mismatch between parasitic capacitances at the amplifier input terminals. In a practical design, the parasitic capacitance of the capacitive pressure sensor, $C_{P-S}$, is likely to be different from that of the reference capacitor, $C_p$, without calling for a special adjustment step.

A correlated double-sampling (CDS) technique is incorporated in this front-end interface design to suppress $1/f$ noise and DC offset from the preamplifier. Figure 3.9b presents the operation timing diagram together with the corresponding signal level at $V_{Preamplifier}$ at the end of each phase. To illustrate the operation, the preamplifier is mod-eled as an ideal amplifier with an input offset voltage of $V_{OS}$ as shown in Figure 3.9a. $1/f$ noise is included as a part of the offset voltage for a sampling frequency much higher than the noise corner frequency. The operation of the front-end circuit in-volves three phases. During phase 1, the preamplifier is reset. The differential output of the preamplifier, $V_{Preamplifier}$, is equal to $V_{OS}$. During phase 2, $V_{Preamplifier}$ becomes $(V_{OS} + V_{SW} + \tilde{V}_n)$, where $V_{SW}$ is an additional induced voltage due to the opening of reset switches, including a charge injection effect and switch thermal noise, and $V_n$ is the preamplifier output thermal noise. $V_{Preamplifier}$ is sampled onto $C_H$ at the end of this phase. During phase 3, the left plates of the capacitive pressure sensor, $C_s$, and the reference capacitor, $C_{ref}$, are connected to the simulation voltage, $V_s$. $V_{Preamplifier}$ thus becomes $\left(V_{OS} + V_{SW} + \tilde{V}_n\right) - \dfrac{C_s - C_{ref}}{C_I} V_s$. The second term of the expression car-ries the sensor information. Thus, the difference of the preamplifier output voltage between phase 2 and phase 3, $\Delta V$, is amplified by the gain stage with the gain factor set as $-C_H/C_2$. $V_{OS}$ and $V_{SW}$ are nearly constant due to a relatively high sampling fre-quency in the two phases and thus can be eliminated to the first order. The thermal noise of the preamplifier, $V_n$, is uncorrelated in the two phases. Therefore, the total thermal noise power is doubled, resulting in a differential voltage at the output of the gain stage expressed as $V_{out} = \left(\dfrac{C_H}{C_2}\right)\left(\dfrac{C_s - C_{ref}}{C_I} V_s + \sqrt{2}\,\tilde{V}_n\right)$, where $C_2$ is an integra-tion capacitor connected between input and output nodes of the gain stage.

Besides employing the CDS-based front-end topology described above, a chopper-stabilized technique can be applied for the capacitive pressure sensor inter-face circuit as shown in Figure 3.10. The same design approach and consideration presented in this chapter can be applied here.

## 3.4 SIMULATION CIRCUIT DESIGN

Simulation circuit is responsible for delivering well-controlled biphasic electrical stimulation pulses across a targeted nerve or muscle. The biphasic operation is criti-cal to ensure charge balance of the stimulating waveforms. Ideally, all the charge that is conducted through a selected pair of stimulation electrodes should be canceled

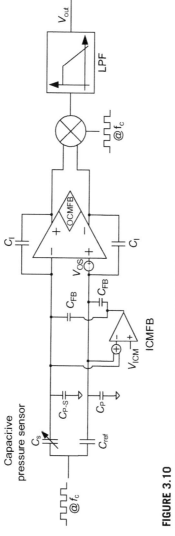

**FIGURE 3.10**

Chopper-stabilized interface design for capacitive pressure sensor.

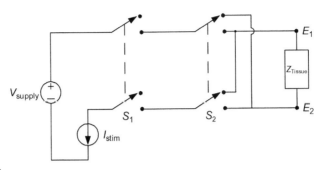

**FIGURE 3.11**

General electrical stimulation circuit topology.

exactly over time by periodic reversal of the current flow direction. As a result, the electrodes and their environment are left in a similar electrochemical state after the stimulation is completed as that existed before the simulation was applied. Figure 3.11 presents a general stimulation circuit topology illustrating the method of stimulation construction, where a DC voltage supply, $V_{supply}$, and a controlled simulation current, $I_{stim}$, are connected to a pair of stimulation electrodes, $E_1$ and $E_2$, by a set of control switches, $S_1$ and $S_2$. Switch $S_1$ is closed when the simulation current is to be delivered to the electrode pair while switch $S_2$ determines the direction of the current flow. Toggling $S_2$ between $E_1$ and $E_2$ produces a biphasic stimulation over the targeted tissue impedance modeled as $Z_{Tissue}$.

Figure 3.12 shows a corresponding stimulation circuit design schematic, which achieves the biphasic operation by $S_2$ and its complementary switch, $\bar{S}_2$. The

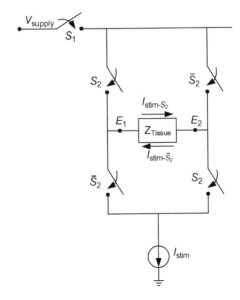

**FIGURE 3.12**

Simulation circuit design schematic.

stimulation current source can be digitally controlled to produce various required amplitude and waveform characteristics. To further ensure there is no any residual DC current flowing between the electrodes, a blocking capacitor is typically inserted between each electrode and an output node of the stimulation circuit, which is not show here for simplicity.

## 3.5 TELEMETRY SYSTEM INTRODUCTION

Telemetry system is a critical part of a bioimplantable systems design. The system not only needs to receive external command to control the operation of an implant but also receives external power to energize an implant in a battery-less manner or to recharge an implanted battery. Various techniques based on RF and ultrasound have been demonstrated. The RF approach can couple an appreciable amount of power to an implanted antenna positioned relatively close to the skin surface, while the ultrasound means can transfer power deeper into the tissue but deserves a careful consideration of impedance matching between an ultrasound transceiver and the surrounding tissue. In this chapter, the RF approach will be chosen as the focus for discussion. Due to implant size and weight constraints, the telemetry system needs to be highly miniaturized, thus presenting a design challenge for achieving an efficient operation. Besides receiving external command and power, the system is also responsible for wirelessly transmitting data to an external receiver. Various design requirements, for example, transmission distance, data rate, and power dissipation, call for an optimized design in terms of telemetry topology and implementation trade-offs.

## 3.6 RF POWER TRANSFER CIRCUIT

Inductively coupled coils have been widely employed to transfer RF power for bioimplantable systems applications. Figure 3.13 presents an inductively coupled RF power transfer architecture, where an external RF power source, $V_{in}$, drives a tuned series resonator consisting of $L_1$ and $C_1$. The resistor, $R_1$, represents the overall series resistance associated with the resonator including the output resistance of the power source. The resonator is tuned to an optimal frequency to maximize the power coupling efficiency.

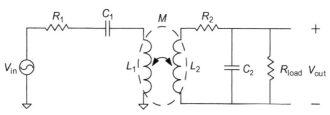

**FIGURE 3.13**

Inductively coupled RF power transfer architecture.

The operating frequency can be determined by the coil's geometry, separation distance, and loading requirement but is limited by the self-resonance of the inductors, which is typically on the order of a few megahertz (MHz).

The RF signal is coupled to a parallel resonator, consisting of $L_2$ and $C_2$ tuned to the same frequency with a total loop resistance of $R_2$. As a result, an AC voltage signal, $V_{out}$, is developed across $R_{load}$, which models the loading from the proceeding stages, for example, voltage rectifier, regulator, and other associated implant electronics. The voltage gain across the tuned resonators network can be expressed as

$$V_{out} / V_{in} = \frac{\omega^2 L_2 M}{\left( R_1 R_2 + (\omega M)^2 + R_1 (\omega L_2)^2 / R_{load} \right)}, \text{ where } M \text{ represents the mutual induc-}$$

tance between the two coil inductors, $L_1$ and $L_2$, and $\omega$ is the operating frequency. It can be shown that reducing $R_1$ and $R_2$, hence lowering loss for $L_1$ and $L_2$, and enhancing $M$ can increase the voltage gain in practical designs. Using metal wires with larger diameter to construct the coils can minimize the resistive loss, but with a penalty of increased size and reduced self-resonant frequency. A large mutual inductance can be obtained with large coil dimension and small separation distance, but often constrained by physical requirements of an implant. It can be shown that an optimal power coupling efficiency can be achieved when the loading resistor, $R_{load}$, matches the equivalent loading of the parallel resonator at the desired operating frequency. To further improve the efficiency, nonlinear driver circuits such as class E amplifiers have been used to deliver power into the series resonator. Recently, multicoil configurations have also been demonstrated to achieve substantially enhanced power coupling efficiency. The induced AC voltage signal across $R_{load}$ will be further rectified and regulated to produce a stable DC voltage to power the implant system. Devices with low rectification voltage such as Schottky diodes and threshold voltage cancellation techniques should be considered in the rectifier design to improve overall system efficiency.

## 3.7 DATA TELEMETRY CIRCUIT

The inductively coupled RF power transfer architecture shown in Figure 3.13 can be used to transfer command data from an external control unit to the implant and to telemeter the monitored physiological data from the implant to an external receiver. Amplitude-shift keying (ASK) modulation is typically employed to impose the command data onto a carrier frequency as depicted in Figure 3.14. This signal drives the series resonator shown in Figure 3.13 to develop an induced voltage swing across $R_{load}$, which exhibits a similar waveform shape, thus allowing a peak detector to readily retrieve the command data. The received ASK signal can also be processed to produce a digital clock for the implant synchronization.

The physiological data can be passively back-telemetered to an external receiver by using the circuit topology shown in Figure 3.13. Figure 3.15 shows that a telemetry capacitor, $C_{Tele}$, can be digitally controlled by a data stream to achieve a phase-shift keying (PSK) telemetry signal, while a telemetry resistor, $R_{Tele}$, can be

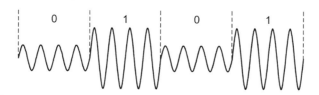

**FIGURE 3.14**

ASK command data waveform.

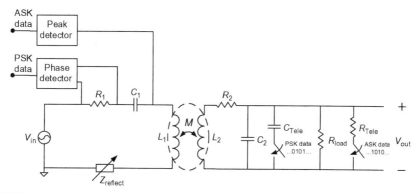

**FIGURE 3.15**

Passive data back telemeter design architecture.

used to realize an ASK telemetry signal. The switching of $C_{Tele}$ and $R_{Tele}$ introduces a reflected impedance, $Z_{reflect}$, in the serial resonator, which results in a voltage signal across $R_1$ to exhibit phase-shift characteristics. Consequently, a phase detector can be applied to retrieve the PSK data from the implant. It can be shown that the voltage across $L_1$ and $Z_{reflect}$ can be processed by a peak detector to obtain the corresponding ASK data.

In most bioimplantable systems applications, a single-channel telemetry will be sufficient. The two-channel telemetry scheme presented here is attractive for applications with signals exhibiting different data rate, for example, a high data rate neural activity signal with a low data rate body temperature. The signals with different data rate can be separated into two different channels and Manchester-encoded for spectral separation to minimize cross-channel interference. Time multiplexing should be considered if the same telemetry topology is used for transmitting command data and back telemetering the implant data in order to avoid potential data collision and loss. The passive data telemetry scheme consumes little power, which is highly desirable for bioimplantable systems applications, but suffers from a limited data rate and telemetry distance. Furthermore, it should be noted that the PSK telemetry causes a resonant frequency detuning of the parallel resonator, thus reducing the power coupling efficiency. The ASK telemetry, although does not detune the resonant

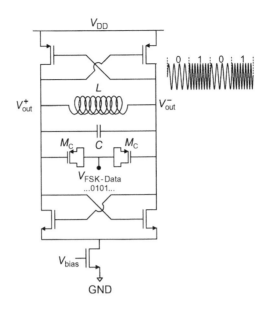

**FIGURE 3.16**

Active FSK data transmitter based on RF VCO design.

frequency, alternates the load resistance presented to the resonators network, resulting in an efficiency degradation.

To wirelessly transmit implant data exhibiting a relatively high data rate to an external receiver over an appreciable distance, active telemetry circuits are required. Figure 3.16 presents an active transmitter circuit based on an RF voltage-controlled oscillator (VCO) to achieve a frequency-shift keying (FSK) modulation. Cross-coupled MOSFETs are employed to form a positive feedback loop to establish a steady-state oscillation with a frequency determined by the resonance of the LC tank. MOSFET-based voltage-controlled capacitors, $M_C$, are used to digitally switch the oscillation frequency between two designed states, thus achieving an FSK operation as shown by the oscillator output waveform in Figure 3.16. The inductor, $L$, is used as an antenna to radiate the RF signal. Typical transmission frequencies within the IMS band, for example, 433 and 915 MHz, are chosen for the oscillator's design. The inductor can be integrated on chip to reduce the overall system size but suffers from a low-quality ($Q$) factor, thus resulting in a high power dissipation. High-$Q$ discrete coil inductors have been considered in a number of bioimplantable systems designs to minimize system power dissipation, however, requiring a careful attention to its parasitic capacitance and trace inductance in order to achieve an accurate oscillation frequency. Depending on applications and system requirements, active telemetry circuits can be duty-cycled to minimize power on-time to save power. Other modulation schemes such as ASK and on–off keying can also be considered for the transmitter design.

## 3.8 SUMMARY

This chapter presents general design methods and considerations for analog front-end and telemetry systems with a focus on bioimplantable systems applications. A number of analog front-end topologies are illustrated with a description of their operating principle, frequency response, noise analysis, noise suppression techniques, and relevant design trade-offs. An inductively coupled power transfer configuration is also presented together with various built-in passive data telemetry schemes. In addition, an active data transmitter design based on an RF VCO is described to achieve high data rate telemetry over an appreciable distance. Various bioimplantable systems applications call for different design requirements and constraints. The design approaches and considerations presented in this chapter should serve as a basis to achieve an optimized system design.

## ACKNOWLEDGMENT

The author would like to thank Qingbo Guo from the University of Utah for preparing the figures used in this chapter.

# Signal processing hardware

# 4

**Arnaldo Mendez, Mohamad Sawan**

*Polystim Neurotechnologies Lab, Department of Electrical Engineering, Polytechnique, Montreal, Quebec, Canada*

## CHAPTER CONTENTS

## 4.1 INTRODUCTION

Biomedical implantable microchips and biosensors are used mainly for prosthetic purposes and for the monitoring or enhancement of the body functions. These devices record biomedical signals of different nature: biopotentials such as the electrograms recorded from the heart (electrocardiogram, ECG), the brain (electroencephalogram, EEG), the retina (electroretinogram, ERN), the muscles (electromyogram, EMG), the nerves (electroneurogram, ENG), etc.; biomechanical such as pressure, volume, flux, displacement, acceleration, angle, etc.; electrochemical using biosensing principles such as potentiometric, conductometric, and voltammetric; and other types of variables such as heat and temperature.

Bhunia et al. Implantable Biomedical Microsystems. http://dx.doi.org/10.1016/B978-0-323-26208-8.00004-2

The signals recorded by these bioimplantable systems are transduced into electronic currents by specialized sensors forming part of the device. These sensors typically generate weak analog signals that are also contaminated by noise produced by themselves and internal and external sources to the body. Therefore, a signal conditioning to amplify and filter out undesired noise and artifacts is needed before starting any signal processing. The analog signal conditioning is not a signal processing that allows interpreting the information conveyed by the recorded signal but an improvement of the signal. An additional signal processing is required to detect, extract, modify, or decode the relevant information carried by the signal for its use in the envisaged application.

The design of the hardware deploying mathematical algorithms used to process the signal is driven by the particular requirements of the application. Although the signal processing may use general-purpose hardware to perform analyses in the time or frequency domains, most of the time, bioimplantable systems require a specialized hardware that maximizes the system performance while employing a very limited amount of energy and fitting to a highly restrained space inside the body. As part of the implanted device, the signal processing hardware has to ensure not only clinical effectiveness but also safety. Power required for the implant operation plays an important role in both. Low power consumption ensures long duration of the stored energy, lower heat dissipation, and lower electromagnetic flux associated with the implant operation and communication, which are required to prevent damage of the tissue surrounding the implant. Power has also a direct impact on the size, weight, and cost of the implant. On the other hand, a small implant size allows for easier and safer chirurgic implantation procedures and reduces adverse immunologic responses, that is, an improvement in its biocompatibility, among other advantages.

The signal processing can be performed on the analog domain or in the digital domain using the corresponding hardware technology. Both of them provide not only advantages but also drawbacks that the hardware designer should trade-off with the system specifications to find the best approach that meets the end user's needs.

This chapter introduces general principles that a designer of signal processing hardware should consider to improve the effectiveness and safety of the electronic-based implantable devices. It starts with the analysis that should be done to define the hardware architecture partitioning between analog and digital domains, followed by an introduction to design principles of analog, digital, and mixed-signal processors. Some circuits implementing processors of each category are presented to bring out the advantages of each approach in real applications.

## 4.2 HARDWARE ARCHITECTURE OF THE SIGNAL PROCESSING SYSTEMS

The definition of the hardware architecture for signal processing system is crucial to achieve the expected results. Since the early beginning of the design of the signal processing algorithms, a concurrent analysis should be performed. This is not a trivial task and many factors should be considered carefully throughout the design process. Without pretending, make an exhaustive list all of the possible factors; below, you will find some that you should consider when defining the hardware architecture:

- Target specifications that prioritize low power, low noise, and small size
- Optimal system partitioning (analog, digital, or mixed)
- Circuit topologies that take advantage of the technology degrees of freedom
- Both fundamental and incidental interactions among the system components, for example, among analog and digital circuits and low- and high-frequency subsystems
- Trade-off between reusability of some modules and overall system performance, that is, an analysis of the pros and cons of choosing a modular architecture
- Hardware flexibility or the ability to update or redesign the system to meet ongoing requirements, for instance, as the biomedical research process advances, new users' needs are identified, or some improvement need to be implemented
- The system implementation and exploitation costs
- Development easiness and time of development allowed

Although bioimplantable systems are used in applications with different purposes, all of them have to deal with specifications that frequently show a nonsynergistic relationship; that is, improving one specification can lead to deterioration of others. Often, high-performance systems need relatively high power levels to function properly. However, a common concern in the implantable electronic systems is a limited power source either from a nonrechargeable battery, like in current cardiac pacemakers, or from wireless energy provided by an external unit outside the body, like in current cochlear implants. Power dissipation of the implanted hardware should be also limited to prevent heating and damage of the surrounding tissue. The upper limit suggested by most of the long-term studies is a 2 °C temperature increase of 40 mW/cm$^2$ heat flux, with 1.6 mW/g of the specific absorption rate (SAR), which is a measure of heating produced by the electromagnetic field in tissue [1]. Thus, both power consumption and power dissipation should be at the top of the priorities when choosing the optimal architecture for the implantable system.

Noise and interference need also to be carefully considered when defining the hardware architecture and choosing the proper technology. The signal processing hardware of bioimplantable systems need to process signals from artificial sensors operating with very low level of voltages and currents or from natural sensors (neural receptors) producing biopotentials ranging from few microvolts to few millivolts. These low-power signals are more likely to be affected by internal noise, mainly the thermal and 1/$f$ noise, as well as by interferences from other external sources inside and outside the body. A comprehensive analysis of noise in CMOS-based circuits, which are ubiquitous in current bioimplantable systems, can be found in Refs. [2,3]. There are also several approaches to prevent or attenuate interference in biomedical devices that are introduced in Ref. [4].

Another factor to consider is the amount of information to process by the implantable hardware. More information to process in a limited time frame turns out in more power consumption. A proper selection of the optimal architecture and hardware technology to implement technology-optimized algorithms can reduce significantly the amount of information to process and ultimately the total power consumption. The technology-optimized algorithms should be conceived during an iterative design process that implies successive changes in the architecture, the circuit technology, and the algorithms until the desired system performance is achieved. Additionally, the reduction in the amount of information to process helps to release analog-to-digital converter (ADC) specifications, the processing time, and the computing precision required to obtain the same information with similar precession at the output without the need of increasing the processing speed.

Throughout the design process, the designer should ensure that the system just meets the specifications with a proper safety margin that cover nonpredictable operating conditions. Indeed, a hardware performance that excessively exceeds one or few target specifications, for instance, processing speed, signal-to-noise ratio (SNR), and the number of bits required to achieve the process precision or the target output error, will not necessarily improve the overall system performance but may impact negatively other system specifications. Outdoing the optimal values found during the identification of system specifications will not allow an efficient use of the limited power available and likely will increase the hardware implementation costs.

The information contained in raw signal acquired from implanted sensors or electrodes can be processed using one of the two typical architectures shown in Figure 4.1. In the architecture shown in Figure 4.1a, the raw signal previously conditioned (typically preamplified, amplified, and filtered) by front-end analog circuits

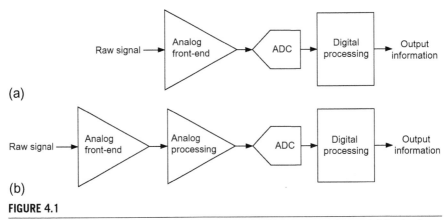

(a)

(b)

**FIGURE 4.1**

General architecture of signal processing systems. (a) Digital processing and (b) mixed-signal processing.

is digitized by an ADC and finally processed using a digital system, which output the information sought. Figure 4.1b shows an alternative architecture where the conditioned analog signal is preprocessed using analog circuits that reduce the amount of information to be processed subsequently by the digital system.

To achieve the best system performance considering concurrently precision, efficiency, and reliability, a suitable partitioning of the analog signal processing and the digital signal processing is required. The goal is to use the optimal architecture, technology, and circuit topology to process the information such that the overall power consumed and dissipated is reduced without sacrificing the precision of the output information and the system reliability over time.

Both analog computation and digital computation present advantages and disadvantages over its counterpart for signal processing, which is useful to keep in mind when deciding the best hardware partitioning for analog signal processing and digital signal processing. Both alternatives are briefly discussed below. The information can be represented by analog (continuous) variables without any loss, whereas digital variables are subject to intrinsic information loss due to discretization that produces quantization errors (distortions). Analog signal processing can exploit the physical laws that drive the behavior of the bipolar or field-effect transistors, the circuits' topology (Kirchhoff's voltage and current laws), and passive component properties to perform computation based on those laws, this way maximizing the amount of information that can be processed on average by each transistor. In contrast, the transistor in digital systems switches between two states and only logical (Boolean) functions can be executed by basic gates that are composed of several transistors. However, digital system parameters are less sensitive to offsets, power supply noise, and thermal drift than analog systems are, and the resulting errors do not accumulate through multiple processing stages. Analog computation is overall more power-efficient than digital computation, but this latter is intrinsically more robust. Also keeping the precision and reliability in analog system is much more difficult and costly when highly complex algorithms are executed. On the other hand, analog implementation of complex algorithms would take a nonpermissible development time for most of beyond-the-lab applications.

Bearing in mind all of this factors, the designer has the hard task of choosing the optimal partitioning between the analog signal processing and the digital signal processing. Excessive analog preprocessing will increase development and implementation costs due to the extra time and hardware required to keep the target output precision and circuit robustness, that is, insensitivity to manufacturing process variations, power supply noise and other external noise, and cross talk among channels. Conversely, excessive digital processing would be inefficient because the degrees of freedom provided by the analog circuits are ignored and a lot of meaningless information would needlessly be processed.

There are general principles that can be considered to design mixed-signal systems to reduce power consumption without deteriorating the system performance. Some of them are summarized below (a comprehensive analysis of these general principles can be found in Ref. [5]):

**(1)** Find the circuit system topology that encodes computation efficiently considering the physics laws of the technology chosen to enhance SNR in analog system and to minimize switching and leakage power in digital systems.

**(2)** Preprocess the analog signal before its digitization to reduce meaningless information considering the specific task to be executed, the required precision, technology, and circuit topology, for example, by using automatic-gain-control (AGC) circuits in analog preprocessing to reduce the internal dynamic range (DR) as in Ref. [6].

**(3)** Use compression techniques, for example, compressed sampling [7] for digital signal processing, to reduce the amount of the information to be subsequently processed.

**(4)** In CMOS-based circuits, not only operate the transistor in the subthreshold regime (weak inversion) to optimize power consumption but also consider the known limitations of working in this zone to minimize them by using special circuit topologies, such as source degeneration and bump linearization techniques, as shown in Ref. [8].

**(5)** Operate the system as slow as possible to reduce power dissipated during transistor switching and, if possible, try to separate high-speed from high-precision circuits.

**(6)** Favor the use of N-parallel structures of lower complexity and speed rather than a single complex structure operating at high speed.

**(7)** Use feedback and feedforward techniques to improve robustness and energy efficiency by reducing correspondingly the unpredictable and predictable errors in closed-loop systems, which in turn reduce the average closed-loop gain over time, as shown in Ref. [14].

## 4.3 ANALOG, DIGITAL, AND MIXED-SIGNAL PROCESSORS

Besides the optimal point of hardware partitioning between analog and digital circuits, another important choice has to be made when designing signal processors: the degree of technology integration to be employed in the hardware implementation. To implement the signal processor, the system designer can choose among commercially available miniaturized integrated circuits (ICs) (general purpose and application-specific integrated circuits (ASICs)), full-custom ASICs, or a combination of both. This choice will certainly influence the design at all system levels and may also compel some modifications of the original signal processing algorithms.

To select the right technology, the system designer should consider on the one hand the constringent specifications of chronically implanted medical devices concerning the effectiveness, power consumption, safety, and size of the implant and on the other hand the hardware implementation cost, the overall development time, and the ability to redefine or improve the signal processing algorithms executed by the hardware during the R&D process (i.e., flexibility), among other factors mentioned earlier in this chapter. Once again, this choice is not a trivial task, and frequently, several iterations during the system design are needed to find the technology that best fits the proposed algorithm precision requirements and best solves the trade-off among these two groups of nonsynergistic specifications.

The use of commercially available miniaturized ICs allows for shorter development time and lower costs for low- to medium-scale production. Most of the electronic-based medical implants fall into this category of production volume, except for cardiac pacemakers [9]. Additionally, by using off-the-shelf ICs, some system modifications can be achieved by changing specific circuits or modules without incurring in costly and long redesign processes. However, power consumption and size of systems designed using off-the-shelf ICs do not generally meet the stringent medical implant specifications, especially for the analog processors. In contrast, custom ASICs allow considerable cost reductions in large-scale productions, and the specifications can be met when the right algorithms are designed and the hardware partitioning point is carefully chosen. Nevertheless, as drawbacks, the development time of custom ASICs is longer and more expensive. Moreover, flexibility to accommodate modifications is very limited, and the cost for low- to medium-scale productions can be significantly higher.

In spite of the current limitations due to the high complexity of the design process, the lack of flexibility, and the higher costs for low- to medium-scale production, custom ASIC processors are chosen frequently in current works to implement signal processing hardware due to their advantages to achieve lower power consumption and smaller size when compared to processors built using off-the-shelf ICs. In the following sections, some examples of signal processor designs illustrating the current tendency of choosing custom ASICs for signal processing in bioimplantable systems are presented.

### 4.3.1 SIGNAL PROCESSING USING ANALOG CIRCUITS

Analog signal preprocessing beyond the initial front-end conditioning stages can contribute to reduce significantly the amount of information to be subsequently processed and the power consumption. Different approaches are shown in this section to demonstrate some advantages of the analog signal preprocessing such as in neural signal processing implants used in both neuroscience research and to control prosthetic devices, and also in cochlear implants and implantable cardiac monitors.

#### 4.3.1.1 Analog signal processing of neural signals

Neuroprosthetic devices used in neurological research and in brain–machine interfaces (BMIs) have to record simultaneously in real time an increasing number of signals produced by neurons in the central or peripheral neural systems. Neural signals are typically detected by microelectrode arrays (MEAs) of large number of channels. Depending on the application, MEAs along with analog signal front-end conditioning circuits can record many channels at time [10–17]. The sensory information is encoded in the firing-rate patterns of action potentials (voltage spikes) produced by neurons over time, as described among other in Refs. [18,19]. Typical sampling rate of neural signals can vary from 15 to 30 kS/s, which means that with an 8-bit analog-to-digital conversion, the resulting data bandwidth would fall in the range of 12 and 24 Mb/s. This amount of data requires an excessive power consumption to be processed, and the power dissipation would heat the surrounding tissue above the allowed limits ($<2\,°C$). Furthermore, wireless transmission of the recorded data to be further processed by an external unit outside the body would not meet either the system specifications for a safe operation of the implant (power density $<40\,mW/cm^2$ and SAR $<1.6\,mW/g$). Therefore, data reduction approaches have been used in the past to address this problem in neural implants using ultralow-power analog signal processing circuits.

One of these approaches is to detect and extract specific snips of the signal, that is, the spikes generated by firing neurons, and remove the rest of the recorded signal that contains noise and meaningless information to reduce the data bandwidth, as in Ref. [20]. The extracted signal can be fed to spike classification systems to identify the firing neurons and decode the neural information generated by a single neuron or by a cooperative network of neurons. These processes are known as spike sorting and neural decoding, respectively. The use of analog processors for learning and decoding tasks that allow data compression is another approach proposed to reduce power consumption when processing neural signals in BMI applications [14]. Both approaches are presented below.

### Low-power analog processor for automatic neural spike detection

The analog signal processor for automatic neural spike detection of Figure 4.2 allows reducing the amount of data to be processed by extracting the spikes from the recorded signal of each channel and removing noise and meaningless data [20].

In Figure 4.2, the amplified and filtered neural signal ($V_{in}(t)$) is fed to both the preprocessor block (PPB) and the time delay block (TDB). The PPB implements the Teager energy operator (TEO) that improves the SNR of the signal by increasing

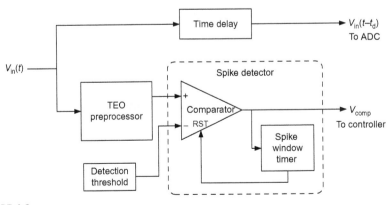

**FIGURE 4.2**

Simplified block diagram of spike detection technique using analog processing.

the amplitude of the spikes and reducing the amplitude of the background noise and interfering signals (artifacts) arising from, for example, distant neurons, nearby muscles (EMG), and breathing and cardiac activity (ECG). The TDB delays the signal the time ($t_d$) elapsed between the onset of the spike and the time when signal crosses the detection threshold ($t_d = 0.6\,\text{ms}$, determined experimentally). The TBD output signal is fed to the spike detector block (SDB) that compares it with the detection threshold. When the spike is detected, $V_{\text{comp}}$ is asserted to signal to the implant controller to capture the whole spike waveform from the delayed signal output ($V_{\text{in}}(t-t_d)$), this way preventing the loss of the first segment of the spike below the threshold. The detection threshold should be set over the background noise so that the probability of true detections is maximized and the probability of false detections is minimized, for example, by using a scaled value of the signal standard deviation as described in Ref. [21] or by using an adaptive threshold circuits as proposed in Ref. [22].

The SDB of Figure 4.2 uses the nonlinear energy operator (NEO) known as TEO, which was introduced for digital signal processing by Kaiser [23]. This operator estimates the instantaneous energy of the signal taking into account both time and frequency domain properties. The TEO analog version is

$$\psi\left[V_{\text{in}}(t)\right] = \left(\frac{dV_{\text{in}}(t)}{dt}\right)^2 - x(t)\left(\frac{d^2V_{\text{in}}(t)}{dt^2}\right) \tag{4.1}$$

where $\psi[x(t)]$ is the estimated instantaneous energy and $V_{\text{in}}(t)$ is the input neural signal. This low-complexity but effective energy operator that improves significantly the SNR is well adapted for analog circuit implementation as shown in Figure 4.3. The TEO preprocessor implementation, fully described in Ref. [20], uses two differentiators, a four-quadrant multiplier and a summing circuit. The quadratic term in Equation (4.1) is obtained using the four-quadrant multiplier with the same signal at the inputs (Diff$_1$). This circuit is implemented using a compact version of a Gilbert multiplier [24]. The second-order derivative term in Equation (4.2) is

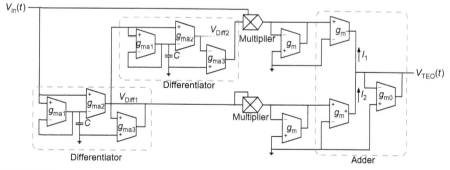

**FIGURE 4.3**

Simplified schematic of the analog processor implementing the TEO operator in Equation (4.1).

achieved by cascading the two differentiators (Diff$_2$). The transfer function of each differentiator is

$$\frac{V_{\text{Diff}}}{V_{\text{in}}} = \frac{j\omega C}{j\omega C + g_{\text{ma1}}} \cdot \frac{g_{\text{ma2}}}{g_{\text{ma3}}} \tag{4.2}$$

where $g_{\text{m}}$ is the gain of the operational transconductance amplifiers (OTAs). Two current-summing OTAs with gain $g_{\text{m}}$ and opposite current output perform the difference between the two terms. Finally, the $g_{\text{m0}}$-OTA converts the output current to voltage, $V_{\text{TEO}}(t) = (I_2 - I_1)/g_{\text{m}}$.

The TDB in Figure 4.2 is implemented by a 9th-order linear-phase OTA $G_{\text{m}}$–$C$ filter with a constant group delay response to prevent spike distortion. This delay filter allows for a proper time shifting of the input signal to capture the entire waveform when the SDB indicates the detection of a spike.

The decision block in Figure 4.2 is implemented by a latched comparator and a 5-bit binary counter clocked at 16 kHz to generate the spike time window of 2 ms. The comparator consists of a kickback noise-free preamplifier followed by a track-and-latch stage operated with low drain currents ($I_{\text{D}}$).

All of the OTAs used in the abovementioned circuits were implemented using CMOS transistors operating in subthreshold regime (week inversion) with currents of few nA. As mentioned earlier in this chapter, this operation regime allows for reducing the power consumption but at the cost of known limitations. In this case, the DR can be significantly reduced when the transistor is operated in the subthreshold regime. Therefore, the source degeneration and bump linearization techniques were used to improve the DR and to optimize the overall transconductance $g_{\text{m}}$ while keeping a small size of the circuit. It can be demonstrated [25] that in this conditions, $g_{\text{m}}$ is linearly related to the $I_{\text{D}}$ according to

$$g_{\text{m}} \cong \frac{I_{\text{D}}}{nV_{\text{T}}} \tag{4.3}$$

where $n$ is the technology slope factor and $V_{\text{T}}$ is the thermal voltage ($V_{\text{T}} = kT/q$).

The IC of this automatic spike detector was fabricated using the CMOS 0.18 μm process. The total area per channel (0.07 mm$^2$) was reduced 44% compared to previous works using similar-scale technology. The main building blocks in Figure 4.2, that is, the TEO preprocessor, the spike detector, and the delay filter consumed 170, 250, and 356 nW, respectively, which amount to 776 nW for the whole analog processor. This analog preprocessing allowed a 98% of data reduction while preserving intact the spike waveforms for subsequently processing.

### Low-power analog processor for decoding of neural signals from the motor cortex

The analog signal processor shown in Figure 4.4 presented in Ref. [14] allows reducing power consumption significantly by adopting a slow-and-parallel architecture

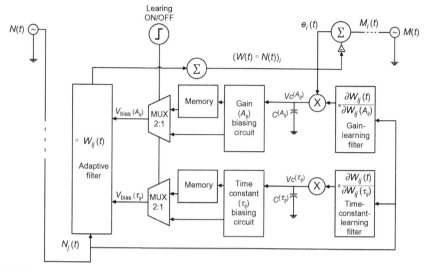

**FIGURE 4.4**

Architecture of the analog processor implementing a continuous-time, adaptive linear filtering algorithms for learning and decoding neural signals from the motor cortex ($N(t)$) and to output the corresponding motor commands ($M(t)$) (from Ref. [14]).

that performs on-chip learning and decoding algorithms. The amount of data to be digitized and transmitted by the implant is reduced to a minimum because only the results are wirelessly transmitted to an external unit, which turns out in a reduction of the overall power consumption throughout the signal processing and the energy emitted during the data transmission.

Preprocessed neural signals recorded from monkey's posterior parietal cortex ($N(t)$) are fed to the analog processor in Figure 4.4 that runs an $n$ to $m$ convolutional decoding algorithm. The processor consists of the following building blocks: continuous-time transconductor–capacitor ($G_m$–$C$) filters and multiplier, adder, and subtractor circuits, all operated in the subthreshold regime.

The architecture is capable of learning how to decode monkey's arm movements using a gradient descent-based algorithm that allows tuning the parameters used for the convolution kernels of the adaptive decoding filter to minimize the mean-squared error between the intended output and the prediction. Once the learning process is finished, the parameters are stored, the circuit is switched to the decoding mode, and then, the filters are optimized and configured for decoding the neural signals.

Results from simulation of this circuit in a $0.18\,\mu m$ process have shown a power consumption of $17\,\mu W$ for 100 channel inputs with three motor outputs using 1 V power supply on a reduced chip size. The data bandwidth is significantly reduced (1000×) from 24 Mb/s (100 channels sampled at 20 Kb/s with 12-bit resolution ADC) to 2.4 Kb/s (three motor parameters at 100 Kb/s with 8-bit resolution).

### Other analog processors of neural signals

Analog signal preprocessing has been incorporated by an increasing number of works to perform some tasks that were usually executed at the back-end stages using digital signal processing such as on-chip spike detection, spike features extraction, and wireless transmission of the isolated spike, the spike features, or both [26–32]. These works also demonstrate that analog signal preprocessing using relatively simple but effective analog architectures allows reducing the bandwidth and the power required to process and transmit the data by adopting the design principles presented in Section 4.2.

#### 4.3.1.2 Low-power analog processor for cochlear implants

Cochlear processors used in deeply deft patient implants transform sound into spatial electrode stimulation patterns for the auditory nerves. These processors can be implemented using either an analog preprocessing (Figure 4.1b) or only a digital signal processing (Figure 4.1a). However, in this application, the analog signal preprocessing has shown to be more power-efficient than the digital processing in more than an order of magnitude [33] by compressing incoming data and using low-resolution and low-speed ADC at later stages of computation. The analog circuits also ensure robustness, that is, rejection to power supply noise, thermal noise, temperature variation, cross talk, and immunity to process variation and artifacts.

In Figure 4.5, is depicted the architecture of an analog cochlear implant processor described in Ref. [33] that runs a low-power algorithm termed as asynchronous interleaved sampling (AIS), originally presented in Ref. [34]. This AIS-based processor, conceived for next generation of fully implantable cochlear prostheses, will allow deaf patients not only speech hearing but also music hearing by providing fine-phase-timing encoding for auditory nerve stimulation.

The AIS algorithm allows detecting the input channel (neuron) with the highest intensity to perform channel sampling at a higher rate than in the lower-intensity channels. This asynchronous sampling approach prevents simultaneous channel

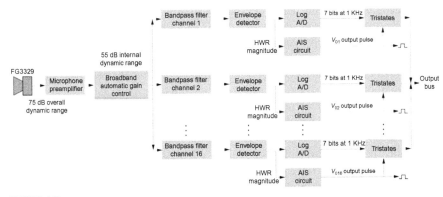

**FIGURE 4.5**

Architecture of the AIS-based processor (from Ref. [33]).

stimulation and spectral smearing in the output stimulation patterns. The working principle is based on an array of neural capacitors that are charged by halfway rectified current outputs (envelope detection). The first charged capacitor that crosses a predefined threshold wins the race among all channels, and a pulse (spike) is fired to signal event detection and also to reset all other capacitors. Then, a negative current is applied to the firing neuron capacitor to inhibit subsequent firing in the same channel during the neuron refractory period. The adaptive sampling rate, which also determines the output stimulation rate, considers the time and spectral signal content of the signal in each channel to avoid power consumption during quite periods and to sample at higher rates the high-intensity predominant channels.

The AIS-based processor in Figure 4.5 works as follows [33,35]. The microphone detects sound and outputs a signal with a wide DR of 75 dB. Subsequently, the microphone signal is preamplified and then compressed using an AGC amplifier, which reduces the internal DR to 55 dB. The AGC circuit is implemented by varying $g_m$ in a transconductance-resistance variable gain amplifier. The compressed signal is fed to a full processing chain of 16 parallel channels. Each channel performs band-pass filtering in the suitable frequency range using two cascaded blocks of second-order OTA $G_m$–$C$ circuits. Next, an envelope detector halfway rectifies the compressed signal and outputs peaks (spikes) that are fed to the AIS processor implementing the algorithms introduced above. The compressed signal is used also as the input of the logarithmic ADC to extract the logarithm of the spectral energy in the corresponding spectrum band. A dual-slope ADC circuit is implemented using a proportional-to-absolute temperature (PTAT) circuit at the input, which converts the output current from the envelope detector into a logarithmic voltage at the input of a wide-linear-range transductor circuit followed by a standard comparator. The PTAT circuit allows for the temperature and offset compensation required to ensure analog circuits robustness. The spike produced by the enveloped detector also enables the tristate buffer of the winning channel connected to a common bus to output the 7-bit digitized envelope amplitude. This output provides in a single event both the amplitude information and the fine phase timing required for generating the electrode stimulation pattern of the auditory nerves.

The analog preprocessing approach adopted for this cochlear implant allowed reducing the power consumption of the whole processor to 357 µW using a 1.5 µm CMOS process compared to the 5 mW estimated by the authors for the digital signal processing approach.

### 4.3.1.3 Ultralow-power analog processor for ECG acquisition and feature extraction

Cardiac rhythm disorder needs to be continuously monitored through ECG signal recording to feedback rate-responsive pacemakers, implantable cardioverters, and defibrillators [36]. Similarly to other chronically implanted devices, minimal power consumption and small implant size are required for these devices, which are powered by an internal battery that should last 10 years or more. The analog processor described here extracts the QRS complex from the ECG to detect abnormal heart rate. The circuit is representative of the state of the art of such ECG signal processing using ultralow-power circuits that minimize the computational complexity of subsequent digital signal processing without compromising other performances significantly.

The architecture of the ECG analog signal processor proposed in Ref. [37] and shown in Figure 4.6 does not require any external passive component to operate. The intracardiac ECG signal is amplified (20 dB) by an instrumentation amplifier with high DR and on-chip rail-to-rail AC coupling. Next, the preamplified differential signal is fed to a single ECG recording channel and a QRS feature extraction (FE) channel. The FE channel, which allows monitoring of the ECG signal in a selected frequency band, consists of a programmable gain amplifier (PGA), a high-pass switched-capacitor filter (SC-HPF) with a selectable corner frequency (1, 2, 5, 10, or 20 Hz) and a low-pass switched-capacitor filter (SC-LPF), which allows a bandpass from 10 Hz (for removing T-waves as well as electrode motion artifacts) to 10, 15, or 25 Hz. The ECG recording channel is similar to the FE channel but does not require the low-pass filter (LPF), and the HPF (high-pass filter) frequency corner is set to 1 Hz instead. The high-pass filtered differential signal is amplified by the PGA implemented by a differential-to-difference amplifier (DDA) with gain set through capacitors $C_g$ and $C_f$ as shown in Equation (4.4), which provides 0–24 dB programmable gain with 6 dB/step. The single-ended output signal of the PGA is then passed through a current multiplexed buffer (CMPX) and subsequently digitized by an ADC of programmable resolution (7–10 bits).

$$G_{DDA} = 1 + \frac{C_g}{C_f} \tag{4.4}$$

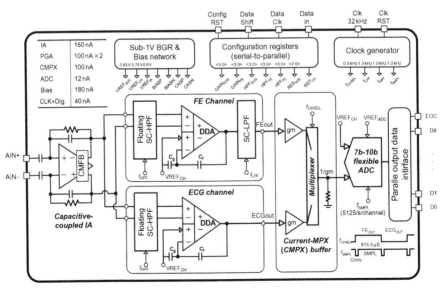

**FIGURE 4.6**

Ultralow-power analog signal processors for ECG signal recording and FE (form Ref. [37]).

The analog processor in Figure 4.6, which was implemented using the 0.18 μm CMOS process, consumes less than 1 μW (680 nA) while features a SNR greater than 70 dB, a common-mode rejection ratio greater than 90 dB and a power supply rejection ratio greater than 80 dB.

## 4.3.2 DIGITAL SIGNAL PROCESSING

Digital signal processing (DSP) has been the preferred processing method for the back-end stages since the arrival of microprocessors. However, microprocessor-based systems can achieve lower signal processing speeds and generally consume more power when compared to analog signal processing counterparts performing similar tasks. However, due to the continuous improvement in microprocessors' processing speed and power consumption per instruction executed, we have to look at microprocessor technology advances before deciding on the optimal signal processing architecture for the envisaged application. Another reason that may explain the preference for digital over analog signal processing is the fact that DSP allows designers to choose the algorithm best suited for the target application considering system specifications without the constraints imposed by analog implementation concerning the complexity of the circuit topology, the cost of keeping the requiring precision and robustness, and the development cost and time.

As mentioned previously, the power consumption and the size are the main challenging constraints that an implantable hardware has to address. To ensure the overall system performance taking advantage of the digital processing implementations, we should consider the DSP system architecture, the type of processor, the hardware–software partitioning, the type of arithmetic (floating or fixed point), the flexibility, and the technology available. In the following sections, these factors will be analyzed, and some examples of DSP hardware used in bioimplantable systems will be given.

### 4.3.2.1 Choosing the right processor for digital signal processing

There is always more than one possible solution when it comes to choosing the central processor unit (CPU) that best delivers the performance required for a particular application. The CPU traditionally has used one of the typical architectures shown in Figure 4.7, depending on the type of processor. The von Neumann architecture shown in Figure 4.7a is commonly used in general-purpose processors (GPPs). This architecture features one bus for program codes and data and shares a common memory space to store both. In contrast, the Harvard architecture and the Super Harvard architecture (SHARC) shown in Figure 4.7b and c, respectively, are typically used in digital signal processors (DSPs). This architecture uses separated buses and memories to store instruction codes and data, which allows higher processing speeds. The SHARC also includes an instruction cache and a dedicated I/O controller that improve the overall DSP processors performance.

The selection of the best processor is greatly dependent on the targeted application. Thus, a careful selection of the processor considering the trade-offs among

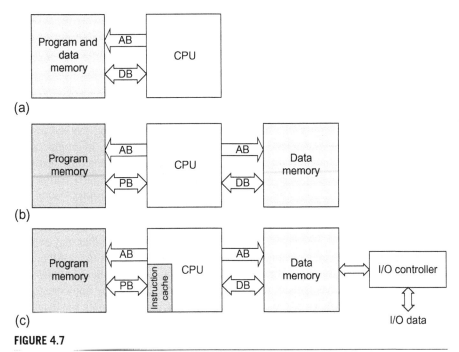

**FIGURE 4.7**

Architectures used in different CPUs. (a) The von Neumann architecture typically used in general-purpose microprocessors. (b) The Harvard and (c) SHARCs typically used in digital signal processors (DSPs). AB: address bus, DB: data bus, and PB: program memory bus.

the application requirements and the hardware performance should be carried out to optimally implement the DSP system. A comparison considering the performance, on one side, and the flexibility, area, and power, on the other side, of the available CPU choices to implement DSP systems is shown in Figure 4.8, followed by a summarized description of the different CPU choices.

### General-purpose processors (GPPs)

GPPs provide great programmability using a particular instruction set, either a reduced instruction set computing (RISC) or a complex instruction set computing (CISC). GPPs require external circuitry to function but offer rich human–machine interfaces running advanced operating systems that support a wide variety of applications and development tools. The GPPs are moderately efficient performing the typical multiply–add–accumulate operations greatly used in DSP functions. They can be combined with DSP processors for computationally intensive signal processing. However, signal processing systems based on GPPs are bulky and power-hungry, which render these systems unsuitable for implantable hardware but useful for implementing base stations linked with the external unit of the implanted hardware.

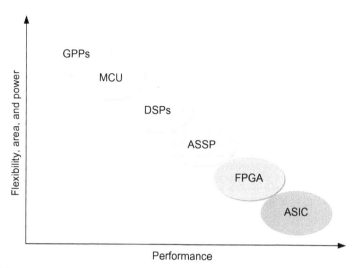

**FIGURE 4.8**

Comparison among processors considering performance, flexibility, area, and power. GPPs: general-purpose processors, DSPs: digital signal processors, ASSPs: application-specific standard products, FPGAs: field-programmable gate arrays, ASICs: application-specific integrated circuits.

The main GPP providers currently are Intel with its x86 series (e.g., Atom, Celeron, Pentium, and Core i3/i5/i7), Advanced Micro Devices (e.g., AMD series of Athlon, Sempron, Turion, and Phenom), and several manufacturers and fabless providers that offer ARM-based GPPs and intellectual property (IP) cores, respectively. The ARM-based processors are able to run a wide variety of operating systems (OS). Examples of OS supported by ARM-based processor in embedded systems are Linux, Windows CE, Symbian, FreeRTOS, MicroC/OS-II, QNX, and VxWorks; in mobile devices are Apple iOS, Google Android, Windows Phone, Windows RT, Samsung Bada, and BlackBerry OS; and in desktops or servers are Unix-based OS (e.g., NetBSD and FreeBSD), Linux, OpenSolaris, Ubuntu, and Chrome OS.

## Microcontrollers (MCU)

MCUs have an on-chip programmable processor with the von Neumann architecture, and frequently, they contain GPPs based on ARM Cortex or 8051 cores. They also integrate on the same chip data and program memories several peripherals such as timers, interrupt and direct memory access (DMA) control circuitry, and programmable serial and parallel interfaces, among other programmable peripherals, and even analog and mixed-signal circuitry depending on the manufacturer and application series. The MCUs are less power-efficient running signal processing functions than the DSPs and dedicated processors built in FPGAs or ASICs, mainly due to their general-purpose architecture and many integrated peripherals that a particular application may not need. Therefore, they may be more suitable for modest DSP demands.

Generally, the MCUs provide a compact set of instructions and lower mathematical processing resources than RISC-GPPs. However, MCU-based systems show smaller sizes, significantly lower costs due to high volume of sales of this processor (the highest among all processors), and faster design time supported by a high variety of development tools.

Currently, there are a considerable number of MCUs providers of both off-the-shelf ICs and IP cores that are often customizable. Among the manufactures that offer IC-MCUs are Atmel (e.g., AVR-8/32 picoPower series for battery-powered devices), Analog Devices (e.g., precision analog microcontroller ADuC7xxx series), Freescale (e.g., ultralow-power S08/CF51 series), Maxim (e.g., ultralow-power high-performance MAXQ 16-bit RISC series), Microchip (e.g., nanoWatt eXtreme Low Power PIC series), Texas Instruments (MSP430 16-bit RISC mixed-signal series), and STMicroelectronics (e.g., the ultralow-power STM32L0 series, claimed as the lowest power consumption as of May 2014).

Some MCU manufacturers like Freescale have added DSP functions to their MCUs to create the so-called digital signal controllers (DSCs) to enhance MCU signal processing capabilities while keeping process control capabilities and lower prices compared to traditional DSPs (e.g., Freescale DSC 56800/E series).

## Digital signal processors (DSPs)

DSPs are specialized processors optimized to perform in real-time digital signal processing (DSP) functions based on repetitive multiply–add–accumulate (MAC) operations, which are heavily used in digital filtering (e.g., FIR filters), fast Fourier transform (FFT), and many other DSP functions. The DSPs integrate a programmable microprocessor typically with a Harvard-based architecture as shown in Figure 4.7b and c. They provide high-speed data processing by implementing single instruction, multiple data (SIMD) operations, special instructions for superscalar architectures cores, single-cycle MAC or fused multiply–add (FMA) computation, parallel computation in several MAC units, fast data streaming using DMA, circular buffering capabilities, zero-overhead hardware-controlled looping, and fast and extended precision computations in fixed or floating point and DSP-optimized features such as single instruction FIR filtering, or FFT, among other specialized DSP functions (e.g., audio and video codecs). They also integrate on-chip some peripherals and memory for a stand-alone operation.

There are many DSPs in the current market that delivers high performance due to the semiconductor technology advancements and improvements in the DSP architectures. Prices and performance vary greatly among the main IC-DSP manufacturers and IP-core providers. However, the high power constraint of biomedical implanted hardware may prevent the use of powerful DSPs despite their high performance. Because DSPs evolve rapidly these days, a comprehensive review of the state of the art should be performed at the beginning of every new project to find the best suited (if any) for the intended application.

Among the main IC-DSP providers are Texas Instruments (e.g., TMS320C5000 Ultra-Low-Power series claimed as the lowest-power 16-bit DSPs as of May

2014), Freescale (e.g., StarCore MSC81xx series), and Analog Devices (e.g., the ADSP-21xx and the SHARC floating-point DSPs series). There are also IP-core developers of low-power DSPs such as NXP (e.g., ultralow-power CoolFlux DSP core), CSEM (ultralow-power icyflex 16/32-bit RISC processor cores), and CEVA (e.g., ultralow-power DSP core CEVA-XC series optimized for advanced wireless communication).

### Application-specific standard products (ASSPs)

The ASSPs are commercially available ASICs that perform specific functions required in different applications, like audio and video coding–decoding, also known as codec, and in communication links such as USB and Bluetooth. They show very high performance when performing the intended functions but lack flexibility to make changes or improvements. ASSPs are usually combined with other processors that perform signal processing tasks, overall system control, and user interfaces.

### Field programmable gate array (FPGA)-based processors

FPGAs allow the implementation of customized DSPs capable of approaching the performance shown by ASICs performing DSP tasks but providing shorter development time and costs. They can be reconfigured within a system, which is a major advantage during the R&D of new applications and in the updating of the existing ones. They can also be reprogrammed at run time to implement self-reconfigurable systems that dynamically adapt their functions to the running tasks. Parallel computing is another benefit of FPGA-dedicated processors. This can be a major advantage when used in systems demanding high sampling rates and high processing speed and when multiple channels need to be processed at time. FPGAs are also very useful in ASIC system prototyping. As drawbacks, FPGA-based DSPs are more expensive than the ASICs for high scale production, and depending on the application, they can consume more power than off-the-shelf DSPs performing the same task. To improve the overall power consumption, some FPGA manufacturers and IP-core providers offer embedded hard-core or soft-core processors that are integrated with FPGA cells in a single chip to take advantage of both technologies.

Currently, there are two major manufacturers of FPGAs, Xilinx and Altera, that provide a wide range of FPGAs well suited for many applications. There are also other manufacturers like Microsemi (former Actel), Lattice Semiconductor, and Atmel that provide interesting products to be considered during the search of the best-suited FPGA-based DSPs. Once again, the power and size restrictions may suggest the use of the FPGAs series exhibiting the best figures for these specifications. Among the products offered by these manufacturers, the following can be found: Xilinx offers the low-power Artix-7 series, Altera proposes the Cyclone V series, Microsemi offers the ultralow-power IGLOO/ProASIC series claimed as the industry's lowest-power FPGAs (as of May 2014), and Lattice Semiconductor offers the ultralow-power iCE40 series.

### Application-specific integrated circuit (ASIC)-based processors

ASICs allow implementation of DSP algorithms in dedicated, fixed-function logic to minimize power consumption and hardware size at ultimate levels. However, high-complexity algorithms can be difficult to deploy in dedicated logic within a reasonable time. In such cases, IP cores of powerful processors can be used to execute the DSP functions. A particular class of IP core, known as application-specific instruction-set processors (ASIPs), is used to further improve power efficiency by customizing the set of instructions and the hardware of the processor core according to the intended application requirements. The customization of the processor core can be made hassle-free with the help of automated tools.

Dataplane processors units (DPUs) is a new category of processor IP core recently proposed by Cadence Tensilica to bring together the strengths of GPPs, DSPs, and FPGAs by integrating all of these technologies in a single chip. The IP Xtensa customizable processor core can improve by 10–100 times the processing speed of data-intensive functions.

As mentioned before, ASIC-based processor design is longer and more expensive than other signal processor designs that use more flexible and cost-effective technologies. Thus, their use in low- to medium-scale production will turn out in a cost penalty to the final device. However, due to the high demanding requirement of power efficiency and reduced size, ASIC-based processor is often selected as the preferred method for implementing DSP processor in current R&D of new bioimplantable systems. The continuous improvement in the processor technology may eventually encourage designers to choose more flexible and cost-effective approaches in the development of new implantable DSP processors.

### 4.3.2.2 *Choosing the right numeric format for DSP processors*

The numeric format for the arithmetic operations in fixed or floating point is another important factor to consider during the design and implementation of DSP algorithms. The choice of the numeric format will impact the overall system performance, the computational precision, the power consumption, the development process easiness, and the cost of the final hardware implementation. A careful selection should be performed, trading off the system specifications and the pros and cons of fixed- and floating-point arithmetic for each particular application.

Computational precision is one of the key factors to consider when deciding the numeric format between fixed point and floating point. Digitization of analog signals introduces the first quantization error to the signal processing chain. Subsequently, the numeric format adds to this error but in a different degree depending on the notation and the number of bits chosen for the registers involved in the arithmetic operations. Additionally, a quantization error (also known as quantization noise) is produced by rounding or truncating the arithmetic operation results to adjust the data to the registry lengths. This quantization noise will accumulate through successive operations, eventually deteriorating the output precision.

To decide on the best-suited numeric format, the designer should consider the differences between fixed- and floating-point formats, the DR of the input data, and the allowed output error. Fixed-point arithmetic handles positive and negative integer numbers of $n$-bits, which yields $2^n$ possible bit patterns. For example, in 16-bit DSP-based processors, 65,536 patterns are possible, but one bit is commonly reserved for the sign; thus, the integer range covered is −32,768 to 32,767. The DSP designer has to decide in advance how many bits will be allocated to the integer and to the fractional part, of both the I/O and the internal (intermediate operation) registers, based on the DR of the input data. Because the number of bit is fixed, an increase of the DR will increase the number of bits that need to be assigned to the integer part to meet the DR, but in turn, the number of bits of the fractional part will decrease, and the other way around. The DR is determined by the largest value and the smallest value that can be presented to the input of the system at all times, in either typical or atypical conditions of operation. If it is not possible to predict the data DR, then the fixed-point notation is not the best choice.

In the floating-point format, the number is handled similarly to the scientific notation where a number ($N$) is represented using a mantissa ($A$) and an exponent ($B$), so that $N = A \times 2^n$. For example, in 32-bit DSP-based systems, 24 bits can be allocated to mantissa and 8 bits to the exponent. This way, the point can be moved throughout the mantissa allowing the representation of very small and very large numbers. Consequently, the DR is significantly improved. Therefore, floating-point format is the best choice for computationally intensive applications where high precision matters and for data sets of unpredictable or high DR. However, floating-point processors can consume more power due to the higher number of bits used to store and process data, and in spite of the cost differences between DSPs of both numeric formats that have decreased significantly, floating-point DSPs are still more expensive. Otherwise, the floating-point numeric format would be the unbeatable choice for all applications.

Nowadays, the DSP design process is highly automatized and it is not much more difficult for fixed-point arithmetic as it used to be. There are many development and simulation tools, like the specialized toolboxes available in MATLAB/Simulink for digital signal processing using fixed-point arithmetic, and other electronic design automation tools offered by the processors providers, which allow estimating the DR, the quantization errors, and the power consumption, through simulation using test vectors made of synthetic and real data. The final decision on the numeric format should be made upon the results obtained from these simulations and validated in the final hardware using similar test vectors.

### 4.3.3 DESIGN EXAMPLES OF DSP PROCESSORS FOR BIOIMPLANTABLE SYSTEMS

In the following sections, examples of signal processors used in bioimplantable systems showing some of the design principles and design trends examined before are presented. We start with a DSP processor for monitoring the bladder volume followed by a mixed-signal DSP processor used to detect epileptic seizures.

### 4.3.3.1 Custom DSP processor for bladder monitoring through afferent neural signals

There are millions of people around the world suffering from bladder dysfunction that can benefit from neuroprosthetic implants. These implants can restore urinary functions in patients who suffer from spinal cord injury or neurological conditions. The effectiveness of implantable neuroprosthesis can be improved by sensing the bladder volume and pressure to adapt the functional electrical stimulation of the bladder nerves to the ongoing bladder state [38].

Signal processing algorithms have been proposed for estimating the bladder volume or pressure from the recording of neural signals at specific spinal cord roots [39]. These algorithms were specially designed to be deployed in an electronic sensor that provides feedback to an implantable neuroprosthetic device aimed to restore autonomously the bladder storing and voiding functions. Such bladder sensor (BS) should be able to record, detect, and decode sensory neural activity in real time in less than 2.5 ms (time window of a neural spike), with low-power consumption to ensure safety and long battery endurance. The power consumption target value was estimated at less than 1 mW.

The BS is based on the digital processing of the neural signal recorded through MEAs. The MEAs placed on selected spinal cord roots record the afferent activity produced by peripheral nerves carrying sensory information from the bladder [40]. The raw neural signal of few microvolts recorded from one channel properly chosen is preamplified, filtered, and digitized using a low-noise, low-power neural amplifier and a sigma-delta ADC, as in Ref. [12].

A custom DSP processor (CDSP) shown in Figure 4.9, which is fully described in Ref. [40], was deployed in an ultralow-power FPGA from the IGLOO series of Microsemi that provides ultralow static and dynamic power consumptions and a small packaging area ($13 \times 13\,\text{mm}^2$). This architecture meets the flexibility requirements for eventual modifications of the BS and also satisfies the target specifications of power, size, latency, and the computational demands of the algorithms designed for detecting, classifying, and decoding bladder sensory information. The numeric format for the arithmetic operation was chosen as 32-bit fixed point. This decision was made upon the results of simulations performed in MATLAB/Simulink to assess the DR, the output precision, and the number of basic arithmetic operations, which ultimately determine the hardware burden, the DSP latency, and the power efficiency. The top-level architecture of the CDSP is shown in Figure 4.9. At the top of this figure, representative outputs of the signal processing blocks that will be described below are depicted.

The preconditioned neural signal is fed to the SDB. This block detects and aligns the spikes by their maximum absolute value using a NEO to improve the SNR, as shown in Equation (4.5):

$$\psi\left[x(n)\right] = x^2(n) - x(n+k)x(n-k) \tag{4.5}$$

where $\psi[x(n)]$ is the estimated instantaneous energy, $x(n)$ is the input neural signal, and $k$ was chosen experimentally as 2.0.

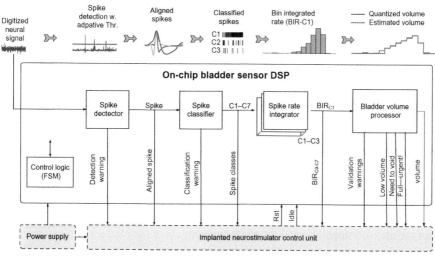

**FIGURE 4.9**

Top-level architecture of the custom DSP processor used to detect, classify, and decode the bladder sensory information recorded through a MEA and preconditioned using analog front-end circuit (not shown) (from Ref. [40]).

A first-order exponential average filter defined by Equation (4.6) is used to extract the estimated noise level $y(n)$ in order to adapt the detection threshold of the spikes to the ongoing background noise. The filter coefficient a is computed off-line as shown in Equation (4.7) using a time window $(t)$ of 100 ms with a sampling frequency $(f_s)$ of 24 kHz. The detection threshold (Thr) is set to a scaled filter output using a scaling factor $(C)$ as shown in Equation (4.8):

$$y(n) = \psi(n) + a\left[y(n-1) - \psi(n)\right] \tag{4.6}$$

$$a = e^{-\frac{1}{f_s t}} \tag{4.7}$$

$$\text{Thr}(n) = Cy(n) \tag{4.8}$$

The spikes captured by the SDB are then fed to spike classifier block (SCB) that discriminates the spike classes by computing the weighted Euclidean distance to the stored templates of each class using Equation (4.9):

$$\Delta_{Ci}^2 = \sum_{j=1}^{N} w_j \left(S_j - P_{i,j}\right)^2 \tag{4.9}$$

where $S_j$ is the $N$-dimensional vector of the spike to be classified $(j = 1,...,N)$, $P_{i,j}$ is the template matrix $(i = 1,...,p$, $p$: number of templates/patterns), which is computed during an off-line training phase, $w_j$ is the vector of weights, and $\Delta_{Ci}^2$ is the squared WED computed between the spike $S$ and the template $P$ of class $i$. The vector of weights ($w_j$) emphasize the dimensions where the template differences are significant (usually in the vicinity of the main peak) and minimize the contribution of the dimensions where the differences are nonsignificant and contaminated by background noise (usually towards both ends of the spike). The SCB outputs the spike raster (pulse trains) of the spike classes discriminated, as shown at the top of Figure 4.9.

Next, the spike rate integrator (SRI) block counts the number of spike events occurring within a bin (time window) of a properly chosen duration ($t_{bw}$) and outputs the bin integrated rate (BIR$_i$) corresponding to the channel identified off-line as the one carrying the bladder sensory information.

The last signal processing stage is performed by the bladder volume decoder (BVD) block. This block can operate in three modes: a qualitative volume estimation output (mode 1), a quantitative volume estimation output (mode 2), and a combined mode, in which both estimated volume outputs are computed (mode 3). In mode 1, the BVD outputs three qualitative values of bladder fullness that can be used to trigger the neurostimulation or to inform the patient about the bladder fullness. In mode 2, the BVD computes the bladder volume using a regression model of programmable order [39]. Mode 2 is also suitable for feedback purposes and for monitoring applications where an accurate value is required. The qualitative volume estimation is achieved by finding the arg min of the absolute difference among the BIR$_i$ and each BIR$_{ref}$ determined during the training phase for each level of detection (low, medium, and high), as shown as follows:

$$\arg\min_{i \in [0,2]} \left\{ \left| \mathrm{BIR}_{C1} - \mathrm{BIR}_i \right| \right\} \tag{4.10}$$

The quantitative volume estimation is performed by a regression model based on the polynomial of order $n$ shown in Equation (4.11). However, to reduce the number of operations, the hardware burden, and the errors arising from fixed-point arithmetic with numbers that differ in magnitude, Horner's method, as shown in Equation (4.12), was used to implement this circuit using a customized MAC unit. As a result, $n$ additions and $n$ multiplications without exponentiation are required:

$$\hat{y} = \sum_{i=0}^{n} a_i x^i \tag{4.11}$$

$$\hat{y} = x\left(x\left(x...\left(a_n x + a_{n-1}\right) + a_{n-2}\right)\cdots + a_2\right) + a_1\right) + a_0 \tag{4.12}$$

The CDSP was deployed in the low-power IGLOO FPGA AGL1000V2 from Microsemi. To maximize power efficiency, the CDSP was supplied by the lowest voltage available of 1.2 V and, taking advantage of the freedom of choice of the clock frequency ($f_{clk}$), $f_{clk}$ was set to 333 kHz, the lowest value that allowed meeting the maximum latency including a safety margin of 20%. The volume estimation circuits, which were tested with real signals, reproduced accuracies achieved by off-line simulations in MATLAB, that is, 94% and 97% for quantitative and qualitative estimations, respectively. The measured power consumption during the CDSP operation was $485\,\mu W$. The static power consumed during the IGLOO FPGA Flash*Freeze mode (idle) was $55.6\,\mu W$. The power density achieved, considering single-side area of the package and the measured power, was $287\,\mu W/cm^2$. The CDSP latency was $2.1\,ms$ and used 68% of the FPGA cells.

### 4.3.3.2 Mixed-signal processor to detect epileptic seizures

The epileptic seizure detectors can be used in drug-refractory patients who suffer from recurrent and multifocal epileptogenic seizures. These detectors can provide feedback to closed-loop self-triggered implantable microsystems that deliver focal treatment upon early detection of seizure onsets. These devices record intracerebral EEG (icEEG) signals of microvolt-level voltages usually contaminated by noise and artifacts. The preconditioned icEEG signal is processed in order to detect in advance patterns of an incoming seizure to trigger the treatment-delivering subsystem, for instance, to start vagus nerve stimulation (VNS), deep brain stimulation (DBS), or local drug infusion through miniaturized infusion pumps as in Ref. [41], which allows preventing or reducing the handicapping symptoms of the epileptic seizures (e.g., body shaking, disability of motor functions, and loss of consciousness).

Two signal processing approaches have been used to implement these devices: synchronous seizure detection (SSD) and asynchronous seizure detection (ASD). The latter has shown to be more power-efficient due to minimization of the transistor switching by performing data-depending analyses [41]. These analyses are performed in time–frequency and time–amplitude domains to detect seizure onsets while minimizing the false detection of unrelated activity in the brain.

The architecture of an epileptic seizure detector implementing the ASD algorithm in a mixed-signal processor, which is fully described in Ref. [42], is shown in Figure 4.10. In this ASD processor, the recorded icEEG ($V_{In}$) signal is amplified using a low-power, low-noise bioamplifier. Subsequently, artifacts and noise are filtered out with a LPF. The preconditioned signal ($V_a$) is then passed through voltage window detectors (VWDs) to discriminate different amplitudes within predefined thresholds ($V_{THi}$ and $V_{TLi}$). The corresponding $V_{Wi}$ output is asserted to indicate that the signal amplitude falls into the corresponding voltage window. Next, frequency analyzers measure the $V_{Wi}$ signal frequency ($F_{Hi}$) in a defined time frame ($T_f$), as shown in Equation (4.13), to compare their outputs ($F_{Hi}$) to a seizure detection threshold ($F_{Sz}$). If $F_{Hi}$ is greater than $F_{Sz}$ within the time frame $T_f$, then the comparator output ($V_{Di}$) is asserted to signal an abnormally fast activity. A dedicated logic analyzes the signal $V_{Di}$ and quantifies specific features indicating a progressive increase of the amplitude and the presence of high-frequency activity to finally assert the seizure onset. All of the circuit parameters can be adjusted to patients' specific conditions to maximize the detection of the seizure onsets and minimize the false alarms.

$$F_{Hi} = \frac{1}{T_f} \sum_{t=0}^{T_f} V_{Wi}(t) \qquad (4.13)$$

The analog front end of the epileptic seizure detector of Figure 4.10 consists of the following building blocks: a low-power, low-noise, continuous-time, common-mode

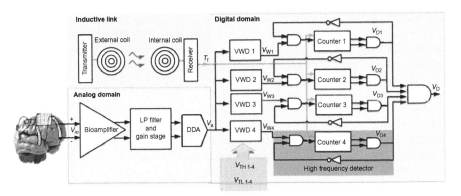

**FIGURE 4.10**

Top-level architecture of the implantable device implementing the asynchronous epileptic seizure detection algorithm (DDA: differential difference amplifier, VWD: voltage window detector).

feedback (CMFB) differential amplifier, as described in Ref. [43]; a LPF implemented with OTA $G_m$–$C$ circuit followed by an amplification stage that provides an overall gain of 60 dB; a DDA used to convert the differential filter output to a single-ended signal ($V_a$); four VWD circuits implemented with standard VWDs, two of which are used for detecting positive amplitudes and two for negative amplitudes, as described in Ref. [41]; and four high-frequency detectors implemented by $D$ flip-flops and dedicated logic.

The ASD processor was initially validated using miniaturized off-the-shelf components (Figure 4.11a), which yielded a power consumption of 47.2 mW in a PCB area of 1967 mm². This processor showed a power reduction of 45% compared to SSD prototyped architecture [41]. Nevertheless, the ASIC implementation of this

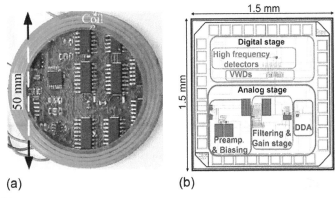

(a)                              (b)

**FIGURE 4.11**

(a) Asynchronous seizure detection prototype made up of discrete components. (b) ASIC implementation of the asynchronous seizure detector.

ASD processor (Figure 4.11b), using a 0.13 µm CMOS process, occupied a die area of 2.3 mm² and achieved a power consumption of only 9 µW, with 13.7 s of average latency, which allowed 100% of the seizure detection well before the onset of clinical manifestations.

## 4.4 CONCLUSIONS

In this chapter, we have presented some basic principles for designing low-power and small-size signal processors for bioimplantable systems. Several examples of hardware used for signal processing were presented throughout the chapter to show the application of these principles in actual designs. We have also shown that the selection of the architecture and the technology is a crucial step to achieve the best performance of the signal processing hardware.

The examples presented in this chapter helps to understand why it is not possible to design a universal signal processor for bioimplantable systems applications. The hardware should be designed to meet the specifications of each particular application. Furthermore, it has been shown through several author works that an optimal hardware partitioning between analog and digital domains can lead to a significant improvement of the overall signal processor performance.

Digital signal processing hardware deployed in custom circuits (ASICs) has demonstrated to be more power efficient and smaller than in off-the-shelf processors. However, the designer should carefully review the current state-of-the-art of commercially available processors before deciding on the best choice for the envisaged application. The ASICs design process is still long and complex, and ASICs-based applications are only cost-effective in large-scale productions.

## REFERENCES

[1] Wolf PD. Thermal considerations for the design of an implanted cortical brain–machine interface (BMI). In: Reichert WM, editor. Indwelling neural implants: strategies for contending with the in vivo environment. Boca Raton, FL: CRC Press; 2008.

[2] Baker RJ, Li HW, Boyce DE. Electrical noise: an overview. In: Baker RJ, editor. CMOS circuit design, layout, and simulation. 3rd ed. Piscataway, NJ: Wiley-IEEE Press; 2010. p. 269–310.

[3] Sarpeshkar R. Noise in devices. In: Ultra low power bioelectronics: fundamentals, biomedical applications, and bio-inspired systems. Cambridge, UK: Cambridge University Press; 2010. p. 155–211.

[4] Webster JG. Medical instrumentation application and design. 4th ed. Boston: Wiley; 2010.

[5] Sarpeshkar R. Universal principles for ultra low power and energy efficient design. IEEE Trans Circuits Syst II Express Briefs 2012;59:193–8.

[6] Baker MW, Sarpeshkar R. Low-power single-loop and dual-loop AGCs for bionic ears. IEEE J Solid-State Circuits 2006;41:1983–96.

[7] Ganguli S, Sompolinsky H. Compressed sensing, sparsity, and dimensionality in neuronal information processing and data analysis. Annu Rev Neurosci 2012;35:485–508.

[8] Sarpeshkar R, Lyon R, Mead C. A low-power wide-linear-range transconductance amplifier. In: Lande T, editor. Neuromorphic systems engineering, 447. US: Springer; 1998. p. 267–313.

[9] Buch E, Boyle NG, Belott PH. Pacemaker and defibrillator lead extraction. Circulation 2011;123:e378–80.

[10] Harrison R, Watkins P, Kier R, Lovejoy R, Black D, Normann R, et al. A low-power integrated circuit for a wireless 100-electrode neural recording system. Presented at the solid-state circuits conference, 2006. ISSCC 2006. Digest of technical papers. IEEE International; 2006.

[11] Wattanapanitch W, Sarpeshkar R. A low-power 32-channel digitally programmable neural recording integrated circuit. IEEE Trans Biomed Circuits Syst 2011;5:592–602.

[12] Gosselin B, Sawan M. A low-power integrated neural interface with digital spike detection and extraction. Analog Integr Circuits Signal Process 2010;64:3–11.

[13] Majidzadeh Bafar V, Schmid A. Circuits and systems for multi-channel neural recording. In: Wireless cortical implantable systems. New York: Springer; 2013. p. 67–130.

[14] Sarpeshkar R, Wattanapanitch W, Arfin SK, Rapoport BI, Mandal S, Baker MW, et al. Low-power circuits for brain–machine interfaces. IEEE Trans Biomed Circuits Syst 2008;2(September):173–83.

[15] Wang YF, Wang ZG, Lue XY, Gu XS, Li WY, Wang HL, et al. A multichannel neural signal detecting module: its design and test in animal experiments. Prog Nat Sci Jun 2007;17:675–80.

[16] Rodriguez-Perez A, Ruiz-Amaya J, Delgado-Restituto M, Rodriguez-Vazquez A. A low-power programmable neural spike detection channel with embedded calibration and data compression. IEEE Trans Biomed Circuits Syst 2012;6:87–100.

[17] Clark GA, Ledbetter NM, Warren DJ, Harrison RR. Recording sensory and motor information from peripheral nerves with Utah Slanted Electrode Arrays. In: 2011 annual international conference of the IEEE engineering in medicine and biology society, EMBC; 2011. p. 4641–4.

[18] Brown EN, Kass RE, Mitra PP. Multiple neural spike train data analysis: state-of-the-art and future challenges. Nat Neurosci 2004;7(May):456–61.

[19] Rolls ET, Treves A. The neuronal encoding of information in the brain. Prog Neurobiol 2011;95:448–90.

[20] Gosselin B, Sawan M. An ultra low-power CMOS automatic action potential detector. IEEE Trans Neural Syst Rehabil Eng 2009;17:346–53.

[21] Donoho DL. De-noising by soft-thresholding. IEEE Trans Inf Theory 1995;41:613–27.

[22] Watkins PT, Santhanam G, Shenoy KV, Harrison RR. Validation of adaptive threshold spike detector for neural recording. In: 26th annual international conference of the IEEE engineering in medicine and biology society, 2004. IEMBS '04; 2004. p. 4079–82.

[23] Kaiser JF. On a simple algorithm to calculate the 'energy' of a signal. 1990 international conference on acoustics, speech, and signal processing, 1990. ICASSP-90 1990; vol. 1:381–4.

[24] Gunhee H, Sanchez-Sinencio E. CMOS transconductance multipliers: a tutorial. IEEE Trans Circuits Syst II Analog Digital Signal Process 1998;45:1550–63.

[25] Gosselin B, Sawan M, Chapman CA. A low-power integrated bioamplifier with active low-frequency suppression. IEEE Trans Biomed Circuits Syst 2007;1:184–92.

[26] Loi D, Carboni C, Angius G, Angotzi GN, Barbaro M, Raffo L, et al. Peripheral neural activity recording and stimulation system. IEEE Trans Biomed Circuits Syst 2011;5:368–79.

[27] Hiseni S, Sawigun C, Ngamkham W, Serdijn WA. A compact, nano-power CMOS action potential detector. Presented at the biomedical circuits and systems conference, BioCAS IEEE; 2009.

[28] Chae MS, Yang Z, Yuce MR, Hoang L, Liu WT. A 128-channel 6 mW wireless neural recording IC with spike feature extraction and UWB transmitter. IEEE Trans Neural Syst Rehabil Eng 2009;17:312–21.

[29] Chen T-C, Liu W, Chen L-G. 128-Channel spike sorting processor with a parallel-folding structure in 90 nm process. Presented at the ISCAS, 2009.

[30] Perelman Y, Ginosar R. An integrated system for multichannel neuronal recording with spike/LFP separation, integrated A/D conversion and threshold detection. IEEE Trans Biomed Eng 2007;54(January):130–7.

[31] Harrison RR, Kier RJ, Chestek CA, Gilja V, Nuyujukian P, Ryu S, et al. Wireless neural recording with single low-power integrated circuit. IEEE Trans Neural Syst Rehabil Eng 2009;17:322–9.

[32] Olsson RH, Wise KD. A three-dimensional neural recording microsystem with implantable data compression circuitry. IEEE J Solid-State Circuits 2005;40:2796–804.

[33] Ji-Jon S, Sarpeshkar R. A cochlear-implant processor for encoding music and lowering stimulation power. IEEE Pervasive Comput 2008;7:40–8.

[34] Ji-Jon S, Simonson AM, Oxenham AJ, Faltys MA, Sarpeshkar R. A low-power asynchronous interleaved sampling algorithm for cochlear implants that encodes envelope and phase information. IEEE Trans Biomed Eng 2007;54:138–49.

[35] Sarpeshkar R, Salthouse C, Ji-Jon S, Baker MW, Zhak SM, Lu TKT, et al. An ultra-low-power programmable analog bionic ear processor. IEEE Trans Biomed Eng 2005;52:711–27.

[36] Dell'Orto S, Valli P, Greco EM. Sensors for rate responsive pacing. Indian Pacing Electrophysiol J 2004;4:137–45.

[37] Long Y, Harpe P, Osawa M, Harada Y, Tamiya K, Van Hoof C, et al. A 680 nA fully integrated implantable ECG-acquisition IC with analog feature extraction. In: 2014 IEEE international solid-state circuits conference digest of technical papers (ISSCC); 2014. p. 418–9.

[38] Sinkjaer T, Haugland M, Inmann A, Hansen M, Nielsen KD. Biopotentials as command and feedback signals in functional electrical stimulation systems. Med Eng Phys 2003;25:29–40.

[39] Mendez A, Sawan M, Minagawa T, Wyndaele JJ. Estimation of bladder volume from afferent neural activity. IEEE Trans Neural Syst Rehabil Eng 2013;21:704–15.

[40] Mendez A, Belghith A, Sawan M. A DSP for sensing the bladder volume through afferent neural pathways. IEEE Trans Biomed Circuits Syst 2014;8(4):552–564.

[41] Salam MT, Mirzaei M, Ly MS, Dang Khoa N, Sawan M. An implantable closedloop asynchronous drug delivery system for the treatment of refractory epilepsy. IEEE Trans Neural Syst Rehabil Eng 2012;20:432–42.

[42] Mirzaei M, Salam MT, Nguyen DK, Sawan M. A fully-asynchronous low-power implantable seizure detector for self-triggering treatment. IEEE Trans Biomed Circuits Syst 2013;7:563–72.

[43] Charles CT, Harrison RR. A floating gate common mode feedback circuit for low noise amplifiers. In: Southwest symposium on mixed-signal design, 2003; 2003. p. 180–5.

# Energy management integrated circuits for wireless power transmission

# 5

**Hyung-Min Lee, Maysam Ghovanloo**

*School of Electrical and Computer Engineering, Georgia Institute of Technology, Atlanta, GA, USA*

## CHAPTER CONTENTS

## 5.1 INTRODUCTION

Wireless power transmission is one of the few viable techniques to power up implantable medical devices (IMDs) across the skin without any direct electrical contact between the energy source and the IMD. There are also other wirelessly powered applications with various levels of power requirements from nanowatts in wireless sensors and radiofrequency identification (RFID) tags, milliwatts in near-field communication (NFC), watts in mobile electronics, and kilowatts in electric vehicles. Figure 5.1 shows some of the state-of-the-art applications for wireless power transmission [1–5]. High power transfer efficiency (PTE), robustness against nearby objects and coil misalignments, and extended power transfer range are highly desired in all of these applications.

For example, IMDs have already been used successfully in the form of cochlear implants to substitute a sensory modality (hearing) that might be lost due to diseases or injuries [1]. More recent IMD applications demand higher performance and power

Bhunia et al. Implantable Biomedical Microsystems. http://dx.doi.org/10.1016/B978-0-323-26208-8.00005-4

(a)                (b)                (c)                (d)                (e)

**FIGURE 5.1**

Various applications for wireless power transmission. (a) A cochlear implant [1], (b) a visual prosthesis [2], (c) a mobile device charger [3], (d) an NFC device [4], and (e) wirelessly charging electric vehicles [5]

efficiency to enable sophisticated treatment paradigms, such as retinal implants for the blind or bidirectional cortical brain–computer interfaces with sensory feedback for amputees or those suffering from severe paralysis [2,6–8]. These IMDs require more power to handle more functions on a larger scale under variable environmental and loading conditions. Particularly, when stimulation is needed through a large number of electrodes at high rates, power level is inherently high and less dependent on the power consumption of the internal circuits [9,10]. Therefore, the power consumption of the new IMDs is going to be one or more orders of magnitude higher than the traditional IMDs, such as pacemakers [11]. In most cases, supplying the IMDs with primary batteries will not be an option because of their large volume, limited lifetime, difficult replacement, and cost [12–15]. Moreover, considering that the temperature at the outer surface of the IMD should not increase more than $2\,^{\circ}\text{C}$ for the surrounding tissue to survive [16], it is of utmost importance for the inductive link and the IMD power management circuitry to maintain very high PTE without generating excessive heat.

Since the transferred power level through the inductive link is typically limited to the size constraints of the secondary power receiver coil, various energy management schemes should be adopted in every stage of the path of power flow to efficiently deliver as much energy as possible to the target device. Maximizing PTE from the power transmitter (Tx) to the receiver (Rx) through an optimized inductive link, power management units such as AC–DC and DC–DC converters, and energy storage systems is key in improving the overall system power efficiency, wireless power transfer range, and low-temperature operation. With higher PTE, the Rx can operate with smaller transmitted power from a larger coil distance while reducing the risk of overheating. There are also other energy management applications, such as energy harvesters, in which arbitrary AC signals from transducers need to be efficiently converted to stable DC supply through similar conversion, charging, and storage building blocks. Table 5.1 summarizes power requirements and energy sources of various IMDs.

This chapter reviews various mechanisms for wireless power transmission while referring to structures and circuit techniques for inductively powered IMDs as a challenging example. Design and optimization of the inductive power transmission are described to improve overall efficiency through the power delivery path from

**Table 5.1** Power Requirements and Energy Sources of Various IMDs [10]

| IMD | Power Requirement | Lifetime | Energy Source |
|---|---|---|---|
| Pacemaker | <100 µW | 10 years | Primary battery |
| Hearing aid | 100–2000 µW | 1 week | Rechargeable battery |
| Cochlear implant | 20–100 mW | N/A | Inductive power |
| Retinal implant | 40–250 mW | N/A | Inductive power |
| Neural recorder/stimulator | 1–100 mW | N/A | Rechargeable battery/ inductive power |
| Artificial heart | 10–100 W | N/A | Inductive power |

the external power source to the target device (IMD) while maintaining robustness against nearby conductive objects and coil misalignments. Key energy management units for providing the regulated DC supply on the IMD side are presented by receiving the AC input through the inductive link and passing it through power-efficient AC–DC and DC–DC converters. Power storage options, such as rechargeable batteries and supercapacitors, are also discussed in terms of their characteristics, charging mechanisms, and circuit techniques.

## 5.2 WIRELESS POWER TRANSMISSION MECHANISMS

Several wireless power transmission mechanisms have been utilized in various applications: (1) high-frequency electromagnetic wave transmission, (2) ultrasound power transmission, (3) inductive power transmission, and (4) coupled-mode magnetic resonance-based power transmission. Figure 5.2 describes basic structures of the wireless power transmission mechanisms.

The high-frequency electromagnetic wave transmission can offer reliable near-field or far-field power links, but the received power through a high-frequency antenna is typically limited to microwatts, which is not sufficient for most applications [17,18]. For IMD applications, the specific absorption rate of the human body significantly increases at higher frequency, absorbing more energy in the tissue, which reduces the electromagnetic field penetration, increases the tissue temperature, and leads to safety issues, such as tissue damage from overheating.

The ultrasound has been considered to wirelessly transfer power through the human body, which contains approximately 60% water [19]. However, low PTE, in the order of 40% at 10 mm separation, has severely limited its usage to high-power IMD applications [20]. In addition, the size of typical ultrasonic systems is still too large to be implanted in the body. More recently, a group of researchers have proposed using ultrasound to deliver power across a short distance of a few mm to ultraminiaturized implants distributed in the neural tissue for brain interfacing application [21]. However, few functional devices have been demonstrated at this point.

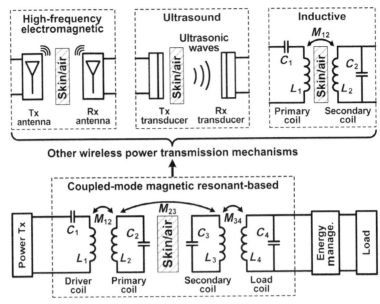

**FIGURE 5.2**

Basic structures used in wireless power transmission.

The inductive power transmission is a viable method to transfer wireless power with high PTE at frequencies in kHz to MHz bands [22–24]. In a conventional inductive link, a primary LC circuit generates a time-varying magnetic field that passes through the secondary loop and induces current according to Faraday's law of induction. In order to provide sufficient power delivered to the load (PDL) with high PTE, both primary and secondary $LC$ circuits should be tuned to match the designated carrier frequency, $f_c$. The coils' geometry, which includes their relative distance, orientation, and number of turns, also significantly changes coils' inductance, quality factor, and mutual coupling, affecting both PDL and PTE.

The coupled-mode magnetic resonant-based power transmission is a similar technique that utilizes three or more coils and significantly improves the PTE particularly at large coupling distances compared to the conventional two-coil inductive links [25,26]. This method operates based on the resonant coupling principle, which states that two same-frequency resonant objects tend to couple. Multicoil links can be utilized for various applications to improve both PDL and PTE while matching impedance values on the source (Tx) and load (Rx) sides of the inductive link [27–29].

In the following sections, the inductive power transmission with two coils will be utilized to describe various power management integrated circuits that are needed in wireless power transfer applications.

## 5.3 OVERALL STRUCTURE OF INDUCTIVELY POWERED DEVICES

A generic inductively powered device consists of three main components: a power transmitter (Tx), an inductive link, and a power receiver (Rx), as shown in Figure 5.3. On the Tx side, a power amplifier (PA) drives the primary coil, $L_1$, at the carrier frequency, $f_c$. This signal is induced on to the secondary coil, $L_2$, through the electromagnetic flux coupling and generates an AC voltage, $V_{COIL}$, across the Rx resonance circuit, $L_2$ and $C_2$. The $L_2C_2$ tank is followed by an AC–DC converter to provide the rest of the receiver with a DC output voltage, $V_{OUT}$.

Various AC–DC converter topologies can be used following the inductive link: passive rectifiers, passive voltage doublers or multipliers, active rectifiers, and active voltage doublers, which will be discussed in Section 5.4. The choice of AC–DC converter structure depends on the desired specifications, such as peak input voltage, power conversion efficiency (PCE), dropout voltage, operating frequency, delivered power capacity, and size. The PCE of the AC–DC converter is a key factor in improving the overall power efficiency of the system because the entire received power from the inductive link needs to pass through this block. Since $V_{OUT}$ may vary significantly with the changes in the coils' relative distance and alignment, a low-dropout regulator (LDO) often follows the AC–DC converter to provide a constant supply voltage, $V_{DD}$, to the load.

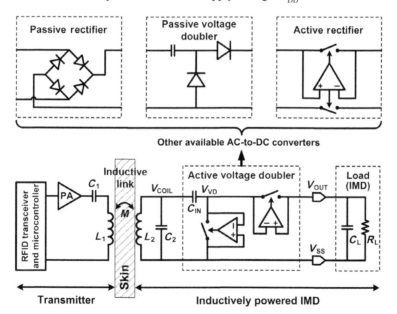

**FIGURE 5.3**

Block diagram of an inductively powered IMD with emphasis on the power transmission through the AC–DC converter.

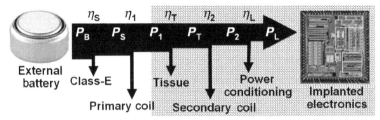

**FIGURE 5.4**

Inductive power transmission flow from the external power source to the IMD electronics [24].

Figure 5.4 shows the power flow from the external energy source to the IMD electronics through the inductive link [24]. Considering power losses at each stage of power delivery path, the total PTE, $\eta_{\text{Total}}$, from the external Tx to the IMD can be calculated from

$$\eta_{\text{Total}} = \eta_S \times \eta_1 \times \eta_T \times \eta_2 \times \eta_L, \tag{5.1}$$

where $\eta_S$, $\eta_1$, $\eta_T$, $\eta_2$, and $\eta_L$ are the efficiencies of the PA, primary $LC$ circuit, transcutaneous powering through the tissue, secondary $LC$ circuit, and power management units. The inductive link efficiency, $\eta_{\text{Link}}$, can be represented as $\eta_1 \times \eta_T \times \eta_2$, while $\eta_L$ typically includes efficiencies of the AC–DC converter and the regulator, $\eta_{\text{ACDC}}$ and $\eta_{\text{Reg}}$, respectively.

Achieving higher overall PTE ($\eta_{\text{Total}}$) is very important in inductively powered applications because it allows them to operate with smaller received power from a larger distance. Alternatively, lower transmitted power also reduces the risk of interference with other devices and damage from overheating. In the IMD applications, the transmitted power level and link efficiency ($\eta_{\text{Link}}$) are typically limited due to the size constraint of the implanted secondary coil. Therefore, improving PCEs of the AC–DC converter ($\eta_{\text{ACDC}}$) and the regulator ($\eta_{\text{Reg}}$) is key for safe operation. Other energy management methods such as rechargeable battery and supercapacitor charging can be applied in the system architecture to utilize the stored energy when the inductive powering is interrupted, insufficient, or unavailable. Nonetheless, high charging efficiency through the inductive link is still highly desired.

## 5.4 AC–DC CONVERSION UNITS

Various AC–DC converters have been utilized to improve PCE over an extended range: (1) passive AC–DC converters, (2) comparator-based active AC–DC converters, (3) adaptive reconfigurable AC–DC converters, and (4) regulated AC–DC converters. In addition, several voltage regulator structures following AC–DC converters can be used to provide constant supply voltages to the load while achieving high efficiency and fast regulation capability.

## 5.4.1 PASSIVE AC–DC CONVERTERS

Passive rectifiers and voltage doublers have simple structures using diodes or diode-connected transistors. However, they suffer from large voltage dropout and power loss because of the large MOSFET threshold voltages, resulting in low PCE [12,30–33]. A bridge rectifier using Schottky diodes has low dropout voltage [34], at the cost of high leakage current and low reverse breakdown voltage. It is not available in a standard CMOS process without extra fabrication steps either. Several threshold voltage ($V_{Th}$) compensation techniques have been proposed to reduce the forward voltage drop by adjusting the effective $V_{Th}$ in passive rectifiers, voltage doublers, and multipliers [35–37]. However, they still need to deal with issues such as sensitivity to process variations, leakage currents, and reverse currents.

Figure 5.5 shows the reduced-$V_{Th}$ diode scheme for passive rectifiers in Ref. [35], as an example of $V_{Th}$ compensation techniques. The conventional half-wave passive rectifier in Figure 5.5a delivers power through $D_1$ (diode-connected MOS) when $V_{IN} > V_{OUT} + V_{Th(D1)}$. Thus, $V_{OUT} = V_{IN} - V_{Th(D1)}$ results in large power dissipation in $D_1$. To reduce $V_{Th(D1)}$, $D_1$ is replaced with a reduced-$V_{Th}$ diode circuit, shown in the dashed box in Figure 5.5b. In this scheme, initially $V_{OUT}$ reaches $V_{IN} - V_{Th(D2)}$ through $D_2$, leading to $V_C = V_{OUT} - V_{Th(D3)} = V_{IN} - 2V_{Th(D2,3)}$. Since $V_{SG(M1)} = V_{IN} - V_C = 2V_{Th(D2,3)} > V_{Th(M1)}$, $M_1$ is pushed into triode, and $V_{OUT}$ is charged up until $V_C$ becomes $V_{IN} - V_{Th(M1)}$. Therefore, $V_{OUT}$ finally reaches $V_{IN} - V_{Th(M1)} + V_{Th(D3)}$, which means that the $V_{Th}$ of $D_1$ has been reduced to $V_{Th(M1)} - V_{Th(D3)}$ with the help of $D_3$ and $C_c$. This simple method reduces the diode voltage drop and increases the PCE, but it has several limitations. First, threshold voltages are sensitive to process variations, and it may be difficult to achieve the desired $\Delta V_{Th}$ ($= V_{Th(M1)} - V_{Th(D3)}$). Therefore, to avoid reverse current, $\Delta V_{Th}$

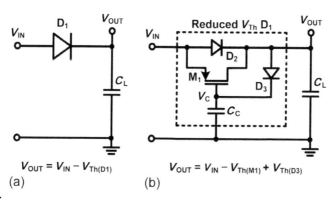

$$V_{OUT} = V_{IN} - V_{Th(D1)}$$

(a)

$$V_{OUT} = V_{IN} - V_{Th(M1)} + V_{Th(D3)}$$

(b)

**FIGURE 5.5**

Schematic diagram of the reduced-$V_{Th}$ diode rectifier scheme: (a) conventional half-wave diode rectifier and (b) reduced-$V_{Th}$ diode rectifier [35].

needs to be designed more than 100–200 mV, which may lead to insufficient PCE. In addition, the body terminals of NMOS and PMOS transistors are connected to the lowest and highest potentials in the system, respectively, and their threshold voltages may vary depending on $V_{IN}$ or $V_{OUT}$.

The passive rectifier described in Ref. [38] utilized the $V_{Th}$ compensation with ferroelectric capacitors, which were connected between the gate and the source of the diode-connected transistors to reduce $V_{Th}$. However, ferroelectric capacitors are not available in standard CMOS processes and impose extra cost in fabrication. A differential-drive CMOS rectifier for UHF RFID applications has been introduced in Ref. [39], which has a cross coupled bridge configuration driven by differential RF inputs. This method aims for higher-frequency RF applications, for example, $f_c = 953$ MHz, with very small input power, as opposed to higher-power biomedical applications, for example, $f_c < 20$ MHz, or charging consumer electronics, $f_c < 1$ MHz.

## 5.4.2 ACTIVE AC–DC CONVERTERS

In order to increase the PCE further by decreasing the rectifier dropout voltage, active rectifiers with synchronous switches are considered the most promising solutions [40–45]. Figure 5.6 shows the schematic diagram of a full-wave active rectifier employing a pair of comparators ($CMP_{1A}$ and $CMP_{2A}$) to drive the rectifying pass transistors ($P_{1A}$ and $P_{2A}$). Since the rectifier input voltage, $V_{IN} = V_{IN1} - V_{IN2}$, ideally has

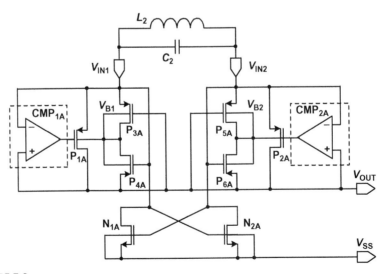

**FIGURE 5.6**

Schematic diagram of the comparator-based active rectifier in which the rectifying pass transistors ($P_{1A}$ and $P_{2A}$) operate in the triode region as switches with low on-resistance [43].

a sinusoidal waveform, $P_{1A}$ and $P_{2A}$ turn on alternatively by the comparators depending on the polarity and amplitude of $V_{IN}$.

When $V_{OUT} > V_{IN}$ $(=V_{IN1} - V_{IN2}) > V_{Th(N2A)}$, the positive feedback operation of the cross coupled NMOS pair ($N_{1A}$ and $N_{2A}$) connects $V_{IN2}$ to $V_{SS}$ through $N_{2A}$ and turns off $N_{1A}$. In this case, $CMP_{2A}$ output goes high because $V_{OUT} > V_{SS}$, and $P_{2A}$ is turned off. $P_{1A}$ also remains off as long as $V_{IN} < V_{OUT}$. However, when $V_{IN} > V_{OUT}$, $CMP_{1A}$ output goes low and turns $P_{1A}$ on. Therefore, current flows from $V_{IN1}$ to $V_{OUT}$ and supplies the rectifier's output load. In the next half cycle, when $|V_{IN}| > V_{Th(N1A)}$, in the opposite direction, $V_{IN1}$ is connected to $V_{SS}$ through $N_{1A}$, and $N_{2A}$ turns off. Both $P_{1A}$ and $P_{2A}$ are also initially off when $|V_{IN}| < V_{OUT}$. However, when $|V_{IN}| > V_{OUT}$, $CMP_{2A}$ turns $P_{2A}$ on and current flows from $V_{IN2}$ to $V_{OUT}$ to supply the output load again. Since $V_{IN1}$ and $V_{IN2}$ undergo large and sharp voltage transitions, it is important to prevent latch-up and substrate leakage currents, particularly among $P_{1A}$ and $P_{2A}$. Dynamic body biasing technique from Ref. [30] has been commonly used by adding auxiliary PMOS transistors, $P_{3A}$ to $P_{6A}$, to automatically connect $V_{B1}$ and $V_{B2}$ to the highest potential among the input and output voltages, that is, $\max(V_{IN1}, V_{OUT})$ and $\max(V_{IN2}, V_{OUT})$, respectively.

In this active rectifier, voltage drop across the rectifying pass transistors is much lower than the diode voltage drop because the pass transistors operate in the triode region as switches with low on-resistance. The dropout voltage of the active rectifier can be as low as $V_{SD(P1,2)} + V_{DS(N1,2)}$, resulting in less power dissipation in the rectifying transistors than passive rectifiers. Moreover, active rectifiers are less sensitive to process variation by not depending on the $V_{Th}$ values. On the downside, the comparators' power consumption and delay can result in PCE degradation, which should be dealt with.

To maximize the PCE in active rectifiers, the pass transistors, $P_{1A}$ and $P_{2A}$, need to turn on and off at proper times. If they turn on too late, the forward current conduction time, during which power is delivered from the $LC$ circuit to the load, will be wasted. On the other hand, if the pass transistors turn off too late, reverse current flows from the load back to the $LC$ circuit when $|V_{IN}| < V_{OUT}$, severely degrading the PCE. Comparators play a more critical role in achieving high PCE when the active rectifier needs to operate at high frequencies, such as 13.56 MHz in the industrial, scientific, and medical band.

Various comparator topologies have been utilized to optimize the timing of active rectifier switching and maximize the PCE. In Ref. [40], a negative feedback loop was added to a 4-input common-gate comparator (CG CMP) in order to turn off the pass transistors early and prevent the reverse current. Even though this comparator can operate at high frequencies in the order of 13.56 MHz, the turn-off time may change with process variations along with the feedback loop. Moreover, the turn-on delay compensation was not considered in this design, and as such, there was room to further improve the PCE. An active rectifier using a phase lead comparator was introduced in Ref. [41], which compensated for both turn-on and turn-off delays. However, its maximum operating frequency was limited to ~2 MHz, and an additional supply voltage was required for the comparators from an auxiliary passive rectifier that further complicated the design. The active rectifier in Ref. [42] was equipped with unbalanced-biased comparators, which generated the intended offset inside the comparator to reduce the turn-off delay. However, this rectifier also operated at low frequencies <1.5 MHz, and the turn-on delay was increased because of the comparator's unbalanced biasing condition.

To switch the large pass transistors at the right times while operating at high frequencies, an active rectifier using high-speed comparators with offset-control functions was proposed in Ref. [43]. Figure 5.7 shows the conceptual block diagram of the comparator used in this rectifier, which consists of a CG CMP, two offset-control blocks (Offset$_F$ and Offset$_R$), and two current-starved (CS) inverters. In this design, offset-control blocks alternately inject a programmable offset current, OS$_F$ and OS$_R$, into the inputs of the CG comparator depending on the state of the $V_{CMP}$ feedback signals, FB$_F$ and FB$_R$. Therefore, offset-control functions expedite the falling or rising transition by sensing them ahead of time to compensate for both turn-on and turn-off delays in comparators. These functions improve the PCE by maximizing the forward current delivered to the load and minimizing the reverse current.

More recently, an active rectifier with cross coupled latched comparators was proposed in Ref. [44], which utilized a pair of four-input CG CMPs that are capacitively cross coupled to drive the rectifying switches, controlling the reverse leakage current and maximizing the PCE at 13.56 MHz. The active rectifier in Ref. [45] used comparators with switched-offset biasing to compensate for the delays at 13.56 MHz while eliminating multiple pulsing and reverse current. Moreover, the comparator utilized a modified peaking current source to reliably control the reverse current over a wide AC input range.

Even though the aforementioned active rectifiers can achieve a high PCE at 13.56 MHz, they still require higher input voltage than the output voltage due to the inevitable rectifier dropout voltage across the on-resistance of the pass transistors. To address this limitation, active voltage doublers in which pass transistors are driven by synchronous comparators have been proposed [46–48]. Active voltage doublers in Refs. [46,47] were designed for energy scavenging from mechanical vibrations

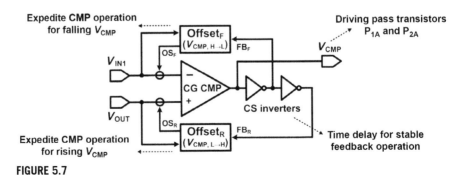

**FIGURE 5.7**

Block diagram of the high-speed comparator employing offset-control functions for both falling and rising $V_{CMP}$ transitions [43].

through piezoelectric transducers, while they operate at low frequencies in the order of 100 Hz. Even though these active voltage doubles offer high PCE, they are not suitable for inductively powered biomedical applications, which operate at much higher frequencies through near-field inductive links. Moreover, active voltage doubler in Ref. [47] was implemented with discrete components. Recently, a high-frequency active voltage doubler was reported in Ref. [48], which used offset-control functions in its comparators to achieve high PCE and high output voltage at 13.56 MHz.

### 5.4.3 ADAPTIVE RECONFIGURABLE AC–DC CONVERTERS

Active rectifiers require higher peak inputs than the desired outputs, which may be temporarily unavailable due to the weak coupling of the inductive links. On the other hand, active voltage doublers are capable of generating higher output voltages than their inputs, but their PCEs are generally lower than active rectifiers with similar size. In order to address such limitations, adaptive reconfigurable AC–DC converters have been proposed for robust wireless power transmission through inductive links over an extended range [49,50].

In Ref. [49], both voltage doubler (VD) and rectifier (REC) modes are integrated into a single structure, employing low-dropout active synchronous switches, leading to high PCE. Moreover, by adding an output voltage sensing circuit, reconfigurable VD/REC can automatically change its operating mode to either VD or REC depending on which one is a better choice for generating the desired output voltage to accommodate with a wider range of mutual coil couplings.

Figure 5.8 shows the block diagram of a wireless power transmission link that includes the active VD/REC. A PA drives the primary coil, $L_1$, at the designated carrier frequency (here $f_c = 13.56$ MHz), which improves the coils' quality factors ($Q$) and increases the overall power transmission efficiency while maintaining the sizes of $LC$ components small for implantable applications. Coupled signal across the secondary coil, $L_2$, creates an AC voltage, $V_{IN}$ ($= V_{INP} - V_{INN}$), across $L_2 C_2$, which is tuned at $f_c$. VD/REC, which follows the $L_2 C_2$ tank, converts $V_{IN}$ to an automatically adjusted DC voltage, $V_{OUT}$, for supplying the load after regulation. If $V_{IN}$ falls below a certain level, which is determined by comparing a portion of $V_{OUT}$ with a reference voltage, $V_{REF}$, using a hysteresis comparator, then mode = 1 and VD/REC operates in VD mode. Since the voltage doubler can generate the desired $V_{OUT}$ with much lower $V_{IN}$ than the rectifier, VD/REC can still provide sufficient $V_{OUT}$ to the load even with decreased $V_{IN}$. On the other hand, if $V_{IN}$ increases above $V_{REF}$ + hysteresis window, then mode = 0 and VD/REC will operate in the REC mode, which can achieve higher PCE than the VD mode while generating the desired $V_{OUT}$.

The active VD/REC can be a combination of two separate AC–DC converters, a rectifier and a voltage doubler, in which the operating mode, REC or VD, can be selected based on the input or output voltages. Figure 5.9 shows the conceptual diagram of the active VD/REC converter that consists of the full-wave rectifier and the voltage doubler with active diodes. The full-wave rectifier requires two diodes, $D_1$ and $D_2$, and a cross coupled NMOS pair, $N_1$ and $N_2$. Either $D_1$–$N_2$ path or $D_2$–$N_1$ path

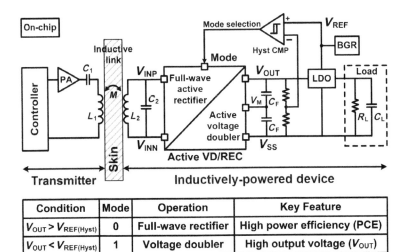

| Condition | Mode | Operation | Key Feature |
|---|---|---|---|
| $V_{OUT} > V_{REF(Hyst)}$ | 0 | Full-wave rectifier | High power efficiency (PCE) |
| $V_{OUT} < V_{REF(Hyst)}$ | 1 | Voltage doubler | High output voltage ($V_{OUT}$) |

**FIGURE 5.8**

Block diagram of an inductively powered device with emphasis on the wireless power transmission through the active VD/REC converter [49].

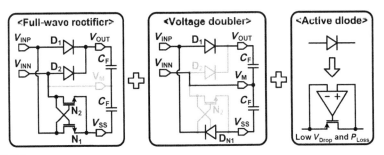

**FIGURE 5.9**

Conceptual diagram of the active VD/REC converter in which a full-wave rectifier and a voltage doubler are combined using active diodes [49].

is activated depending on the amplitude of $V_{INP}$ and $V_{INN}$ to transfer the input power to the output filtering capacitors, $C_F/2$. The voltage doubler requires only two diodes, $D_1$ and $D_{N1}$, charging one $C_F$ per half cycle depending on the polarity and amplitude of $V_{IN}$ ($=V_{INP}-V_{INN}$). Therefore, $V_{OUT}$ becomes almost doubled compared to the peak voltage of $V_{IN}$. In order for VD/REC to include both structures, $D_1$ is shared, and $D_2$ and $N_2$ have enable functions to turn them on/off in the REC and VD modes, respectively. $N_1$ operates as part of a cross coupled pair in the REC mode, while it is reconfigured as an NMOS diode, $D_{N1}$, in the VD mode. $V_{INN}$ and $V_M$ are also shorted through a switch, $N_3$, in the VD mode. VD/REC utilizes active diodes, $D_1$, $D_2$, and $D_{N1}$, in which rectifying pass transistors are driven by fast comparators to operate as switches in the deep triode region with low dropout voltages. Therefore, these active diodes dissipate less power compared to passive diodes, leading to higher PCE in both operating modes.

### 5.4.4 REGULATED AC–DC CONVERTERS

While the active AC–DC converters are capable of providing the DC voltage with high PCE, their output voltages easily vary because of coupling variations and coil misalignments, which significantly affect the input amplitude. Therefore, there should be a voltage regulator to generate the constant supply voltage from the variable output voltage of the AC–DC converter. The voltage regulator results in additional power loss, especially when its input voltage is much higher than the desired output voltage. Recently, several designers have proposed combinations of the AC–DC converter and voltage regulator, which can generate the regulated output voltage through one-step AC–DC conversion, to increase overall power efficiency, reduce the chip area, and decrease the number of off-chip filtering capacitors [51–54].

Figure 5.10 shows the simplified voltage waveforms of the conventional and regulated AC–DC converters (rectifiers) depending on their turn-on topologies. Conventional rectifiers aim to generate the maximum output voltage, $V_{REC}$, from the AC input, $V_{IN}$, at high PCE. Thus, they turn on as long as $V_{IN} > V_{REC}$, as shown in Figure 5.10a. Consequently, $V_{REC}$ becomes dependent on the $V_{IN}$ amplitude, and it is not internally adjustable. In Figure 5.10b, $V_{REC}$ can be adjusted by controlling the turn-on time around the peak of $V_{IN}$. If the turn-on period is reduced, the lower forward current reduces $V_{REC}$ as well. However, the large voltage drop between $V_{IN}$ and $V_{REC}$ during the turn-on period results in large power loss across the rectifying transistors, resulting in low PCE. To adjust $V_{REC}$ while maintaining high PCE, the rectifier turn-on phase can be controlled as shown in Figure 5.10c. In this method, the rectifier turns on when $V_{IN} > V_{REC}$, similar to the conventional rectifiers. However, its turn-off timing is controlled to limit the forward current. Therefore, $V_{REC}$ is adjustable depending on the rectifier turn-on phase, while the small dropout voltage between $V_{IN}$ and $V_{REC}$ during the on period provides high PCE.

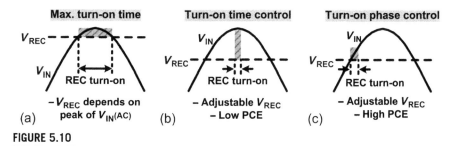

**FIGURE 5.10**

Simplified voltage waveforms of the rectifiers with (a) the maximum turn-on time, (b) the turn-on time control, and (c) the turn-on phase control.

A timing-controlled AC–DC converter in Ref. [51] utilized two converter cores that detect the output voltage and adjust the rectifier turn-off timing, as shown in Figure 5.10c, through the comparators, digital controller, and boosted switch driver. Therefore, the turn-off timing is adaptively controlled by target output voltage, input amplitude, and load current. It also included an additional full-wave rectifier to supply the converter cores. This regulated AC–DC converter generates 1.25–2 V output voltages from 3.5 V peak input amplitude at 1 MHz with 1–10 mW output power while achieving 60–90% PCE. The hybrid combination of a rectifier and a regulator in Ref. [52] used two negative feedbacks through error amplifiers and rectifying switches, so that the regulated rectifier turns on and charges the output capacitor when the output voltage was lower than the reference voltage. It can provide the regulated 2.4 V DC output when the input amplitude varies from 4 to 7 V while requiring smaller on-chip and off-chip area for size-constraint applications.

The adaptive regulated rectifier in Ref. [53] utilized the phase control feedback loop that can be added to conventional comparators to control the turn-off timing and the output voltage level while performing active AC–DC conversion with high PCE. Figure 5.11 shows the schematic diagrams of the adaptive regulated rectifier with active switches and one of its phase control comparators. In Figure 5.11a, a pair of comparators, $CMP_1$ and $CMP_2$, which are equipped with the phase control feedback, drives the rectifying switches, $P_1$ and $P_2$, respectively, for low dropout voltage and high PCE. The reference voltage, $V_{REF}$, controls the transition times of the comparator output voltages, $V_{O1}$ and $V_{O2}$, in a way that the rectifier turn-off timing can be adjusted to change the turn-on phase and consequently the $V_{REC}$ level. $P_1$ and $P_2$ turn on alternatively depending on $V_{INP,N}$ polarity, while a cross coupled NMOS pair, $N_1$ and $N_2$, closes the rectifier current path.

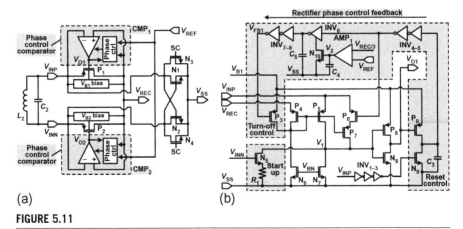

(a)                    (b)

**FIGURE 5.11**

Schematic diagrams of (a) the adaptive regulated rectifier with active switches and (b) one of its phase control comparators, $CMP_1$ [53].

In the phase control comparator (CMP$_1$), shown in Figure 5.11b, P$_4$, P$_5$, N$_6$, N$_7$, P$_8$, and N$_8$ form a CG CMP with input voltages, $V_{REC}$ and $V_{INP}$, while the current source, P$_7$, injects additional current when $V_{O1}$ is high and P$_6$ turns on, forcing $V_1$ to increase earlier and expedite the turn-on transition of P$_1$. The phase control feedback loop consists of inverter chains along with the CS inverter, INV$_6$ and N$_{10}$, in which bias current is controlled through AMP$_1$ by comparing $V_{REC}/3$ and $V_{REF}$ to generate the corresponding time delay. INV$_6$ output is further delayed before affecting the turn-off control transistor, P$_3$, which forces the rectifier to turn off adaptively even before $V_{INP} < V_{REC}$ to generate the desired $V_{REC}$. Therefore, unlike conventional AC–DC converters, in which output levels are dependent on the $V_{INP,N}$ amplitude, the adaptive rectifier is capable of generating variable supply voltages regardless of the $V_{INP,N}$ amplitude, thanks to the phase control feedback.

In Figure 5.11b, a start-up circuit with $R_1$ and N$_5$ driven by $V_{INN}$ guarantees the rectifier operation before $V_{REC}$ is charged up without requiring additional start-up circuits while not affecting the normal rectifier operation after start-up. The reset control circuit on the lower right resets the phase control feedback loop to turn off P$_3$ and P$_6$ after P$_1$ turns off and $V_{INP}$ goes low. Here, the timing of the reset signal depends on $V_{INP}$, which is independent of process variations.

Figure 5.12 shows the timing diagram of the adaptive regulated rectifier depending on the actual $V_{REC}$ level versus the target $V_{REC}$, which is $3 \times V_{REF}$. For example, when $V_{REC} > 3V_{REF}$ in Figure 5.12a, AMP$_1$ increases $V_2$, decreasing the delay of INV$_6$. Once $V_{O1}$ drops to turn on the rectifier, P$_3$ also turns on by $V_{FB1}$ after a small delay, $T_D$, limiting the charging period of the load and decreasing $V_{REC}$. On the other hand, when $V_{REC} < 3V_{REF}$ in Figure 5.12b, the delay of INV$_6$ increases as $V_2$ decreases, and P$_3$ turns on after a longer $T_D$ or even remains off, allowing more forward current to increase $V_{REC}$. When $V_{REC} = 3V_{REF}$ in Figure 5.12c, $V_2$ results in a $T_D$ that can maintain $V_{REC}$ at the desired value. Since the turn-off timing is controlled in every rectifier cycle, the ripple on $V_{REC}$ can be reduced to that of conventional rectifiers once it is settled on

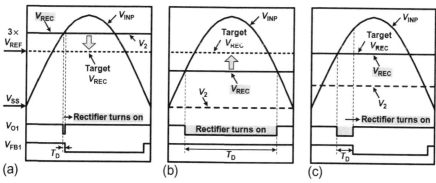

(a)     (b)     (c)

**FIGURE 5.12**

Timing diagram of the adaptive regulated rectifier when (a) $V_{REC} > 3V_{REF}$, (b) $V_{REC} < 3V_{REF}$, and (c) $V_{REC} = 3V_{REF}$ [53].

the desired $V_{REC}$ value. This adaptive regulated rectifier can generate adjustable DC output between 2.5 and 4.6 V with 3-bit resolution from 5 V peak AC input at 2 MHz while achieving 72–87% PCE at 2.8 mA loading. It also results in small output voltage change (<3 mV) against input amplitude variations between 3 and 5 V, ensuring the line regulation capability.

More recently, a resonant regulating rectifier has been proposed in Ref. [54] for resonant wireless power transfer at 6.78 MHz. By sharing the secondary resonant coil in the inductive link, this regulating rectifier can perform switching-mode regulation without requiring any additional inductor. It employed both continuous and discontinuous conduction modes for different output power levels between 0 and 6 W while achieving peak PCE of 86% for the switching regulator and 55% for the overall system, including full-bridge diode rectifier.

## 5.4.5 VOLTAGE REGULATORS

Voltage regulators can generate a constant supply voltage from the AC–DC converter output, which can vary due to coupling and load variations. The linear voltage regulators, including LDOs, have been widely used because of their compact structure and fast operation with no switching noise and low ripple voltage. Figure 5.13 shows a typical LDO structure, which consists of a voltage reference, error amplifier, pass transistor, and resistive divider. A portion of the output voltage, $V_{DD}$, is sensed through the resistive dividers, $R_1$ and $R_2$, and compared to the reference voltage, $V_{REF}$. The error amplifier generates a corresponding error signal, $V_{EA}$, to adaptively control the voltage drop across the pass transistor by adjusting its gate voltage. Thus, the feedback control loop ensures that $V_{DD}$ is close to the target supply voltage, that is, $V_{REF} \times (1 + R_1/R_2)$, when the loop gain is sufficiently large.

The error amplifier used in the linear regulator should consume very low power while achieving high slew rate to drive the large parasitic gate capacitance, $C_G$, of the pass transistor in response to rapid loading and input voltage variations. Moreover, the feedback loop should have high gain and bandwidth to accommodate with wide variations in the input voltage (line regulation) and load current (load regulation) in

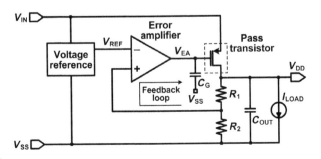

**FIGURE 5.13**

Schematic diagram of a typical LDO.

wirelessly powered biomedical applications. Several linear regulator topologies have been reported to achieve these features.

The linear regulator in Ref. [55] utilized a high-slew-rate amplifier with a push–pull output stage to improve transient response while consuming low quiescent current to increase light-load efficiency. This amplifier is capable of providing the instantaneous push–pull transient output current, which can reduce the required time to charge or discharge the gate capacitance of the large pass transistor, leading to improved load and line regulation. The slew-rate enhancement block in Ref. [56] injected extra instantaneous current to the gate of the pass transistor, and the active feedback compensation loop reduced the required compensation capacitance, leading to high slew rate for fast transient responses. In addition, the dynamic bias-current boosting technique in Ref. [57] instantaneously increased the bias current of the error amplifier when the output transient occurs, which improves both slew rate and bandwidth of the linear regulator. It consumes low power while improving line and load regulation without requiring an output capacitor.

More recently, a linear regulator with embedded voltage reference was proposed in Ref. [58] to generate a low regulated 0.6 V supply from 0.65 to 0.9 V input voltage for low-power biomedical and energy-harvesting applications while occupying small chip area with no additional reference generator circuits. The LDO regulator in Ref. [59] utilized the flipped voltage follower and buffer impedance attenuation techniques to push the internal poles to be higher than the unity-gain frequency, leading to high power supply rejection ratio at high frequency ranges (<−12 dB from DC to 20 GHz), which is suitable for wideband communication applications. This fully integrated regulator also achieved 0.6 ns transient response when the load current changed between 0 and 10 mA.

Power efficiency in these linear regulators drops when the output voltage is far below the input voltage because of the large voltage drop across the pass transistor. To improve the regulator efficiency, switched-capacitor regulators offer a suitable alternative. However, they require large chip area for their capacitors and suffer from switching noise [60–62]. Switched-capacitor regulators can generate output voltages that can be much higher or lower than their input voltages while maintaining their high efficiency by charging capacitors and reconfiguring capacitor connections through switches. The switched-capacitor regulators can be useful especially when the input voltage needs to generate much lower or higher output voltage because they can minimize the power losses across the regulator.

## 5.5 RECHARGEABLE BATTERY AND SUPERCAPACITOR CHARGING UNITS

Since the wireless power transmission through the inductive link can be affected by coupling variations or coil misalignments, the received power of the IMDs can be interrupted, leading to system supply failure. Alternatively, it might be advantageous for the IMD to be able to operate independently for certain periods of time without requiring the inductive link and its associated external power transmitter, for example, when taking a shower. To address this issue, secondary energy sources can be utilized in IMD applications, which supply the low-power IMDs or augment the inductive power when it is unavailable or insufficient. These energy sources should have small size, high energy capacity, and long lifetime. Moreover, fast and efficient charging schemes through the inductive link would be highly desired.

### 5.5.1 SECONDARY ENERGY SOURCES

As secondary energy sources, rechargeable batteries and supercapacitors can be employed to store the harvested energy from the inductive link because of their small size and high power density. In general, the rechargeable battery can achieve high capacity and low leakage rate. However, the cycle life, which is defined as the number of charge–discharge cycle before its capacity falls below 80% of its initial rated capacity, is relatively short, limiting the lifetime in IMDs. For example, the cycle life of Li-ion battery is 1000–1200 cycles [63]. On the other hand, a supercapacitor has more than 1,000,000 cycle life and also benefits from high charge–discharge efficiency and fast charge–discharge characteristics [64]. However, the supercapacitor suffers from lower energy density and higher leakage rate. To optimize the secondary energy sources, several systems have been proposed, using either supercapacitors alone or a combination of supercapacitors and rechargeable batteries [65], while requiring the development of power-efficient wireless charging units for rechargeable batteries and supercapacitors.

### 5.5.2 LI-ION BATTERY CHARGER

Li-ion batteries have been widely used in size-constrained portable systems because of their high energy density, high full-charge voltage, and absence of memory effects. However, battery longevity and capacity are highly sensitive to the optimum full-charge voltage, $V_{FC}$. For example, undercharging the battery by 1.2% of its full-charge voltage leads to 9% reduction in battery capacity [66]. On the other hand, overcharging the battery results in electrolyte oxidation and decomposition, leading to the risk of thermal runaway [67]. In addition, deeply discharging the battery below the minimum discharge voltage, $V_{Min}$, can permanently reduce the battery capacity [68]. Therefore, to maximize the battery capacity and cycle life, the Li-ion battery should be charged to within 1% of its optimum full-charge voltage, which is typically 4.2 V, through the well-known constant current (CC) and constant voltage (CV) charging techniques [69]. Moreover, reliable and safe operation of the charger under varying supply voltage needs to be guaranteed especially for wirelessly charged IMD applications.

Figure 5.14 shows the CC–CV charge profile with four distinct regions: (1) trickle current charge, (2) CC charge, (3) CV charge, and (4) end of charge. When deeply discharged, the battery should be charged with a small amount of trickle current, $I_{TC}$, which is typically no more than 0.1 times the rated battery capacity, $C$ (mAh), up to

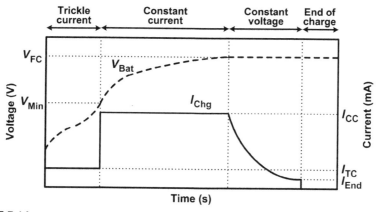

**FIGURE 5.14**

Li-ion battery charging profile with the CC–CV technique.

$V_{Min}$, for example, 3 V, to prevent heat generation and capacity loss. Once the battery voltage, $V_{Bat}$, reaches $V_{Min}$, the battery can be charged with higher CC, $I_{CC}$, which is typically 1–1.5C, for CC charging. Then, when $V_{Bat}$ nears $V_{FC}$, for example, 4.2 V, it enters the CV charging mode, which charges the battery with CV to gradually decrease the charge current and smoothly reach to $V_{FC}$. The charger stops its operation when the charging current falls below end-of-charge current, $I_{End}$.

Several Li-ion battery chargers have been proposed to achieve precise full-charge battery voltage with accurate CC–CV charging while reducing the chip area and improving the charging efficiency. Figure 5.15 shows various Li-ion battery charging techniques with an emphasis on the CC charging mode, which dominates the overall charging efficiency, when they are inductively charged. To supply the battery charger, the AC–DC converters and regulators typically convert an AC input voltage from an inductive link to a DC supply voltage, $V_{DD}$, resulting in AC–DC power loss. In Figure 5.15a, the constant current source ($I_S$) directly charges the battery by

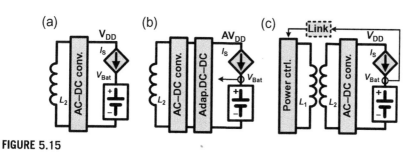

**FIGURE 5.15**

Various inductively powered Li-ion battery charging techniques in the current source ($I_S$) charging mode with (a) a fixed supply voltage [70,71], (b) an adaptive supply voltage [72], and (c) a supply voltage adjusted by an external control loop [73].

controlling the gate voltage of the $I_S$ [70,71]. However, the current source still wastes energy because of the voltage drop between supply and battery voltages, $V_{DD} - V_{Bat}$. Generating an adaptive supply voltage, $AV_{DD}$, in Figure 5.15b keeps the dropout voltage of the current source small, $AV_{DD} - V_{Bat}$, while suffering from the additional DC–DC power loss [72]. The charging system in Figure 5.15c utilizes a back telemetry link to control the inductive power, adjusting $V_{DD}$ depending on the $V_{Bat}$ level to reduce the voltage drop across the current source [73]. However, it requires additional sensing and control circuits and an external feedback loop through an optical link.

### 5.5.3 WIRELESS CAPACITOR CHARGER

Large capacitors can also be utilized as temporary energy sources and augment the inductively delivered power when it is interrupted or insufficient [74,75]. Capacitors can also be used in neural stimulation by storing charge and transferring it to the tissue periodically at high efficiency [76,77]. Therefore, it is important to charge capacitors rapidly and efficiently not from batteries but directly through inductive links while reducing the risk of overheating. It can be proven that charging capacitors from a voltage source through a switch achieves less than 50% efficiency, wasting at least half of the input energy in the switch. On the other hand, charging capacitors with a current source can minimize the switching loss as the fixed charging current becomes smaller [78]. However, it still requires the DC supply voltage through AC–DC conversion to generate the current source, while the voltage drop across the current source also wastes significant power during charging. Unlike Li-ion batteries, capacitor charging is not constrained by the specific charging profile such as CC or CV modes. Therefore, various circuit techniques can be used to improve the capacitor charging efficiency through wireless power transfer.

Figure 5.16 shows the wireless capacitor charging system in Ref. [79], which efficiently charges positive and negative capacitor banks directly from the AC input voltage through the inductive link, without requiring AC–DC converters, regulators, or current sources. A power transmitter drives the primary coil, $L_1$, at the designated carrier frequency, $f_c$. The secondary coil, $L_2$, and its parallel resonant capacitor, $C_2$, which generate a coil voltage, $V_{COIL}$, are followed by a series charge injection capacitor, $C_S$, which provides an input voltage, $V_{IN}$, to a capacitor charger. The capacitor charger

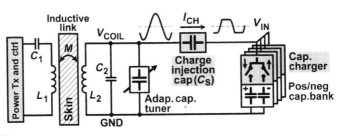

**FIGURE 5.16**

Wireless capacitor charging system through an inductive link [79].

consists of switches driven by high-speed active drivers to charge a bank of positive and negative capacitors. When charging, the capacitor charger connects $V_{IN}$ to either positive or negative capacitors to hold $V_{IN}$ relatively constant, while voltage variation across $C_S$ generates a fixed amount of charging current, $I_{CH}$. For example, when $V_{COIL}$ increases and $V_{IN}$ is less than the positive capacitor voltage, the capacitor charger turns off, and $V_{IN}$ becomes floating. Thus, $V_{IN}$ also increases along with $V_{COIL}$. When $V_{IN}$ exceeds the positive capacitor voltage, the capacitor charger connects $V_{IN}$ to the positive capacitor voltage, which remains relatively constant during one carrier cycle, while $V_{COIL}$ continues increasing. Therefore, the voltage variation across $C_S$ generates a positive charging current, $+I_{CH}$, to charge the positive capacitor. The amount of $I_{CH}$ can be controlled by adjusting $C_S$, $V_{COIL}$ amplitude, and $f_c$. Charging the capacitors with $I_{CH}$ can minimize the switching losses in the capacitor charger. Moreover, the voltage drop across $C_S$, which operates like an ideal current source, does not dissipate power, maximizing the capacitor charging efficiency from $L_2$ to the capacitor bank. An adaptive capacitor tuner compensates for the $L_2C_2$ tank resonance capacitance variations during charging and maintains $V_{COIL}$ amplitude constant at its peak.

## REFERENCES

[1] Wilson BS, Dorman MF. Cochlear implants: a remarkable past and a brilliant future. Hear Res 2008;242(1–2):3–21.

[2] Cruz LD, Coley BF, Dorn J, Merlini F, Filley E, Christopher P, et al. The Argus II epiretinal prosthesis system allows letter and word reading and long-term function in patients with profound vision loss. Br J Ophthalmol 2013;97(5):632–6.

[3] Zhong WX, Liu X, Hui SYR. A novel single-layer winding array and receiver coil structure for contactless battery charging systems with free-positioning and localized charging features. IEEE Trans Ind Electron 2011;58(9):4136–44.

[4] Near field communication (NFC) forum. Available: http://www.nfc-forum.org [accessed 19.02.14].

[5] Wang C, Stielau O, Covic G. Design considerations for a contactless electric vehicle battery charger. IEEE Trans Ind Electron 2005;52:1308–14.

[6] Shire D, Kelly S, Chen J, Doyle P, Gingerich M, Cogan S, et al. Development and implantation of a minimally invasive wireless subretinal neurostimulator. IEEE Trans Biomed Eng 2009;56(10):2502–11.

[7] Ortmanns M, Gehrke M, Tiedtke H. A 232-channel epiretinal stimulator ASIC. IEEE J Solid-State Circuits 2007;42(12):2946–56.

[8] Nurmikko AV, Donoghue JP, Hochberg LR, Patterson WR, Song Y-K, Bull CW, et al. Listening to brain microcircuits for interfacing with external world-progress in wireless implantable microelectronic neuro-engineering devices. Proc IEEE 2010;98:375–88.

[9] Ghovanloo M. Integrated circuits for neural interfacing: neural stimulation. In: Iniewski K, editor. VLSI circuits for biomedical applications. Norwood, MA: Artech House; 2008.

[10] Rasouli M, Phee LS. Energy sources and their development for application in medical devices. Expert Rev Med Devices 2010;7(5):693–709.

[11] Haddad SAP, Houben RPM, Serdijn WA. The evolution of pacemakers. IEEE Eng Med Biol Mag 2006;25(3):38–48.

[12] Ghovanloo M, Atluri S. An integrated full-wave CMOS rectifier with built-in back telemetry for RFID and implantable biomedical applications. IEEE Trans Circuits Syst I Regul Pap 2008;55(10):3328–34.

[13] Catrysse M, Hermans B, Puers R. An inductive power system with integrated bidirectional data-transmission. Sens Actuators A 2004;115:221–9.

[14] Ghovanloo M, Najafi K. A wireless implantable multichannel microstimulating system-on-a-chip with modular architecture. IEEE Trans Neural Syst Rehabil Eng 2007;15(3):449–57.

[15] Sauer C, Stanacevic M, Cauwenberghs G, Thakor N. Power harvesting and telemetry in CMOS for implanted devices. IEEE Trans Circuits Syst I Regul Pap 2005;52(12):2605–13.

[16] Fujii T, Ibata Y. Effects of heating on electrical activities of guinea pig olfactory cortical slices. Eur J Physiol 1982;392:257–60.

[17] Brown W. The history of power transmission by radio waves. IEEE Trans Microw Theory Tech 1984;32:1230–42.

[18] Poon A, O'Driscoll S, Meng T. Optimal frequency for wireless power transmission into dispersive tissue. IEEE Trans Antennas Propag 2010;58:1739–50.

[19] Ozeria S, Shmilovitza D, Singera S, Wang C. Ultrasonic transcutaneous energy transfer using a continuous wave 650 kHz Gaussian shaded transmitter. Ultrasonics 2010;50:666–74.

[20] Zhu Y, Moheimani S, Yuce M. Ultrasonic energy transmission and conversion using a 2-D MEMS resonator. IEEE Electron Device Lett 2010;31:374–6.

[21] Seo D, Carmena JM, Rabaey JM, Alon E, Maharbiz MM. Neural dust: an ultrasonic, low power solution for chronic brain-machine interfaces, 2013. arXiv:1307.2196.

[22] Baker M, Sarpeshkar R. Feedback analysis and design of RF power links for low-power bionic systems. IEEE Trans Biomed Circuits Syst 2007;1:28–38.

[23] Kendir G, Liu W, Wang G, Sivaprakasam M, Bashirullah R, Humayun M, et al. An optimal design methodology for inductive power link with class-E amplifier. IEEE Trans Circuits Syst I Regul Pap 2005;52:857–66.

[24] Jow U, Ghovanloo M. Design and optimization of printed spiral coils for efficient transcutaneous inductive power transmission. IEEE Trans Biomed Circuits Syst 2007;1:193–202.

[25] Haus H, Huang W. Coupled-mode theory. Proc IEEE 1991;79:1505–18.

[26] Karalis A, Joannopoulos J, Soljacic M. Efficient wireless non-radiative mid-range energy transfer. Ann Phys 2007;323:34–48.

[27] Cannon L, Hoburg J, Stancil D, Goldstein S. Magnetic resonant coupling as a potential means for wireless power transfer to multiple small receivers. IEEE Trans Power Electron 2009;24:1819–25.

[28] RamRakhyani A, Mirabbasi S, Chiao M. Design and optimization of resonance-based efficient wireless power delivery systems for biomedical implants. IEEE Trans Biomed Circuits Syst 2011;5:48–63.

[29] Sample A, Meyer D, Smith J. Analysis, experimental results, and range adaptation of magnetically coupled resonators for wireless power transfer. IEEE Trans Ind Electron 2011;58:544–54.

[30] Ghovanloo M, Najafi K. Fully integrated wideband high-current rectifiers for inductively powered devices. IEEE J Solid-State Circuits 2004;39(11):1976–84.

[31] Ham JV, Puers R. A power and data front-end IC for biomedical monitoring systems. Sens Actuators A 2008;147(2):641–8.

[32] Sauer C, Stanacevic M, Cauwenberghs G, Thakor N. Power harvesting and telemetry in CMOS for implanted devices. IEEE Trans Circuits Syst I Regul Pap 2005;52(12):2605–13.

[33] Sawan M, Hu Y, Coulombe J. Wireless smart implants dedicated to multichannel monitoring and microstimulation. IEEE Circuits Syst Mag 2005;5(1):21–39.

[34] Li P, Bashirullah R. A wireless power interface for rechargeable battery operated medical implants. IEEE Trans Circuits Syst II Express Briefs 2007;54(10):912–6.

[35] Le T, Han J, Jouanne A, Marayam K, Fiez T. Piezoelectric micro-power generation interface circuits. IEEE J Solid-State Circuits 2006;41(6):1411–20.

[36] Mounaim F, Sawan M. Integrated high-voltage inductive power and data-recovery front end dedicated to implantable devices. IEEE Trans Biomed Circuits Syst 2011;5(3):283–91.

[37] Yoo J, Yan L, Lee S, Kim Y, Yoo H. A 5.2 mW self-configured wearable body sensor network controller and a 12 μW 54.9% efficiency wirelessly powered sensor for continuous health monitoring system. IEEE J Solid-State Circuits 2010;45(1):178–88.

[38] Nakamoto H, Yamazaki D, Yamamoto T, Kurata H, Yamada S, Mukaida K, et al. A passive UHF RF identification CMOS tag IC using ferroelectric RAM in 0.35-μm technology. IEEE J Solid-State Circuits 2007;42(1):101–10.

[39] Kotani K, Sasaki A, Ito T. High-efficiency differential-drive CMOS rectifier for UHF RFIDs. IEEE J Solid-State Circuits 2009;44(11):3011–8.

[40] Lam YH, Ki WH, Tsui CY. Integrated low-loss CMOS active rectifier for wirelessly powered devices. IEEE Trans Circuits Syst II Express Briefs 2006;53(12):1378–82.

[41] Bawa G, Ghovanloo M. Active high power conversion efficiency rectifier with built-in dual-mode back telemetry in standard CMOS technology. IEEE Trans Biomed Circuits Syst 2008;2(3):184–92.

[42] Guo S, Lee H. An efficiency-enhanced CMOS rectifier with unbalanced-biased comparators for transcutaneous-powered high-current implants. IEEE J Solid-State Circuits 2009;44(6):1796–804.

[43] Lee H-M, Ghovanloo M. An integrated power-efficient active rectifier with offset-controlled high speed comparators for inductively-powered applications. IEEE Trans Circuits Syst I Regul Pap 2011;58(8):1749–60.

[44] Cha HK, Park WT, Je M. A CMOS rectifier with a cross-coupled latched comparator for wireless power transfer in biomedical applications. IEEE Trans Circuits Syst II Express Briefs 2012;59(7):409–13.

[45] Lu Y, Ki W-H. A 13.56 MHz CMOS active rectifier with switched-offset and compensated biasing for biomedical wireless power transfer systems. IEEE Trans Biomed Circuits Syst 2014;8(3):334–44.

[46] Dallago E, Miatton D, Venchi G, Bottarel V, Frattini G, Ricotti G, et al. Active autonomous AC–DC converter for piezoelectric energy scavenging systems. In: Proceedings of the IEEE custom integrated circuits conference (CICC), September 2008; 2008. p. 555–8.

[47] Cheng S, Jin Y, Arnold D. An active voltage doubling ac/dc converter for low-voltage energy harvesting applications. IEEE Trans Power Electron 2011;26(8):2258–65.

[48] Lee H-M, Ghovanloo M. A high frequency active voltage doubler in standard CMOS using offset-controlled comparators for inductive power transmission. IEEE Trans Biomed Circuits Syst 2013;7(3):213–24.

[49] Lee H-M, Ghovanloo M. An adaptive reconfigurable active voltage doubler/rectifier for extended-range inductive power transmission. IEEE Trans Circuits Syst II Express Briefs 2012;59(8):481–5.

[50] Lu Y, Li X, Ki W-H, Tsui C-Y, Yue CP. A 13.56 MHz fully integrated 1×/2× active rectifier with compensated bias current for inductively powered devices. In: IEEE international solid-state circuits conference (ISSCC), February 2013; 2013. p. 66–7.

[51] Lee KFE. A timing controlled AC–DC converter for biomedical implants. In: IEEE international solid-state circuits conference (ISSCC), February 2010; 2010. p. 128–9.

[52] Sun TJ, Xie X, Li GL, Gu YK, Wang ZH. Rectigulator: a hybrid of rectifiers and regulators for miniature wirelessly powered bio-microsystems. Electron Lett 2012;48(19):1181–2.

[53] Lee H-M, Park H, Ghovanloo M. A power-efficient wireless system with adaptive supply control for deep brain stimulation. IEEE J Solid-State Circuits 2013;48(9):2203–16.

[54] Choi J-H, Yeo S-K, Park S, Lee J-S, Cho G-H. Resonant regulating rectifiers (3R) operating for 6.78 MHz resonant wireless power transfer (RWPT). IEEE J Solid-State Circuits 2013;48(12):2989–3001.

[55] Man TY, Mok PKT, Chan M. A high slew-rate push–pull output amplifier for low-quiescent current low-dropout regulators with transient response improvement. IEEE Trans Circuits Syst II Express Briefs 2007;54(9):755–9.

[56] Ho ENY, Mok PKT. A capacitor-less CMOS active feedback low-dropout regulator with slew-rate enhancement for portable on-chip application. IEEE Trans Circuits Syst II Express Briefs 2010;57(2):80–4.

[57] Ho M, Leung KN. Dynamic bias-current boosting technique for ultralow-power low-dropout regulator in biomedical applications. IEEE Trans Circuits Syst II Express Briefs 2011;58(3):174–8.

[58] Chen W-C, Su Y-P, Lee Y-H, Wey C-L, Chen K-H. 0.65 V-input-voltage 0.6 V-output-voltage 30 ppm/°C low-dropout regulator with embedded voltage reference for low-power biomedical systems. In: IEEE international solid-state circuits conference (ISSCC), February 2014; 2014. p. 304–5.

[59] Lu Y, Ki W-H, Yue CP. A 0.65 ns-response-time 3.01 ps FOM fully-integrated low-dropout regulator with full-spectrum power-supply-rejection for wideband communication systems. In: IEEE international solid-state circuits conference (ISSCC), February 2014; 2014. p. 306–7.

[60] Zhang X, Lee H. An efficiency-enhanced auto-reconfigurable 2×/3× charge pump for transcutaneous power transmission. IEEE J Solid-State Circuits 2010;45(9):1906–22.

[61] Ramadass YK, Fayed AA, Chandrakasan AP. A fully-integrated switched-capacitor step-down DC-DC converter with digital capacitance modulation in 45 nm CMOS. IEEE J Solid-State Circuits 2010;45(12):2557–65.

[62] Ng VW, Sanders SR. A high-efficiency wide-input-voltage range switched capacitor point-of-load DC–DC converter. IEEE Trans Power Electron 2013;28(9):4335–41.

[63] Kularatna N. Modern batteries and their management: Part 1. In: Proceedings of the 36th annual conference on IEEE industrial electronics society, November 2010; 2010. p. 1–103.

[64] Yang H, Zhang Y. Analysis of supercapacitor energy loss for power management in environmentally powered wireless sensor nodes. IEEE Trans Power Electron 2013;28(119):5391–403.

[65] Yang H, Zhang Y. Self-discharge analysis and characterization of supercapacitors for environmentally powered wireless sensor network applications. J Power Sources 2011;196(20):8866–73.

[66] Dearborn S. Charging Li-ion batteries for maximum run times. Power Electron Technol Mag 2005;(April):40–9.

[67] Teofilo VL, Merritt LV, Hollandsworth RP. Advanced lithium ion battery charger. IEEE Aerosp Electron Syst Mag 1997;12(11):30–6.

[68] Hoffart F. Proper care extends Li-ion battery life. Power Electron Technol Mag 2008;(April): 24–8.

[69] Buxton J. Li-Ion battery charging requires accurate voltage sensing. Anal Devices Anal Dialog 1997;31(2):3–4.

[70] Li P, Bashirullah R. A wireless power interface for rechargeable battery operated medical implants. IEEE Trans Circuits Syst II Express Briefs 2007;54(10):912–6.

[71] Valle BD, Wentz CT, Sarpeshkar R. An area and power- efficient analog Li-ion battery charger circuit. IEEE Trans Biomed Circuits Syst 2011;5(2):131–7.

[72] Chen M, Rincon-Mora GA. Accurate, compact, and power-efficient Li-ion battery charger circuit. IEEE Trans Circuits Syst II Express Briefs 2006;53(11):1180–4.

[73] Chen JJ, Yang FC, Lai CC, Hwang YS, Lee RG. A high-efficiency multimode Li-ion battery charger with variable current source and controlling previous-stage supply voltage. IEEE Trans Ind Electron 2009;56(7):2469–78.

[74] Duncan M. Distributed functional electrical stimulation system. U.S. Patent 7127287, October 24, 2006.

[75] Jow U, Kiani M, Huo X, Ghovanloo M. Towards a smart experimental arena for long-term electrophysiology experiments. IEEE Trans Biomed Circuits Syst 2012;6(5):414–23.

[76] Kelly S, Wyatt J. A power-efficient neural tissue stimulator with energy recovery. IEEE Trans Biomed Circuits Syst 2011;5(1):20–9.

[77] Vidal J, Ghovanloo M. Toward a switched-capacitor based stimulator for efficient deep-brain stimulation. In: Proceedings of the IEEE engineering in medicine and biology conference (EMBC), September 2010; 2010. p. 2927–30.

[78] Paul S, Schlaffer AM, Nossek JA. Optimal charging of capacitors. IEEE Trans Circuits Syst I Reg Papers 2000;47(7):1009–16.

[79] Lee H-M, Ghovanloo M. A power-efficient wireless capacitor charging system through an inductive link. IEEE Trans Circuits Syst II Exp Briefs 2013;60(10):707–11.

# System integration and packaging

**Wen Ko\*, Peng Wang\*, Shem Lachhman†**
*\*Case Western Reserve University, Cleveland, Ohio, USA*
*†MediMEMS, LLC, Shaker Heights, Ohio, USA*

## CHAPTER CONTENTS

## 6.1 INTRODUCTION

The human body is a harsh environment; bodily fluids are highly conductive and contain many chemical and biochemical species that are potentially harmful to implanted electronics and MEMS devices and components. At the same time, the body is a complex delicate living system that may be harmed by the material, shape, and size of the implant devices. The package is the outmost part of the implant that interfaces with the body; it has to protect the device from bodily fluids while protecting the body from any harmful effects of the packaged implant device.

Bhunia et al. Implantable Biomedical Microsystems. http://dx.doi.org/10.1016/B978-0-323-26208-8.00006-6
**113**

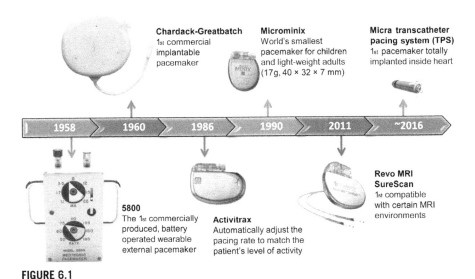

**FIGURE 6.1**

Cardiac pacemaker evolution in 50 years [39].

Implant devices were first used from the 1950s to the early 1960s by biologists for animal tracking, physiological exploration of the body [1], and ingestible devices to record the motion of the intestines [2]. Beeswax, paraffin, epoxy, and silicone were used for packaging, with corresponding lifetimes from days to months. From 1960 to 1970, the progress in microelectronics reduced the implant size and weight and made possible many biological and medical applications of implantable systems. Hermetic box-type packaging was developed for chronic implants, using metal, glass, ceramic, and Macor® (a machinable ceramic) [3,4]. From 1960 to the present, the hermetic box package technology evolved in size and shape. The series of packages of the Medtronic cardiac pacemaker, shown in Figure 6.1, illustrate the evolution of packages through the last 50 years. The early pacemaker used thick epoxy and silicone package with lifetime from months to 1 year. Next, metal (titanium alloy and stainless steel) was used to make the box that housed the implant, with a lifetime beyond 10 years.

All joints of the metal box had to be welded with electron beams, so that the temperature in the package would not be too high to damage the devices, and glass–metal feedthrough for lead wires was used for connection to components outside of the box. The connecting leads and the feedthrough are the most easily damaged parts of the implant devices, which ultimately limits the implant lifetime. A technique to use Macor® for box packages was also developed in the 1960s. Figure 6.2 shows some early ceramic (Macor®) packages used in Ko and Neuman [3] and Ko *et al.* [4]. The Macor ceramic package is hermetic when machined to form a box and sealed with a laser and then coated with thin soft biocompatible material—silicone—to permit many years of implant lifetime.

Most of the hermetic box package surfaces are smooth with no sharp corners and are often coated with soft biocompatible material such as silicone. The coating is designed to protect the implant and to mimic the tissue of the body. The materials

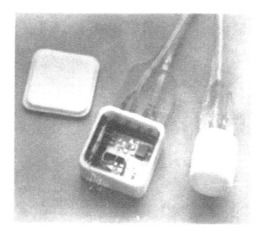

**FIGURE 6.2**

Macor® package samples.

used are nontoxic and biocompatible. More discussion on biocompatibility will be presented in Section 6.3.

As implant electronic and component technology advanced from discrete components to PCBs, to miniature integrated circuit (IC), to large-scale IC and application-specific IC (ASIC) and MEMS, the volume and weight of implantable devices were reduced from $cm^3$ to $mm^3$ and from ounces to fractions of a gram. After 1990, the development of very-large-scale integration (VLSI) further reduced the electronics and some implants to $\mu m^3$ and $\mu g$. At the same time, the implant electronic systems with various microsensors and stimulators progressively took on more complex and precise functions. The currently used box packages can no longer meet the needs of these new microimplants. This unmet need is now a roadblock to the progress of new microimplants in significant biomedical research and clinical areas, such as visual prostheses and brain–machine interfaces.

Since 2003, there has been an increased effort from universities and research institutions to explore new micropackaging technology. Most of the work has focused on finding the materials to act as vapor barrier to protect the implant device from bodily fluids. The purpose is to protect the device from low leakage resistance that would discharge implant batteries, due to malfunction of electronic circuits, and reduce the implant lifetime. It is assumed that implant safety and nontoxicity can be achieved as before by using biocompatible materials as the package outer layer and by pursuing proper system integration and package design. The goal of this new package technology is to reduce the micropackage to the same order of magnitude in size and weight as the implant devices while meeting the requirements of new implant functions. The new packaged implant devices would be small and light and can be integrated with microsensors and actuators and able to perform complex functions with desired levels of precision. Furthermore, new micropackages must be nontoxic and must not cause irritation to surrounding tissues over long periods—that is, the packaged implants must be biocompatible.

## 6.2  BRIEF REVIEW OF IMPLANT PACKAGE TECHNOLOGIES

Unlike the IC technology for which packaging is highly standardized, there is no general packaging method applicable for all implantable devices. Each implantable system requires a specific design to suit the needs for the device operation. However, packaging techniques can be split into two main categories: the traditional box-type hermetic packages and the new nonhermetic micropackages. Both can be used for short-term and chronic implant applications with lifetime from months to years. A short discussion on their technologies and a few published examples are given in each group below, to provide a general view of the present status of package technologies.

### 6.2.1  HERMETIC BOX PACKAGING TECHNOLOGIES FOR LONG-TERM IMPLANTS

Hermetic box package involves the use of a capsule or box made of hermetic materials, metal, glass, and ceramic, to house the implant device. The cover and the body of the capsule need to be hermetically sealed by four main sealing techniques: soldering, brazing, glass sealing, and welding [5,6]. All of these techniques involve high temperatures to form the needed hermetic seal. Hermetic package offers a much higher level of protection against the ingress of water vapor and moisture from device electronics than its nonhermetic counterparts [5–7]. Hermeticity, as it relates to packaging, is generally defined as the condition of being impervious to the diffusion of gas and moisture. By military standard MILSTD883J, hermetic materials must have a measured helium leakage rate lower than $10^8$ atm cm$^3$/s. Because of their effectiveness at protecting sensitive electronic circuitry, hermetic packaging techniques have been widely used in implantable applications since the late 1960s (Medtronic Inc.). Several research groups have utilized hermetic boxes for their specific devices. Herein are a few of the many that are currently in clinical or investigational use. Samples of hermetic box packaged implantable devices are given in Figure 6.3.

The sources of photos in Figure 6.3 and brief description of these devices are given below:

(a)  Medtronic, Inc.'s Micra—The world's smallest pacemaker that is affixed to the inside of the heart with tines [8].
(b)  The Envoy Esteem is the world's first FDA-approved totally implantable hearing device. It is implanted under the skin behind the ear and in the middle ear. It uses the eardrum as the microphone, working at the incus to send current to a sound processor. The electrical current signal is amplified, filtered, compressed, and converted back to mechanical vibrations by a piezoelectric transducer connected to the sound processor. The body of the device is hermetically sealed with titanium, and the packaging materials for the stabilizer include platinum–iridium, tin, stainless steel, and silicone tubing [9].
(c)  A prototype transponder built with a PZT-based piezoelectric cantilever at a mechanical resonance frequency of 435 Hz was reported by Kim *et al.* The

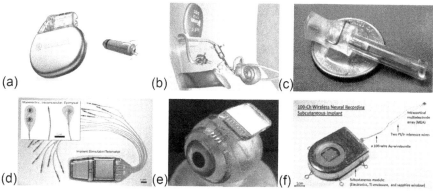

**FIGURE 6.3**

Some box hermetic packages of implantable devices [27]. (a) Implantable pacemaker by Medtronic, Inc.; (b) Envoy Esteem middle ear implantable hearing device; (c) implantable passive transponders; (d) advanced implantable neuroprosthesis; (e) hermetic wireless subretinal neurostimulator for vision prostheses; and (f) 100-channel wireless neuronsensing devices.

device, which was encapsulated in a hermetic glass tube, scavenges musical sound and radiates an radio-frequency (RF) pulse at the passive sensor resonant frequency. The whole device was coated with $10 \mu m$ parylene-C for the purpose of biocompatibility [10].

**(d)** Hart *et al.* developed an advanced implantable neuroprosthesis with myoelectric control. The neuroprosthesis was designed for the restoration of limb function in paralyzed individuals. The stimulator telemeter had 12 channels and was hermetically sealed within a titanium package. The outer layer is epoxy-coated with Silastic for strain relief. The device demonstrated full operational functionality after a 15-month implantation in a dog [11].

**(e)** Ming *et al.* developed a 100-channel wireless neuronsensing device that was hermetically sealed by a titanium enclosure that incorporated a sapphire window for transmission of RF signals. The functionality and reliability were tested in freely moving swine and monkey animals and exhibited a lifetime in excess of 1 year [12].

**(f)** Kelly *et al.* developed a hermetic wireless subretinal neurostimulator for vision prostheses. The 15-channel stimulator chip and discrete circuit components were hermetically sealed in a titanium case with feedthrough for power transmission and data communication. The electrode array insertion was in its own quadrant for the ease of surgical access [13].

## 6.2.2 NONHERMETIC MICROPACKAGE TECHNOLOGIES

Advanced MEMS implants need adequate nonhermetic micropackage technology. The nonhermetic micropackage uses low vapor permeability polymeric material and

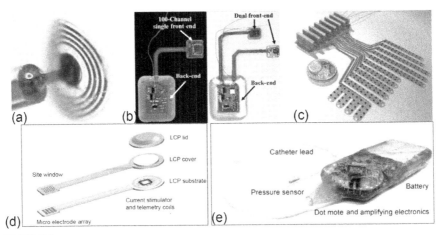

**FIGURE 6.4**

Samples of micropackages developed recently [27]. (a) Parylene-based intraocular pressure sensor [14]; (b) polymeric packaging implantable neural microsensors [15]; (c) MEMS-based flexible multichannel ECoG-electrode array [16]; (d) packaging of LCP-based neuroprosthetic devices [18]; and (e) fully implantable wireless implantable monitoring system with LDPE film and PDMS [4,19].

thin film coating techniques to encapsulate implants. They have advantages of low geometric form factors that occupy smaller volumes, less weight, flexibility with softer surfaces, RF transparency, good biocompatibility. They can be integrated and packaged with MEMS devices (sensors and actuators), with low cost, excellent protection (against ions and moisture), and short fabrication time. When properly packaged, nonhermetic micropackages can have implant lifetime from months to 10 years. Nonhermetic micropackage technologies have been under development for bioMEMS and MEMS implants at research institutes and universities for the past decades. Some examples of current research conducted with nonhermetic packaging are shown in Figure 6.4. The sources of photos in Figure 6.4 and brief description of these devices are given below:

(a) A parylene-based pressure sensor was developed to measure intraocular pressure. The device utilized a passive LC tank resonant circuit whose frequency shift was registered by an external circuit. A 35 µm parylene-C film was applied on the sensing coil to serve as a passivation layer to maintain a high-quality factor even in the environment of body fluid. The device was evaluated in rabbits [14].

(b) Aceros et al. used polymeric materials to package implantable wireless neural microsensors. The packaging was based on a novel plasma-enhanced chemical vapor deposition process that uses hexamethyldisiloxane and oxygen ($O_2$). The package exhibited a maximum lifetime of 508 days in nonhuman primates [15].

**(c)**  252-Channel, flexible multichannel ECoG-electrode array was fabricated. The electrodes were built on thin polyimide foil substrates with the electrodes made from sputtered platinum thin films. The array subtended an area of approximately 35 mm by 60 mm and could cover a large part of one hemisphere of a macaque cortex. Epoxy was applied as a fixture on the pins, solder, polyimide foil, and connector bodies. All the electrodes were functional and exhibited no signs of signal quality degradation after 4.5 months of implantation [16].

**(d)**  A packaging structure using liquid crystal polymer (LCP) in a box was developed. The system consisted of an LCP-based substrate, cover, and lid, sealed with thermal compression bonding. The microelectrode array's package kept its insulation resistance to $10^8 \Omega$ after soaking in 75 °C saline solution for 300 days [17,18].

**(e)**  A fully implantable wireless pressure monitoring system was developed. The system could measure pressures up to 1.5 psi with a resolution of 0.02 psi. The packaging was done by wrapping the sensor in 44.5 μm thick cellophane film sealed with epoxy. An extra polydimethylsiloxane (PDMS) silicone film was coated on the outside of the device for biocompatibility and short-term protection. The device was used *in vivo* in porcine for 3 days [19].

Polymeric materials have been widely used and some are quasihermetic box types using very low vapor permeability materials. Some are simple coated layers of selected polymeric materials. Many other nonhermetic packages for MEMS implantable devices use parylene-C or other vapor barrier films as the packaging material to protect the electronic devices from low surface leakage resistivity caused by vapor permeation and condensation into water. Other micropackage work at Case Western Reserve University (CWRU) used three materials, silicone for biocompatibility, epoxy for mechanical strength, and parylene-C for vapor barrier, to package implantable devices for long lifetimes [20–29]. They studied the leakage resistance ($10^{11}$ to $10^8 \Omega$) over the lifetime ranging from a few months to 5 years and the mechanism of failure for a series of packaging processes. Furthermore, the packaging processes were evaluated in the laboratory and in rats for up to 6 months without failure. Further details of this work will be given in Sections 6.5 and 6.6.

## 6.3 SYSTEM INTEGRATION AND BIOCOMPATIBILITY

The package is the last of the design/fabrication steps of an implantable device or system but is arguably the most important part of the biomedical implementation of implant device. Proper packaging will ultimately affect the success of the implant. Before actually applying the package to the device, the first step of "packaging" is to collaborate with the implant design group to work on system integration—the arrangement of the building blocks and components of the system. The basic building blocks of an implant device are power supply, sensors or actuators, electronic signal processor, and RF links. The system integration includes the design of the substrate and the proper layout of the building blocks or other components on the substrate or in the 3-D position. Each implant device has unique requirements according to its functions and implant location. However, there are some general considerations: (1) The system weight should be as uniformly distributed as possible. (2) Tall parts should be grouped together near the center of the package; there should not be many peaks and valleys. (3) All connections (solder joints and leads) should be strong enough or be reinforced to withstand the maximum acceleration expected in operation. (4) Batteries or other large conductors should be outside of RF antenna loops. (5) Sensors, actuators, and fragile components should be located near well-protected areas or the center of the package. (6) There should be no sharp corners or protruding parts and all corners should be rounded. (7) The overall shape of the package should be such that it is easy to implant and is difficult to move around the implant location. Some stabilization devices may be used such as the spring open hook used in implant electrodes. (8) There should not be hot spots or stress corners when tissue moves around the implant.

For biocompatibility, the packaged implant should not cause toxicity or other long-term harmful reactions such as long-term irritation to the surrounding tissues and the host body. The specific gravity of the package or regions of the package should not be very different from the local tissue to avoid undue stresses generated when the host body moves. The surface of the package should be smooth, with peaks and valleys fully covered by packaging material to form large radius of curvature. The biocompatibility of materials used for package should be tested according to the FDA, the International Organization for Standardization (ISO 10993), and the U.S. Pharmacopeial Convention (USP Class VI).

The sterilization, implant operation, animal/host care and inflammation detection, and treatments should follow the NIH and medical guidelines and be carried out by medical professionals, and biopsy or tissue sample examinations should be performed by experienced pathologists.

## 6.4 PACKAGING MATERIALS AND TECHNOLOGIES

Current approaches to packaging implantable MEMS devices have several limitations in terms of material selection, vapor permeability, biocompatibility, size, robustness, and lifetime. At present, packaging material options for implants are significantly limited. Packages based on metallic or ceramic enclosures guarantee mechanical and hermetic protection, thus long device lifetime, but block RF signals and limited in miniaturization and flexibility. Polymeric packages offer mechanical flexibility, reduced size and weight, and RF transparency, but are not readily suited for long-term implantation. Research directed to increase the lifetime and robustness of a polymer-based packaging technology can provide many potential benefits such as an increased usable lifetime of the implant device and reduction in the size and weight of the implant packaging. Long lifetime packages utilizing polymer films, such as PDMS, are of particular interest because of their ease of use, low cost, and excellent material properties that provide good interface with body environments. In this section, we review the current materials and technologies pursued for packaging of implantable microelectronics systems.

The selection of appropriate materials for the encapsulation of implantable microsystems is one of the most important aspects for the encapsulation design. It is imperative that the packaging engineer knows the characteristics and behavior of the materials in advance and adapts the application method to the particular geometric characteristics of the device to be packaged. Since there may be no single polymeric material that can achieve all of the requirements for successful long-term encapsulation of implantable microdevices (i.e., vapor barrier, tissue interface, and mechanical protection), several materials and/or multiple layers of the same material may be used to provide robust, water-impermeable, and biocompatible microencapsulation. Thin films offer a wide range of advantages: they occupy a small volume, they can be deposited with a variety of techniques, and they can take any shape. Thin film materials for packaging implantable microsystems can be split into two main categories: inorganic and organic [5].

Inorganic materials for packaging purposes include ceramic thin films such as silicon carbide, silicon nitride, and polycrystalline diamond and metal thin films such as gold, aluminum, and silver. Ceramic thin films offer excellent protection against water permeation and allow RF signals to be transmitted throughout the package. Ceramics also bring forth relatively small amounts of tissue response in the human body [30]. Unfortunately, their mechanical properties are not favorable as they can fracture with minimal deformation under stress. Another drawback is the high temperatures needed for a good conformal, pinhole-free deposition. A workable approach is to interlace thin ceramic films with flexible soft polymeric films that have good bonding strength with the ceramic film in order to limit the maximum stress in the ceramic film to a tolerable range.

Metal thin films are also attractive due to the outstanding barrier against moisture they provide. An additional insulating layer is necessary when using metal films for encapsulation to provide electrical isolation from the encapsulated devices [5,7]. Careful selection of the insulating layer is necessary to achieve a successful encapsulation with small parasitic capacitance. Metal films can also be interlaced with selected polymer films (depending on good bonding strength with the metal) to make the package flexible and with insulating surface.

Organic packaging materials primarily used in nonhermetic encapsulations are composed of thin film polymeric materials. Their low cost, RF transparency and nonconductive nature, low-temperature sealing, and simple deposition processes make them extremely attractive as packaging materials for short and medium lifetime (below 2–5 years) implants. The polymers are deposited by special deposition processes, such as spin coating, dam and fill procedures, dip coating, and roller casting. The most popular polymers primarily used for encapsulation of implantable devices include polyurethanes, epoxies, parylene, and polydimethylsiloxane or PDMS (medical-grade silicone) [5].

In order to protect the devices from water that causes low leakage resistance on the substrate and component surfaces, the package materials should (i) have low water vapor permeability, (ii) retain high resistivity after soaking in bodily fluids over the expected lifetime, (iii) bond well with the substrate and all components, (iv) be biocompatible with tissues around the implant site and be nontoxic and have mechanical properties and surface structure compatible with the tissue, and (v) be strong enough to protect the connections and structures of electronic devices and components from possible mechanical stress and deformation in the lifetime. When it is difficult to select one material to satisfy all these requirements, multiple nontoxic materials are used to package an implantable device.

Currently, for nonhermetic packages using polymeric materials, a popular choice for a vapor barrier material is parylene, which can be vapor-deposited with conformal coverage and has low water vapor and oxygen permeability. For biocompatibility, medical-grade silicone is a common choice as it has mechanical properties similar to tissue and has been shown to possess long-term stability and biocompatibility. Moreover, it can be easily applied to an implant with methods such as dipping, spinning, spraying, and roller casting. For mechanical stability and potting, various epoxies are commonly selected.

When multiple materials are used in the package, the temperature coefficients, vapor absorption properties, and bonding strength between adjacent layers are important considerations. If the bonding strength is small or decreases after implantation, small temperature variations or external stresses can cause delamination and decrease the package lifetime. This situation can be alleviated by using thin layers of materials or using alternating layers of hard and soft materials. Preferably, compatible materials would be selected, or the bonding strengths between different thin layers of package materials would be increased.

## 6.5 CWRU NONHERMETIC MICROPACKAGE TECHNOLOGY

The micropackage research at CWRU has two approaches. One is the study of failure modes of polymeric material nonhermetic micropackages, in order to extend the lifetime of single nonhermetic polymer material (PDMS) packages for chronic implants. Furthermore, by understanding the failure mechanisms of nonhermetic micropackages, better substrates and packaging materials may be selected for even longer lifetime micropackages. This approach used interdigitated electrodes (IDEs) as test devices and PDMS as the single coating material.

The second approach is to use low-permeability materials (e.g., parylene-C) as the water vapor barrier material, outer PDMS for biocompatibility, and epoxy for mechanical strength to realize a three-material, multilayer micropackage for three-dimensional implant devices/systems. This approach was developed in the laboratory and later evaluated in rats.

### 6.5.1 CHARACTERIZATION OF PDMS MICROPACKAGE TECHNOLOGY AND EVALUATION

Hypotheses: PDMS has been used to package implant devices for 60 years with demonstrated lifetime varying from a few days to a few weeks. Our attempt to develop nonhermetic micropackage processes to achieve months to years of lifetime was based on Dr. Ko's hypothesis. From 50 years of observations using PDMS and epoxy as packaging materials, Dr. Ko hypothesized that a nonhermetic, polymer-based package technology with long lifetime can be developed if (1) the polymeric materials maintain a high electrical resistivity when saturated with saline solution, (2) the device to be packaged is clean and free of any ionic contamination, (3) the coating has no localized defects, and (4) there are no void-like cavities between the coating materials and the substrate—no unfilled cavities for water to condense from vapor. Guided with these hypotheses, a series of packaging processes were developed with successive improvements to extend the lifetime from months to years. The current lifetime of PDMS roller-casted devices exceeds 2 years, which is more than 100 times longer than that of traditional dip coating.

Laboratory development of packaging processes with PDMS: IDEs were fabricated on printed circuit board as the implantable device to be packaged. The IDE was inspected carefully to eliminate any irregularity before packaging. The process started with thorough cleaning of the IDE in organic solvents and water with ultrasonic agitation. IDEs that then passed a cleaning screening test were advanced for packaging. The devices were packaged with PDMS films using a series of processes of multiple, thin film layers, applied through repeated roller casting with various pressures and number of strokes and directions. The coated PDMS films were then vacuum-cured at the appropriate processing temperatures. The process was repeated to form multiple three thin film coatings. The three thin films of same material are used to greatly reduce or eliminate the defects on the films and microbubbles. All the processes were carried out in a class 100 clean room to reduce the inclusion of foreign materials on/in the films. Processed IDE devices were evaluated in an accelerated lifetime-test saline bath at 85 °C, while the resistance between IDE electrodes was monitored and recorded every 30 min. The IDE resistance after cleaning was around $10^{11}$ to $10^{13}\,\Omega$ initially and dropped in a three-phase paradigm shown in Figure 6.6. When the IDE resistance dropped to $10^{8}\,\Omega$, the packaged IDE was considered failed. The time to IDE failure was recorded as the lifetime of the package at 85 °C saline. The implant lifetime in the 40 °C body environment can be estimated by multiplication with an acceleration factor (10–64), which is calculated from the Arrhenius reaction rate equation [31] and activation energy of the failure mechanism. However, the underlying failure mechanism of nonhermetic packaged devices may be more complicated than just vapor permeation, and major works still need to be made in identifying the mechanism with accuracy [32,33]. The published experimental results are scattered over a large range [25,26,34]. We used a conservative acceleration factor of 10 to extrapolate the mean time to failure (MTTF) lifetime from 85 °C saline solution to 40 °C body environment [24,27].

Laboratory-tested IDE lifetime results: From 2007 to 2014, several modifications of the roller casting processes were made, with successively improved MTTF and yield rate as shown in Table 6.1 and Figure 6.5.

In Figure 6.5, the 2009 and 2011 processes used three 100 μm thick PDMS layers, each with 30 strokes in the lateral (*Y*) direction of the IDE, without pressure. The 2012 processes used three 50 μm thick PDMS layers, each with 200 strokes in the *Y* direction with ~34 psi pressure on the roller. The 2013 and 2014 used three 30–50-μm PDMS layers, each with 200 strokes in the *Y* direction and 200 strokes in the

**Table 6.1** Laboratory Tested IDE Lifetime Results

| Investigator | Testing Saline | Rolling Direction | Longest Lifetime 40°C (days) | MTTF in 40°C (days) | Yield Rate |
|---|---|---|---|---|---|
| Bu et al. [20] | 40°C | x | 38 | 28.7 | 33% above 30 days |
| Zhang [40] | 40°C | x | 52 | 31.3 | 33% above 30 days |
| Lachhman [41] | 40°C +85°C | x | 400[a] (1.1 years) | 223[a] (0.6 years) | 25% above 30 days in 85°C saline |
| Wang et al. [42] | 85°C | x+y | 1300[a] (3.5 years) | 1300[a] (3.5 years) | 100% above 130 days in 85°C saline |
| Sun [43] | 85°C | x+y | 2020[a] (5.5 years) | 1566[a] (4.3 years) | 80% above 100 days in 85°C saline |

[a] Assuming a conservative failure acceleration factor of 10 between 85 and 40 °C saline.

axial ($X$) direction (perpendicular to $Y$) with 37–46 psi pressure. The total number of strokes applied to the surfaces of an IDE was 3600 strokes.

Figure 6.6 shows the resistance changes with time during 85 °C saline bath tests. It has three regions marked I, II, and III. Region I is the period in which water vapor penetrates the PDMS layers. Region II is the major lifetime period, with vapor saturation of the interface to water droplet condensation and spreading throughout the

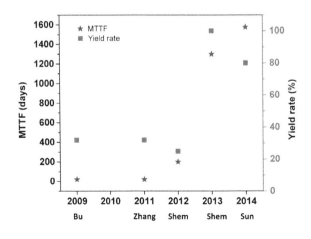

**FIGURE 6.5**

The MTTF and yield versus processes.

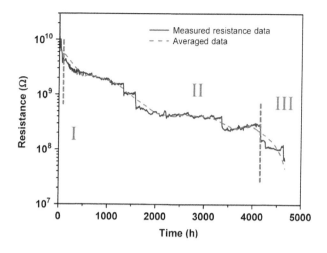

**FIGURE 6.6**

The PDMS resistance versus lifetime test results.

PDMS and substrate interface. Region III is the period in which water spreads over the IDE and reduces the resistance to $10^8\,\Omega$, causing failure of the device.

## 6.5.2 VERIFICATION OF THE HYPOTHESES AND MECHANISM OF PDMS MICROPACKAGE PROCESSES

The successive improvement of lifetime and yield rate can be interpreted according to the hypotheses. For the 2009–2012 processes, the lifetime was extended due to better cleaning processes, increased number of strokes in roller casting, and increased roller pressure during strokes. The improvement from 2012 to 2014 was due to an added screening step after the cleaning process, a doubled total number of strokes in $X$ and $Y$ directions, and higher roller pressure during the strokes so that most of the cavities (voids) in $Y$ and $X$ directions on the PC board were filled with PDMS by the pressured strokes. This is evidenced by the SEM photos in Figure 6.7c–f, of the PCB surface with the valley and the inner surface of the PDMS film showing the peaks of PDMS matching the valley of the substrate surface. They indicate that the PDMS did fill in the valleys of the substrate due to the repeated forced rolling of PDMS. This mechanism likely extended the implant lifetime of the PDMS packaged device.

We also measured the properties of the PDMS material after repeated roller casting and compared them with dip-coated PDMS. The results are listed in Table 6.2. The package processes increased the density of the PDMS material, reduced the vapor permeability, and increased the bonding force with the substrate. These changes also contributed to longer lifetime by delaying the vapor condensation rate.

There is another unexplained observation from the measured PDMS-coated IDE resistance–time characteristics. Referred to Figure 6.6, the resistance–time characteristics have three regions: Region I is about 100 h that is the time for the water

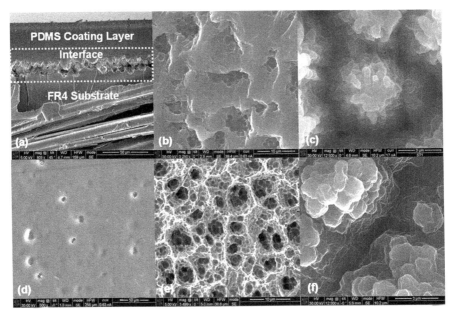

**FIGURE 6.7**

The SEM photos of the FR4 surface and PDMS coating layer interface. (a) Cross-sectional view of the interface between PMDS coating layer and FR4 substrate. (b) Zoomed-in view of FR4-substrate surface facing the PDMS coating layer. (c, f) The PDMS coating layer surface with dip and roller casting method, respectively. (d) Smooth side of FR4 substrate. (e) Rough side of FR4 substrate surface after removing copper by wet etching. (d, e) have the roughness of 0.51 and 1.45 μm, respectively.

vapor to penetrate the PDMS film. Region II is the longest period and lasts about 90% of the lifetime. The water vapor saturated the PDMS and FR4 interface and water starts to form droplets and spread over the interface. And region III is when the water spread over the FR4 surface and lifted the PDMS bonding; thus, the resistance starts to drop below the failure level of $10^8 \, \Omega$. In region II, the vapor saturated the interface at the beginning, and there is a long period before resistance decreases with

**Table 6.2** The Measured PDMS Properties Between the Roller-Casted and Dip-Coated Samples

| Group | Coating Process | WVTR (g mm/ m² day) Ratio | OTR (cc mm/ m² day) Ratio | Density (g/cm³) Ratio | Bonding force to FR4 (psi) Ratio |
|---|---|---|---|---|---|
| A | Dip coated PDMS | 1.00 | 1.00 | 1.000 | 1.00 |
| B | Spin coated PDMS | 0.95 | 0.92 | 1.006 | 1.14 |
| C | Rolling coated PDMS | 0.87 | 0.76 | 1.020 | 1.19 |

steeper slope, which we believe is the time when water starts to form. Then, there is a long period, 80 days for a device with 100 days lifetime; the device has saturated water vapor at the interface and is immersed in 85 °C saline bath with a coating of thin water vapor-permeable PDMS (permeability 2.8 gm/m$^2$ day). What happens during this long period? *More study is needed.*

### 6.5.3 MICROPACKAGE TECHNOLOGY FOR 3-D IMPLANTABLE SYSTEMS

From the aforementioned pilot study, we have developed an approach to 3-D micropackaging. This technique includes the following technical components:

**(1)** A thorough cleaning procedure for the substrate and the implant device assemblies using solvents and deionized water.

**(2)** A screening method to exclude unacceptable devices (including a function/leakage test in vapor environment).

**(3)** The use of a multimaterial design for the package. For the packaging materials, we have selected PDMS (MDX4-4210, Dow Corning Corporation Inc.) for tissue compatibility, epoxy (EB-107LP-2, EpoxySet Inc.) for mechanical protection, and parylene-C (Specialty Coating Systems, Inc.) for water vapor barrier.

**(4)** The use of a multilayer, thin film architecture to greatly reduce the defect density in the encapsulation [20,35]. For example, we estimated that the stack of three 50 μm thick layers of PDMS would reduce the fully penetrating defect density by 10$^6$ [20,29]. Furthermore, all processes were conducted in a class 100 clean room to minimize airborne contamination.

*Cleaning and screening* are guided by quantitative supervision. The devices were rinsed by flux remover (Superior Syberkleen 2000, Superior Flux & Mfg. Co.), isopropyl alcohol, and deionized water in sequence with ultrasonic agitation. The resistivity of each cleaning bath was monitored during rinsing. The solution is changed every 15 min. Completion of the cleaning process was defined when the measured resistivity of the solution stays higher than 70% of its original value for more than 5 min. After the cleaning process, the device function was screened to exclude unacceptable devices. For example, the leakage current of IDE devices was measured with Keithley 4200 SCS semiconductor characterization system ($10^{-15}$ A sensitivity) at 70% RH and 23 °C. Leakage current higher than 1 pA was recognized failed in screening. Voltage applied for leakage current test was 3 V.

*Parylene-C conformal coating process* was realized by chemical vapor deposition polymerization (VDP) with PDS 2010 system (Specialty Coating Systems, Inc.) and parylene-C dimer (diXC, DAISAN KASEI Co., LTD. Japan). The coating process starts with vaporizing solid dimer into a gaseous state (100–175 °C, 1 torr), then pyrolysis (~680 °C, 0.5 torr) into monomers, and finally polymerization (<40 °C, 0.1 torr) into sample surfaces (room temperature) [36].

After VDP, parylene-C film thickness was measured with a Filmetrics F20 thin film measurement system. The film thickness is linearly proportional to the weight of parylene-C dimer used [23] and the total thickness was selected between 10 and 25 μm. Thinner films will have defects and thicker will have cracks, according to our previous observations. We showed that two 5 μm layers of parylene-C are better than a single 10 μm one in a short-term test [20]. Thorough cleaning and dry runs are necessary before each coating process to minimize the parylene-C defects caused by dust contamination as shown in Figure 6.8.

*Molding and spin casting process* was adapted to epoxy and PDMS materials after the parylene VDP process. Injection molding was conducted for casting epoxy. Epoxy was filled into the mold containing devices under vacuum at 60 °C, at which temperature the epoxy viscosity is very low that facilitates the material flow and degas reactions. The mold was kept under vacuum for 30 min to be fully degassed, and 4 h later, the epoxy was fully cured. Next, the device was demolded and coated with three 10–50 μm PDMS layers. For a single-layer PDMS coating process, the device was dipped into the mixed PDMS at room temperature, spun within a centrifuge at 1700 rpm under vacuum for 2 min, and cured for 30 min at 60 °C. The single-layer PDMS casting process was repeated 3 times. Finally, the implantable device was fully cured at 60 °C for at least 12 h. The final curing time and temperature can be adjusted according to the tolerance of packaged devices.

*Bench evaluation* was conducted in 1% saline solution at 85 °C. The package made of parylene-C only showed a shorter lifetime than combination of multimaterials as expected. They are shown in Figure 6.9a and b. with extrapolated lifetimes of 3.3 and 6.6 years, respectively. Implant evaluation is introduced in Section 6.6.

*Failure mechanism of parylene-C film* is observed in the early stage of our parylene-C deposition studies. Two major failure modes were observed. One failure mode is delamination between the parylene-C film and substrate, where water droplets condense and reduce the resistivity of the materials and the measured IDE resistances as shown in Figure 6.10a and d. The delamination defect can be reduced by

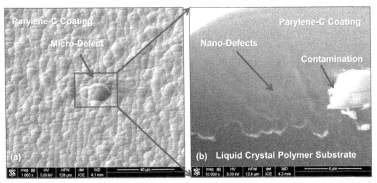

**FIGURE 6.8**

The defect in CDP parylene-C. (a) A microdefect of the CDP parylene-C coating on a LCP substrate. (b) Zoomed-in cross-sectional view of the defect in (a). There are nanodefects on the order of 200 nm width and contamination, which has a different spectrum with respect to parylene-C.

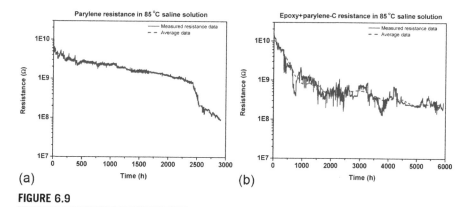

**FIGURE 6.9**

The in-lab evaluation of multimaterial multilayer micropackage technology. (a) The $R$–$T$ lifetime characteristics of a parylene-C-packaged IDE device. The lifetime in 85 °C saline solution is 120 days. (b) The $R$–$T$ lifetime characteristics of IDE device packaged with epoxy and parylene-C. The lifetime in 85 °C saline solution is 240 days.

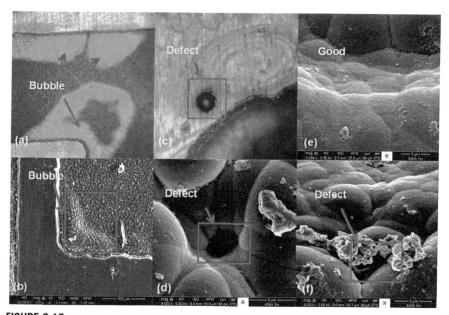

**FIGURE 6.10**

Failure mode of parylene-C. (a, b) Photo and SEM pictures of delamination defect between the parylene-C film and substrate. (c, d, f) Photo and SEM pictures of cross-sectional defect of parylene-C film. (e) SEM picture of a good package with smooth parylene-C film. (f) SEM picture of failed package with rough parylene-C film.

enhancing the bonding force between the parylene-C film and substrate with adhesion promoter (A174 SCS Inc.). After the use of adhesion promoter, the problem seems solved. Another failure mode is leakage across the parylene-C film. The possible reasons are intrinsic leakage path inside parylene-C film (Figures 6.8b and 6.10d, f), copper substrate overetch (Figure 6.10c), and dust contamination (Figure 6.8b) [37]. The suggested and tried solutions include polishing the substrate and annealing (300 °C, 3 h, under inert gas), which will make parylene-C film smoother and thus reduce the leakage-path defects, and cleaning the VDP environment, which will decrease the amount of foreign dust contamination and defects. The failure mode for parylene-C-coated devices with lifetime longer than 100 days in 85 °C saline solution may be different from those reported here and may need more careful study in the future.

## 6.6 IMPLANT EVALUATION OF NONHERMETIC MICROPACKAGE TECHNOLOGIES

Implant evaluations of nonhermetic micropackages in rats were made with a 1-month pilot implant trial first and then 3- and 6-month implants following ANSI/AAMI/ISO 10993-2, 6 standards, using micropackaged implantable pressure telemetry devices. The *in vivo* results indicate that all hand-packaged pressure telemetry devices survived without package failure and demonstrated good biocompatibility and stability.

### 6.6.1 PACKAGE METHODS AND IMPLANT-TELEMETRY-DEVICE DESIGN

The devices were packaged with two 5 μm thick layers of VDP parylene-C (Specialty Coating Systems, Inc.), as well as molded epoxy (EB-107LP-2, EpoxySet Inc.) and roller-casted PDMS (MDX4-4210, Dow Corning Corporation Inc.) as developed in Sun *et al.* [28] and Ko and Wang [22]. The package compensated for the component height variations to form a uniform coating with thickness varying from 250 μm to 1 mm. The packaged device, the circuit diagram of device, and the picture of packaged implant device are shown in Figure 6.11.

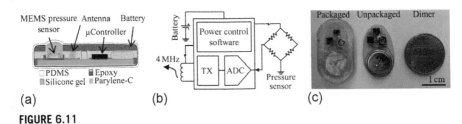

**FIGURE 6.11**

The (a) package, (b) circuit diagram, and (c) picture of implant and sensor.

Sixteen telemetry devices were fabricated for evaluation. As shown in Figure 6.11b, each implant consisted of a commercial MEMS pressure sensor (ASB1200VR, EPCOS Inc.), a microcontroller (PIC12F1822, Microchip Technology Inc.), a lithium battery (CR1225, Renata Batteries Inc.), an SMD inductor antenna, and other passive components. Pressure samples were transmitted at 10 Hz using ASK modulation of a 4 MHz carrier to a receiving antenna placed 10 cm away. The implanted device drew 400 μA while transmitting and 30 nA in sleep mode. To save power, the implant automatically entered sleep after 3 min of usage and was selectively activated by external transmission of a "wake up" signal. With an average usage rate of 15 min/week, the implant could run for longer than 3 years before depleting the battery.

## 6.6.2 IMPLANT EVALUATION RESULTS AND DISCUSSION

**(1)** *One-month preliminary evaluation*: Eight devices were implanted in six rats and preliminary evaluation turned out 100% successful. All devices functioned well during and after recovery from the rats. The tissue around the implant site also showed good biocompatibility, as shown in Figure 6.12.

**(2)** *Long-term evaluation:* The packaged devices were sterilized and implanted in eight rats, with each rat carrying two implants beneath the skin and fat layer near the shoulder and tail. The implanted devices were explanted after

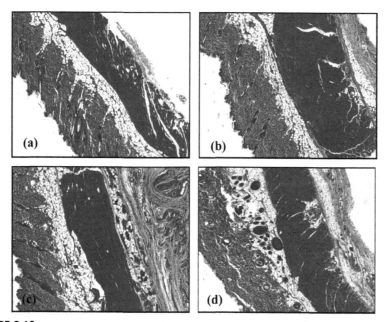

**FIGURE 6.12**

Tissue response images of one-month implants. (a, c) Optical image from the control group. (b, d) Image around the implant. (a, b) are from animal #1. (c, d) are from animal #2.

3- and 6-month periods, and tissue samples around the implant were taken for histopathology analysis. All devices were evaluated for vapor and water penetration, electronic circuit function, and battery condition. Tissue samples were sliced and stained with trichrome and hematoxylin and eosin (H&E) for biopsy.

All 3- and 6-month implants recovered were examined in the laboratory; no trace of water penetration was found inside the packages; all circuits and pressure sensors functioned well, while 7 of 16 devices had overdischarged battery. The overdischarge was determined due to battery damage during the electronics assembly process. Function was restored to all failed devices when the failed battery was charged or replaced.

**(3)** *Biocompatibility evaluation*: Figure 6.13 shows samples of histology slides indicating a completely healed fibrous capsule around the implant site. No acute/chronic inflammation or granulation tissue was found; only a minimal foreign body reaction consisting of macrophages and foreign body giant cells was seen around the implant/tissue interface. This foreign body reaction was a well-formed collagen predominant fibrous capsule indicating good biocompatibility.

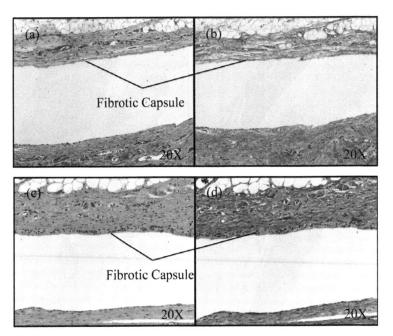

**FIGURE 6.13**

Micrographs of tissue samples from biopsy analysis. (a, b) are results of 3-month, while (c, d) are of 6-month.

### 6.6.3 SUMMARY OF IMPLANT EVALUATION

The 1-, 3-, and 6-month implant evaluations of the packaged telemetry devices were successful. We have developed a practical nonhermetic micropackaging method using three materials and verified package reliability and biocompatibility through 3- and 6- month animal implant evaluations. We are confident that the micropackage will prove to be practical for chronic implant beyond a year. The nonhermetic package is under modification to accommodate a custom ASIC [38] and wire-bonded battery, for a 1-year implant evaluation according to ANSI/AAMI/ISO 10993-6 for chronic implant evaluation procedure. Micropackage equipment is being designed for batch processing of devices in groups for commercial medical applications.

## 6.7 CONCLUSION

Package is the essential part of implantable devices/systems either implanted or surface-attached. It is usually neglected at the beginning phase of the design and only found necessary when system evaluation is needed. The system integration and the nonirritation aspect of the biocompatibility are usually overlooked and are not able to be incorporated in the implant devices. This will then affect the overall performance of the implant device/system. To pay attention to system integration and package considerations early in the design proving stage will avoid unnecessary modifications in layout and improve the performance of implanted device/system.

At present, the micropackage technology is not fully developed. The existing approaches for package of MEMS implants may include (a) box-type hermetic thin ceramic or glass package, (b) quasihermetic polymeric package using low-permeability polymers, and (c) nonhermetic polymeric micropackage. The selection of a suitable package technology will be depending on the application, resource, and individual preference. By understanding the advantages and disadvantages of each technology, it would help to make an intelligent choice. Reviewing the literature to bring up-to-date the current technology will help in choosing the right technology for an implant device.

## ACKNOWLEDGMENTS

The kind guidance and work on pathological evaluation of tissue samples of Professor James Anderson of the Medical School, CWRU, is deeply appreciated. The technical help from the MEMS/NEMS group of EECS Department, CWRU, and APT Center, VA medical Center of Cleveland, and the technical guidance and editorial help of Professor Christian A. Zorman of EECS Department, CWR University, are gratefully acknowledged. The project described was supported by Award Number 1I01RX000443-01A2 from the Rehabilitation Research & Development Service of the VA Office of Research and Development, Case Western Reserve University, and the Cleveland Clinic.

# REFERENCES

[1] Mackay RS. Biomedical telemetry. Sensing and transmitting biological information from animals and man, biomedical telemetry. New York (London & Sydney): John Wiley; 1968.

[2] Nagumo J, Uchiyama A, Kimoto S, Watanuki T, Hori M, Suma K, et al. Echo capsule for medical use (a batteryless endoradiosonde). IRE Trans Bio-Med Electron 1962;9(3):195–9.

[3] Ko WH, Neuman MR. Implant biotelemetry and microelectronics. Science 1967;156(3773):351–60. http://dx.doi.org/10.1126/science.156.3773.351.

[4] Ko WH, Neuman MR, Lin KY. Body reaction of implant packaging materials. In: Stuart L, editor. Biomaterials. NewYork: Plenum Press; 1969. p. 55–65.

[5] Najafi K. Packaging of implantable microsystems. In: Proceedings of IEEE sensors, October 28–31, 2007, Atlanta, GA; 2007.

[6] Ulrich RK, Brown WD. Advanced electronic packaging. 2nd ed. San Francisco, CA: Wiley-Interscience/IEEE; 2006.

[7] Ko WH, Spear TM. Packaging materials and techniques for implantable instruments. IEEE Eng Med Biol Mag 1983;2(1):24–38. http://dx.doi.org/10.1109/emb-m.1983.5005879.

[8] Brian B. How medtronic made the world's smallest pacemaker, http://www.qmed.com/mpmn/medtechpulse/how-medtronic-made-worlds-smallest-pacemaker; 2013 [accessed 20.12.13].

[9] Haynes DS, Young JA, Wanna GB, Glasscock ME. Middle ear implantable hearing devices: an overview. Trends Amplif 2009;13(3):206–14.

[10] Kim A, Maleki T, Ziaie B. A novel electromechanical interrogation scheme for implantable passive transponders. In: 2012 IEEE 25th international conference on micro electro mechanical systems (MEMS), January 29–February 2, 2012; p. 31–4.

[11] Hart RL, Bhadra N, Montague FW, Kilgore KL, Peckham PH. Design and testing of an advanced implantable neuroprosthesis with myoelectric control. IEEE Trans Neural Syst Rehabil Eng 2011;19(1):45–53. http://dx.doi.org/10.1109/TNSRE.2010.2079952.

[12] Ming Y, Borton DA, Aceros J, Patterson WR, Nurmikko AV. A 100-channel hermetically sealed implantable device for chronic wireless neurosensing applications. IEEE Trans Biomed Circuits Syst 2013;7(2):115–28. http://dx.doi.org/10.1109/TBCAS.2013.2255874.

[13] Kelly SK, Shire DB, Chen J, Doyle P, Gingerich MD, Cogan SF, et al. A hermetic wireless subretinal neurostimulator for vision prostheses. IEEE Trans Biomed Eng 2011;58(11):3197–205. http://dx.doi.org/10.1109/TBME.2011.2165713.

[14] Lin JC, Zhao Y, Chen P-J, Humayun M, Tai Y-C. Feeling the pressure: a parylene-based intraocular pressure sensor. IEEE Nanotechnol Mag 2012;6(3):8–16. http://dx.doi.org/10.1109/MNANO.2012.2203876.

[15] Aceros J, Ming Y, Borton DA, Patterson WR, Bull C, Nurmikko AV. Polymeric packaging for fully implantable wireless neural microsensors. In: Annual international conference of the IEEE engineering in medicine and biology society (EMBC), August 28–September 1, 2012; 2012.

[16] Rubehn B, Bosman C, Oostenveld R, Fries P, Stieglitz T. A MEMS-based flexible multichannel Ecog-electrode array. J Neural Eng 2009;6(3):036003.

[17] Min KS, Oh SH, Park MH, Jeong J, Kim SJ. A polymer-based multichannel cochlear electrode array. Otol Neurotol 2014;35(7):1179–86, Official publication of the American Otological Society, American Neurotology Society, and European Academy of Otology and Neurotology.

[18] Lee SW, Min KS, Jeong J, Kim J, Kim SJ. Monolithic encapsulation of implantable neuroprosthetic devices using liquid crystal polymers. IEEE Trans Biomed Eng 2011;58(8):2255–63. http://dx.doi.org/10.1109/TBME.2011.2136341.

[19] Tan R, McClure T, Lin CK, Jea D, Dabiri F, Massey T, et al. Development of a fully implantable wireless pressure monitoring system. Biomed Microdevices 2009;11(1):259–64. http://dx.doi.org/10.1007/s10544-008-9232-1.

[20] Bu LP, Cong P, Hung IK, Ye XS, Ko WH. Micro package of short term wireless implantable microfabricated systems. In: Proceedings of the annual international conference on IEEE engineering in medicine and biology society, September 2–9, 2009, Minneapolis, MN, USA; 2009.

[21] Ko WH, Wang P. Feasibility study on non-hermetic micro-package. In: IMAPS 46th international symposium on microelectronics, May, 2013, Minneapolis, MN, USA; 2013.

[22] Ko WH, Wang P. Nonhermetic micropackage for implant MEMS systems. In: Proceedings of the annual international conferences on IEEE engineering in medicine and biology society, September 28, 2013, Seattle, WA, USA; 2013.

[23] Kuo HI, Rui Z, Ko WH. Development of micropackage technology for biomedical implantable microdevices using parylene-C as water vapor barrier coatings. In: Proceedings of IEEE sensors, Waikaloa, HI, USA; 2010.

[24] Lachhman SB. Roller-cast poly-dimethylsiloxane as a non-hermetic encapsulant for MEMS packaging [M.S. thesis]. Cleveland, OH, USA: Department of Electrical Engineering & Computer Science, Case Western Reserve University; 2011.

[25] Lachhman SB, Ko WH, Zorman CA. Adhesion and moisture barrier characteristics of roller-cast polydimethylsiloxane encapsulants for implantable microsystems. In: Proceedings of IEEE sensors, Taipei, October, 2012; 2012.

[26] Lachhman SB, Zorman CA, Ko WH. Multi-layered poly-dimethylsiloxane as a non-hermetic packaging material for medical MEMS. In: Annual international conferences on IEEE engineering in medicine and biology society, August, 2012, San Diego, CA, USA; 2012.

[27] Sun D. Characterization of medical grade poly-dimethylsiloxane as encapsulation materials for implantable microelectromechanical systems [M.S. thesis]. Cleveland, OH, USA: Department of Electrical Engineering & Computer Science, Case Western Reserve University; 2013.

[28] Sun D, Wang P, Lachhman SB. Characterization of poly-dimethylsiloxane elastomers for non-hermetic implantable microsystem packaging. In: Conference on cellular and molecular bioengineering, September 28, 2013, La Jolla, CA; 2013.

[29] Wang P, Lachhman SB, Sun D, Majerus SJA, Damaser MS, Zorman CA, et al. Non-hermetic micropackage for chronic implantable systems. In: Proceedings of the 46th international symposium on microelectronics, Orlando, FL, USA; 2013.

[30] Davis SD, Gibbons DF, Martin RL, Levitt SR, Smith J, Harrington RV. Biocompatibility of ceramic implants in soft tissue. J Biomed Mater Res 1972;6(5):425–49. http://dx.doi.org/10.1002/jbm.820060509.

[31] Jensen F. Activation energies and the arrhenius equation. Qual Reliab Eng Int 1985;1(1):13–7. http://dx.doi.org/10.1002/qre.4680010104.

[32] Ko WH. Packaging of microfabricated devices and systems. Mater Chem Phys 1995;42(3):169–75.

[33] Vanhoestenberghe A, Donaldson N. Corrosion of silicon integrated circuits and lifetime predictions in implantable electronic devices. J Neural Eng 2013;10:031002.

[34] Chang JHC, Liu Y, Kang DY, Tai YC. Reliable packaging for parylene-based flexible retinal implant. In: Digest of the 17th international conferences on solid-state sensors, actuators, & microsystems (Transducers' 13), June 16–20, 2013; 2013.

[35] Virlich ÉÉ, Bronshtein BS. Probability method of assessing the quality of multilayer coatings. Chem Pet Eng 1989;25(1):42–5. http://dx.doi.org/10.1007/bf01158342.

[36] Fortin JB, Lu T-M. Chemical vapor deposition polymerization: the growth and properties of parylene thin films. Boston: Kluwer Academic Publishers; 2004.

[37] Song B, Azarian MH, Pecht MG. Effect of temperature and relative humidity on the impedance degradation of dust-contaminated electronics. J Electrochem Soc 2013;160(3):C97–105.

[38] Majerus SJA, Garverick SL, Suster MA, Fletter PC, Damaser MS. Wireless, ultra-low-power implantable sensor for chronic bladder pressure monitoring. J Emerg Technol Comput Syst 2012;8(2):1–13. http://dx.doi.org/10.1145/2180878.2180883.

[39] Medtronic Inc. A legacy of improving lives: our history, http://www.medtronic.com/wcm/groups/mdtcom_sg/@masterbrand/documents/documents/contrib_176744.pdf; July, 2013.

[40] Zhang R. The study of MEMS acoustic sensor for totally implantable hearing-aid system and micropackage technology for implantable devices [M.S. thesis]. Cleveland, OH, USA: Department of Electrical Engineering & Computer Science, Case Western Reserve University; 2011.

[41] Lachhman SB. Roller-cast poly-dimethylsiloxane as a non-hermetic encapsulant for MEMS packaging [M.S. thesis]. Cleveland, OH, USA: Department of Electrical Engineering & Computer Science, Case Western Reserve University; 2012.

[42] Wang P, Lachhman SB, Sun D, Majerus SJA, Damaser MS, Zorman CA, et al. Non-hermetic micropackage for chronic implantable systems. In: Proceedings of the 46th international symposium on microelectronics, Orlando, FL, USA; 2013.

[43] Sun D. Characterization of medical grade poly-dimethylsiloxane as encapsulation materials for implantable microelectromechanical systems [M.S. thesis]. Cleveland, OH, USA: Department of Electrical Engineering & Computer Science, Case Western Reserve University; 2014.

# Clinical and regulatory considerations of implantable medical devices

# 7

Iryna Makovey*, Bishoy Gad†, Randy Scherer‡, Elizabeth K. Ferry§,
Grant Hoffman¶, Margot S. Damaser*,#,**

*Glickman Urological and Kidney Institute, Cleveland Clinic, Cleveland, Ohio, USA
†Orthopaedic and Rheumatology Institute, Cleveland Clinic, Cleveland, Ohio, USA
‡DEKRA Certification Incorporate, Chalfont, Pennsylvania, USA
§Urology Institute, University Hospitals Case Medical Center, Cleveland, Ohio, USA
¶Innovations Institute, Cleveland Clinic, Cleveland, Ohio, USA
#Department of Biomedical Engineering, Cleveland Clinic, Cleveland, Ohio, USA
**Advanced Platform Technology Center of Excellence, Louis Stokes Cleveland Department of
Veterans Affairs Medical Center, Cleveland, Ohio, USA

## CHAPTER CONTENTS

Bhunia et al. Implantable Biomedical Microsystems. http://dx.doi.org/10.1016/B978-0-323-26208-8.00007-8

# 7.1 INTRODUCTION

Implantable biomedical systems come in a variety of shapes and sizes. With the evolution of technology in recent years, there is a remarkable opportunity for medical device development. These devices can include simple mesh or prosthetic implants; intravascular devices such as filters, coils, and cardiac stents; and complex implantable multicomponent electrical systems. Advances have included several different packaging or coating possibilities to ensure biocompatibility, drug release systems to promote healing and prevent infection, tailored stress–strain properties to provide desired mechanics, and wireless transmission capability. However, the human body remains the same and the medical considerations of device implantation rely on a number of basic principles. Each medical device system must have an indication for implantation and be biocompatible and technically feasible to implant and explant. Most importantly, the benefits of the device must outweigh the risks.

Benefits and risks of the proposed device must be carefully examined and considered. For almost every medical condition or disease state, there is a treatment that is considered the standard of care, such as a medication or surgical procedure. When a proposed medical system is developed, it has to be at minimum equivalent to the existing treatment options but preferably superior to the current standard of care. Superiority may be simply defined as superior clinical outcomes, such as improved cure rates, compliance, tolerability, and convenience. However, in the approval process, significant consideration is given to financial and social aspects, including cost of the device and alternatives, disease burden and related loss of productivity, and health-care utilization. Additionally, risks associated with the medical system must be carefully examined and compared to current available treatments. There should be a balance between the margin of improved outcomes and the risks a specific medical system poses as compared to alternative treatments or no treatment at all. Risks of a medical device may include risks of complications from implantation or surgery, malfunction, wear over time, allergic reaction, infection, foreign body reaction, and the eventual necessity of a replacement. This chapter will discuss the clinical considerations of medical device development and describe some of the common risks associated with medical device implantation. Furthermore, it will address ways to minimize these risks in specific clinical situations and patient populations. Finally, regulatory considerations to gain clinical approval for device testing and implantation are reviewed.

## 7.2 PATIENT SELECTION AND SPECIAL POPULATIONS

Medical device systems must be created with the target treatment population in mind. For instance, an implant created for a child must be appropriately sized and account for anticipated growth and change in body habitus, while an implant in an elderly individual must account for expected limited mobility and dexterity.

### 7.2.1 GERIATRIC POPULATION

It is anticipated that by 2050, the population of people over age 65 in the United States will double from 2010 levels [1]. Most of the elderly suffer from chronic illnesses and 2/3 of them have more than one chronic disease diagnosis [2]. Elderly patients receive a disproportionately high number of implanted medical devices, as they often suffer from multiple medical conditions. For instance, 40% of new implantable cardiac defibrillators are implanted in patients 70 years or older [3], and the average age of a hip implant recipient is 70 years [4]. Based on population growth, the number of anticipated total knee and hip replacements in the United States is anticipated to grow approximately 6% per year and is not expected to be affected by economic downturns [5].

Each device has a specific set of indications and considerations for this population. For example, an artificial urinary sphincter is indicated for patients with an incompetent urinary sphincter and incontinence. It requires deactivation by the individual prior to urination to allow urine to pass through the sphincter into the urethra and subsequently out of the body. It therefore requires a patient to have the ability to sense a full bladder and the need to void, as well as to have the manual dexterity necessary to squeeze a reservoir to deactivate the device [6]. Thus, while this device can work very well for many patients, an elderly demented individual may not be able to know when it is time to deactivate the device for urination, leading to prolonged periods of urinary retention that may result in infection, hydronephrosis, or kidney failure [7]. It is therefore important to consider the physical and mental limitations of the targeted patient population in the design and development of any implantable medical device.

Specific considerations in the elderly include mobility, dexterity, visual and hearing impairment, mental status, nutritional status and frailty, and medical comorbidities. Up to 50% of patients over the age of 65 suffer from arthritis [8]. One in three patients over age 65 suffers from visual impairment and one in three has a significant hearing impairment [9,10]. Fourteen percent of those over age 71 suffer from dementia [11], and many are unaware of their conditions, making it necessary for physicians to rely on family members for medical information [12]. All of the above can limit a patient's ability to manage the device and should be carefully considered in device development.

Nutritional status and overall health status of an elderly patient may be the limiting factors for any procedure required for device implantation. Cardiovascular and pulmonary chronic illnesses can significantly increase anesthesia risks and postoperative complications. Diabetes or poor nutrition can increase the risk of infection and significantly impair healing [13,14]. As a result, minimizing the time of the procedure and the size of the incision necessary and developing minimally invasive approaches for device implantation could all contribute to maximizing benefit and reducing risk in the elderly population.

### 7.2.2 PATIENTS WITH NEUROLOGICAL DISEASE

Patients with neurological disease, such as those diagnosed with multiple sclerosis, Parkinson's disease, or spinal cord injury are beneficiaries of a number of medical devices aimed at improving functionality and quality of life. However, this group of patients is often immobile and may have limited manual dexterity and need relatively frequent magnetic resonance imaging (MRI) studies [15,127]. Location of the implanted device must be carefully selected, and devices cannot be placed in locations prone to compression, such as the buttock or lower back in patients who are wheelchair-dependent or bedbound [16].

One of the biggest obstacles to many medical devices in patients with neurological disease is their need for future imaging with MRI. Potential risks of MRI in patients with an implant are migration, heating of the device and surrounding tissue, distortion of the device, and image artifact. In recent years, there has been a significant research effort to optimize medical devices and use MRI-safe materials. Safety of orthopedic implants, prosthetic heart valves, neurosurgical fixation, and reconstruction devices has been demonstrated, and most of these are nonferromagnetic and exhibit only very minor temperature changes [17–20]. Recently, MRI-safe leads used in pacemakers have become available [21]; however, most of the devices that have been implanted to date are not considered to be MRI-safe. It is therefore prudent that each medical device system be individually assessed for MRI safety prior to implantation. The American Society for Testing and Materials has developed standard practice guidelines for making medical devices safe for use in an MRI environment and the International Standards Organization along with International Electrical Commission is developing specifications for active implantable medical devices such as pacemakers, neurostimulators, and cochlear implants [22].

### 7.2.3 CHILDREN

Current medical practice is to modify existing medical devices for pediatric patients. However, this approach is not optimal and poses a number of risks. Therefore, there remains a significant unmet medical need for medical device development for children. Barriers to medical device development for pediatric patients include regulatory and ethical concerns, small target population size, and financial disincentives. The US Food and Drug Administration (FDA) has put forth a significant effort to identify and address these barriers associated with pediatric device development [23].

Physiologic and technical factors also make it challenging to develop medical devices for children. Height, weight, growth and development stage, hormonal influences, activity and maturity level, and immune status have to be taken into account. A certain level of manual dexterity and developmental maturity may be necessary to properly use the device, and the expected level of interaction with a device may have to be adjusted based on the pediatric age group and abilities of the population for which the device is intended. Additionally, there are a variety of technical factors involved in implantation of a device as well as the need for long-term functioning and future need for revision with growth of the recipient or wear on the device. An example of this is the use of growth rods for the treatment of infantile and juvenile scoliosis that allows for growth of the spine while maintaining a correction of the spinal curves. New noninvasive techniques for adjusting the length of these growth rods via magnets are being developed [24].

To account for change in size of the child, there has been a shift to biodegradable devices, such as biodegradable skull fixation [25]. However, these innovations cannot be used in more complex permanent devices, such as cardiac pacemaker systems, whose leads often fail in children due to dislodgement. One study examined 497 children with a median follow-up of 6.2 years and noted a 15% rate of lead failure [26]. The most common failure modes were insulation break, conductor failure, high threshold for pacemaker interrogation, and lead dislodgements, and many of these failures were associated with a younger age of the recipient [26].

There is subsequently a greater expectation to explant and replace medical devices in children due to longer lifetimes and requirements due to growth. This is not without risk as multiple surgeries increase the risks for infection, damage to surrounding structures, and scar formation and may result in chronic pain. Consequently, it is important that we consider a device hierarchy. For instance, if there are two implant options, a less invasive device that is biodegradable or easily removed should be considered for the first line of treatment to allow for future interventions.

## 7.3 BIOCOMPATIBILITY

Biocompatibility is a broad term encompassing various interactions between an implanted medical device and the surrounding host tissues. Device considerations include the chemical composition of the material, biodegradation, and changes in physical properties of the device, such as shape, size, and surface morphology, as well as hydrophilic, hydrophobic, and tensile properties. Human factors include the location of the implant and local tissue reaction to the device. When any device is implanted, a collection of proteins is deposited on the surface, which will subsequently dictate the cellular response to the implant [27,28]. This response can range from minimal local inflammation to an extensive and chronic foreign body reaction. The concept of biocompatibility is that the recognition of the implant as a foreign body is diminished or impaired, usually by careful choice of the external packaging layer, minimizing the initial inflammatory cell response and subsequent fibrosis.

Medical devices can be intended for intravascular, epicardial, subdural, peritoneal, or submucosal use, to name a few, and biocompatibility requirements are different for each of these environments. Some of the commonly used materials for coating are metallic alloys, polymers, ceramics, and biologics. Characteristics, advantages, and disadvantages of various biological materials are described by Haschek *et al.*, who also describes toxicological studies to assess various possible toxicities of the biomaterial used [29], and will not be further discussed here.

## 7.3.1 METALS

Metallic alloys such as stainless steel, cobalt–chromium, and titanium alloys are commonly used in orthopedic and neurosurgical applications. While these provide strength and durability necessary for select applications, there are risks of allergy and toxicity. Specifically, allergic reaction consistent with type IV hypersensitivity has been described with metal alloy implants containing nickel, cobalt, and chromium and usually presents as rash, urticaria, and cutaneous vasculitis. For this reason, nickel and chromium-free devices have been developed [30]. If there is concern for a history of hypersensitivity to metals or a postimplant diagnosis is needed, patch testing performed by a dermatologist should be considered [31].

Toxicity of metallic alloys is associated with corrosion and the development of wear debris—particulate matter separated from the primary implant. Metal debris has been found in synovial fluid, lymph nodes, the liver, and the spleen [32,33]. However, the levels of metal particles have been very low and not clearly demonstrated to be a significant clinical risk to the patient. Carcinogenesis remains a concern as both nickel and chromium are used, but clinical data are lacking to demonstrate causality of malignancy due to metal implants [29]. Additionally, in animal experiments, chronic chromium exposure has led to decreased sperm counts, abnormal sperm morphology, and reduced number of viable follicles and ova, posing a potential concern for fertility [34,35].

## 7.3.2 POLYMERS

Polymers have a wide variety of applications in the field of implantable medical devices, and over recent decades, the selection of polymers has become extensive. Some of the commonly used polymers include polyethylene, polypropylene, hyaluronic acid and hydrogels, and silicone. These can be used for device components, drug delivery, and coatings or can comprise the device itself. A polymer surface can be modified to achieve very specific properties desired for the anticipated application. Biological interactions can be tested *ex vivo* with enzyme-linked immunosorbent assays, blood compatibility, and cell cultures [36], and thus, biocompatibility can be carefully tested prior to *in vivo* experiments.

It is not uncommon that polymers are also used when biodegradation is desired. Biodegradable materials have become popular as they eliminate a potential need for a second surgery to remove a device if no longer necessary. While very helpful, biodegradable materials elicit a local foreign body reaction that can vary in severity from mild with minimal local inflammation to extensive, causing destruction of surrounding tissues and requiring explantation. A foreign body reaction is characterized by a nonspecific inflammatory cell reaction with a lack of infectious organisms. The severity and timing depend on the degradation of the material utilized. This has been well described and summarized for orthopedic fixation implants [37]. A recent case series described an inflammatory reaction to a spinal cord stimulator consistent with hypersensitivity reactions seen with endovascular devices and pacemakers. It is important to differentiate an inflammatory reaction such as dermatitis, granuloma formation, and giant cell response from a true infectious process [38] as a missed diagnosis of an implant infection can lead to devastating outcomes.

Conversely, there may be collateral damage caused by biodegradation. The rate of degradation is determined by both the polymer used and the structure of the device, such as porosity and surface area. Some studies have shown that incorporating alkaline salts within the implant helps decrease the pH during degradation and diminishes this reaction [39].

### 7.3.3 LOCATION OF IMPLANT

Biocompatibility depends on the local tissue reaction to the device otherwise known as foreign body reaction that is described in a detailed review by Anderson *et al.* [28]. Local tissue reaction in turn depends both on the properties of the device, such as shape, size, surface, and material used, and on the properties of the anatomical site [40,41]. An important phenomenon in soft tissue implants is encapsulation and capsular contracture [42]. This occurs when an implant generates a severe local reaction characterized by macrophage infiltration and foreign body giant cell accumulation. The resultant formation of a thick capsule around the implant can result in significant discomfort and pain, as was seen with early breast implants [42,43].

Chemical coatings have been used to minimize implant immunogenicity and include collagen, polyethylene glycol, and hyaluronan, to name a few. Additionally, drug loading into the surface of implant materials has been used to prevent or minimize the inflammatory response. Examples of this include drug-eluting cardiac stents and steroid-eluting pediatric pacing leads [44]. Immunogenic responses observed in soft tissues can also be seen in the bone. Stress on the bone by the implant can contribute to corrosion, debris wear, and implant failure. In addition, many orthopedic implants undergo osteointegration, which means very close anchoring of the implant to the bone, which occurs most commonly in cementless total hip arthroplasty where porous coated surfaces allow for bony ingrowth and interdigitation [128]. In many cases, a scaffold can be used to allow bony ingrowth, or osteoconduction. An example of this is the mixing of calcium phosphate croutons [45] with bone marrow to produce bone in areas of bony defects, particularly for treatment of giant cell tumors. Osteoinduction, however, requires a biological or hormonal stimulus for growth of the bony tissues [46,47] and therefore is an example of a desired form of bioactivity of the implant in which the cellular response leads to organized tissue growth and proliferation, rather than a fibrotic reaction.

Tissue and bones are chronically exposed to static or high-motion environments for short periods of time, and thus, design of implants must account for this. Mechanical considerations are very different for intravascular devices that are exposed to continuous blood flow. Various vessels have different degrees of turbulent flow, constant shear stress, and interaction with immunogenic cells and clotting factors. Drug-eluting cardiovascular stents were created to improve the biocompatibility of bare metal stents and to decrease the rate of thrombosis. A monumental research effort has been invested into this field, and now, third-generation biodegradable drug-eluting polymer coatings are in development [48].

## 7.4 IMPLANTATION

### 7.4.1 THE IDEAL IMPLANT

Implantation of a medical device should strive to be a brief, minimally invasive procedure with low risks of infection, bleeding, and postoperative pain, ideally performed without general anesthesia. A new subdermal contraceptive device called Implanon® is an excellent example of this and has rapidly gained popularity [49]. It is a small thin rod measuring 4 cm in length and can be injected into the arm after administration of a local anesthetic. There is no risk of anesthesia, minimal risk of bleeding or pain, and with sterile technique, very low risk of infection. It can be easily removed if desired, is aesthetically optimal, and does not affect or limit daily activity. Unfortunately, many other medical devices require a much more complex design with multiple components and may need to be placed in specific locations through invasive surgical procedures with general anesthesia.

### 7.4.2 PERIOPERATIVE RISKS

The risk of surgery is a common obstacle to a patient's ability to use a medical device [50]. For instance, an elderly patient may benefit significantly from a hip replacement; but, it is not uncommon that due to significant comorbidities and poor functional status, the patient may not be a candidate for the surgical procedure. On the other hand, the benefits of a procedure can sometimes outweigh the high perioperative risks, depending on the indication for the implant. While it can be difficult to justify implantation of an artificial urinary sphincter to improve lifestyle, an aortic valve replacement may be essential to patient survival [51]. There are risk calculators available online that have been developed by the American College of Surgeons and by the Society of Thoracic Surgeons that consider patient preoperative health status and can provide calculated risks for specific surgery or intervention [52,53].

### 7.4.3 SURGICAL APPROACH AND TECHNIQUE

While a minimally invasive approach to implantation is very important, it is also important to consider the tools already available. There is a significant cost benefit if no new tools for implantation need to be developed. Additionally, the simpler the surgical procedure, the more marketable it will be. This was well demonstrated with a study of 54,925 patients receiving different models of knee implants from 1998 to 2007 [54]. Four of the ten implant models had a significant learning curve with outcomes being poor in the introduction phase, while the other six implant models had similar outcomes at the start of implant use as compared to later use.

Standard surgical techniques currently available include open surgery and minimally invasive surgery, using laparoscopic or robotic technology. Open surgical procedures require larger incisions and a longer recovery time, while minimally invasive surgical procedures are performed through a number of small incisions, potentially improving postoperative recovery [55]. Additionally, implantation of some devices can be accomplished through natural orifices with endoscopic tools such as cystoscopy or colonoscopy and as a result do not require a surgical incision. These can often be performed as outpatient procedures. Advances in interventional radiology have allowed for percutaneous access to almost every organ, particularly the vasculature. Vascular access is usually obtained with a small incision at a distant site with easy access to larger vessels, and using fluoroscopic guidance, medical devices such as vascular coils and stents, inferior vena cava filters, and even heart valves can be deployed [56]. These minimally invasive procedures require minimal to no anesthesia and therefore decrease the cardiopulmonary risk burdens, especially in patients with multiple comorbidities.

### 7.4.4 DEVICE DESIGN

Device design in the context of implantation depends on device size, shape, and flexibility. For some devices, size is predetermined anatomically. Implants such as a hernia repair mesh, a heart valve, a hip or knee arthroplasty implant, and a vascular graft necessitate a specific size dimension, and the physician is able to select between a few standardized size options. However, there are a number of medical devices in which the general consensus is that the smaller the device, the better. Examples of such devices include neurostimulators, pacemakers, insulin pumps, peritoneal dialysis catheters, and cochlear implants. With the development of wireless technology, microelectronics, and nanofabrication technology, creation of miniature devices has become possible. Also, the use of low-power devices and wirelessly rechargeable batteries to support these developments is essential to minimizing the size of the implant [57]. Small devices generally cause less discomfort, can be easily concealed, and provide less disruption to local tissues and physiological mechanisms. The shape of the device may also be dictated by the local anatomy; however, implantation technique has to be carefully considered.

While the final shape of the device is usually anatomically determined, a number of devices have been developed to allow for folding prior to implantation and expansion or opening once *in vivo*. For instance, the design of a recently developed percutaneous cardiac valve has allowed many patients who are not good surgical candidates to receive these necessary treatments and avoid significant surgical risks [58]. Prior to development of this technology, the only way to replace a heart valve was to perform open-heart surgery, including opening the length of the sternum, stopping the heart, placing the patient on cardiopulmonary bypass, and opening the cardiac muscle to gain access to the necessary valve. Now, an aortic valve can be placed either through a small incision in the chest or through groin access with comparable outcomes [59].

To enable these techniques, the device coating and structure have to be able to tolerate the implantation without damage to the surface or fracture of internal components. Rigorous testing of mechanical properties of the device is necessary to simulate both the implantation procedure and the expected chronic wear and ensure stability.

## 7.5 EXPLANTATION

Many medical devices are intended for chronic implantation; however, there are a number of reasons a device has to be explanted, such as infection, malfunction, and mechanical failure. One study reviewed 2827 patients with primary cochlear implants and found that 235 patients required revision surgery, most commonly explantation and reimplantation [60]. The indications for revision included failure of the device (58%), device migration/extrusion (23%), and wound infection or wound complication (17%) [60]. Thus, when a device is designed, it is important to consider the possible need for explantation and address feasibility and ease of implant removal.

### 7.5.1 APPROACH TO DEVICE REMOVAL

Some devices will require invasive surgical intervention for removal, which can be associated with even higher morbidity and mortality than the initial implantation, such as with cardiac pacing wires [61]. This is largely due to fibrosis and scar tissue formation around the device, making it difficult to remove. In the past, cardiac lead extraction was performed with simple manual pull with very high risk of lead fracture, myocardial perforation, cardiac tamponade, and death. Techniques of lead extraction have significantly evolved, and most of these involve using a flexible sheath that passes over the lead and breaks up adhesions and scar tissue with either mechanical force or various types of energy, such as heat or laser [62]. With new techniques, the rates of major complications, such as laceration of large thoracic vessels and perforation of the myocardium, were reported to be 1.8% with all-cause 30-day mortality of 2.2 % in a study of over 5000 procedures [63]. On the other hand removal of ureteral stents placed for kidney stone patients only requires an endoscopic procedure performed in the physician's office without any additional sedation and poses minimal risk to the patient [129]. This device has been developed for short term use and therefore design that allows for ease of extraction is critical.

### 7.5.2 REPLACEMENT AND TIMING

Timing of device replacement is an important consideration. When a device is infected, it is not uncommon for immediate reimplantation to be associated with an increased risk of recurrent infection [64]. As a result, patients must wait a given period of time prior to a second implant. However, this can be problematic if the patient is dependent on the device's function and an alternative therapy may need to be used in the mean time [65,66].

A further consideration is change in tissue environment due to prior implant and explants. For instance, when an implantable penile prosthesis fails, the implant is removed there maybe a delay in replacement of the prosthesis. If so, there is scarring and fibrosis of the corporal bodies of the penis that may lead to significant decrease in penile length and limit the size of a future implant [68].

Another example where timing of device removal is critical is in proximal femur implants in children, as these can be left in place or removed in childhood. It is not clear if it is beneficial to remove the implants as there is the risk of additional surgery and removal may increase the risk for postoperative fracture. However, when these are removed at the time of revision total hip arthroplasty, there are an increased operative time, length of hospital stay, and rate of intraoperative fracture [69]. Therefore, the risks and benefits have to be considered when determining the timing of reimplantation of a specific medical device for an individual patient.

### 7.5.3 BIODEGRADABLE DEVICES

Biodegradable devices offer an advantage to permanent implants, as they do not require a reoperation for removal. The rate of degradation and reaction to degradation can differ based on chemical properties of the material and the biological environment. Additionally, the coating or only parts of the device can be designed to be biodegradable, and the biodegradation reaction itself can be used to manipulate the local environment to promote device integration directly or through local drug delivery [70,71].

## 7.6 INFECTION

Infections of implantable medical systems are not uncommon and can be devastating to the host and the device. The material used, implantation technique, location of the device, exposure to the environment, host immune status, use of perioperative antibiotics, and antimicrobial coatings are just a few of the factors that should be considered to reduce the rate of infection of medical devices. For instance, a Foley catheter that is continuously exposed to urine and skin flora is significantly more likely to be colonized with a number of bacteria [72] compared to a sterilely implanted cardiac pacemaker completely excluded from outside environmental pathogens.

The pathogenesis of the infection of an implant is related to biofilm formation, which is initiated by bacterial adherence to a surface, secretion of extracellular matrix, and maturation of the layer to a complete barrier to external environments, including antibiotics. Therefore, if an implant is encased in a biofilm, treatment with antibiotics will likely be insufficient, and there will be a need for explantation of all of the components of the device [73]. However, the presence of a biofilm is possible without apparent clinical infection. In one study, 10 penile implants were explanted due to mechanical failure and 8/10 were found to have biofilm coatings with multiple organisms despite any signs or symptoms of infection [74]. This further demonstrates the complexity of the interaction between the host and the implantable medical device and development of a true implant infection.

### 7.6.1 INFECTION PREVENTION

A number of methods have been developed to minimize the risk of implantable device infection. Device design, specifically the implantable device coating, can be modified to decrease bacterial adhesion and minimize biofilm formation. There are a variety of antimicrobial coatings available. Additionally, heparin and urokinase have been used to decrease bacterial adherence to devices due to their fibrinolytic properties [75]. Preimplant irrigation with antibiotic solutions has also become common. Proper implantation techniques are essential and sterility has to be carefully maintained throughout the surgical procedure. Generally, a systemic antibiotic is given to the patient prior to making a skin incision for the implantation and is often continued for 24 h postoperatively. Specific quality guidelines have been developed to guide a proper choice of antibiotic therapy in various procedures [130]. There is no clear consensus on the benefits of prolonged postoperative antibiotic use in most medical devices, but antibiotic prophylaxis is common after total joint arthroplasty [76].

### 7.6.2 MANAGEMENT OF INFECTED MEDICAL IMPLANTS

Conservative management of infected medical devices is of limited value since the risks are significant. There is increased risk of recurrence and mortality with treatment of device infection with antibiotics alone or removal of only one component of the device [77]. In one study of cardiac pacemaker infections, 42% of patients presented with clinical signs of a pacer pocket infection only, but 2/3 of these patients had infective endocarditis on final diagnosis. The most common bacterial pathogens were *Staphylococcus aureus* and methicillin-resistant *S. aureus* [77]. There was a 9.4% death rate and most of these were associated with antibiotic-resistant organisms [77].

Infections in orthopedic implants are often severe and can be threatening to life and limb. Many studies have been published discussing the treatment of these issues, which often involve total removal of the device and treatment with a temporary prosthesis with antibiotic-impregnated cement [78]. Rates of infection control after single stage procedures range from 73% to 95% when properly selecting patients [79]. The concern for failure is due to prolonged infections secondary to biofilm formation over the implant, which is essentially impenetrable to antibiotics. When this occurs, removal of the prosthesis is necessary before the patient becomes critically ill [80].

## 7.7 DEVICE WEAR AND TEAR

Medical devices are implanted into seemingly benign environments to the untrained eye. However, a local tissue reaction to an implant can create a hostile environment, resulting in device failure, which in turn can cause profound damage to both local tissues and the human body as a whole. It is important to consider all possible modes of failure with implanted devices to properly design devices that minimize both wear and tear of the implant and damage to the patient.

## 7.7.1 CHANGE IN PERFORMANCE OVER TIME: MECHANICAL FAILURE

As time passes, the various demands on medical implants take a toll. These changes result in changes to both the material properties of the implant and the anatomical environment surrounding the implant. An example of this is polyethylene wear in total knee and hip arthroplasty. As the demands on the implant continue over time, abrasions, pitting, burnishing, or scratching may occur [81]. These often can lead to corrosion and eventually metal-on-metal interplay within the joint. As a result, debris may form, causing significant soft tissue damage. In total hip replacements, this can cause pseudotumors, as previously mentioned [82]. In addition, multiple cyclical loads on an implant may result in device fracture, as has been seen with semirigid penile implants [83].

Mechanical effects must be considered with all implants that are loaded. Forces applied to objects cause strain in various areas of the implant, and depending on the type of strain (compressive, tensile, and shear), the mechanical properties are different. When multiple cyclical loads are applied to an object, its elasticity will gradually change. In addition, galvanic changes of a material *in vivo* may change the mechanical properties of the implant [84]. For example, the titanium in intramedullary nails used to treat femur fractures have a very high modulus of elasticity (~100,000 MPa), which is much stronger than that of the bone (~15,000) [85]. Furthermore, mechanical strength is proportional to the radius of the device (if spherical or conical) to the fourth power. These factors play an important role in implant selection and design, as these devices can allow for immediate weight bearing and rarely fail for mechanical reasons prior to fracture resolution [86]. Mechanical properties of implants with mixed materials are more challenging to determine.

Mechanical failures can occur due to a variety of reasons; improper implantation, poor component design, and patient noncompliance are all possible reasons for failure. However, given the severity of some medical conditions, devices that potentially have high risks of failure may still be implanted. In one study evaluating the survivorship of rotating hinge total knee arthroplasty, there was a 45.9% failure rate requiring reoperation over 4.1 years [87]. Regardless, given the severity of the conditions treated by this prosthesis, since use of this device essentially prevented limb amputation for this period of time, it remains a valid treatment option.

Stress shielding is also a valid concern with the design of orthopedic implants since shielding of bone to stress by a stiff implant can impair healing and reduce bone strength. For example, bone surfaces need to experience mechanical stress to heal fractures and promote ingrowth into porous coated total hip arthroplasty components. Stress shielding is related to the modulus of elasticity of the component. Cobalt chrome implants tend to stress shield more than titanium implants because titanium has a lower modulus of elasticity [88].

The body's physiological response to an implant can also result in mechanical failure of implants. For example, mechanical heart valves have been known to fail via pannus (fibrotic tissues), thrombosis, or a combination of the two [89]. Unrecognized failure of a mechanical valve can lead to death. These failures tend to be treated with a combination of surgery (revision valve replacement versus thrombolysis) and medication (anticoagulation) [90].

## 7.7.2 ELECTRICAL AND ELECTROCHEMICAL WEAR

Electrical failure is an uncommon but potentially growing complication of many medical devices. In one study reviewing long-term follow-up of 22 children who had a pacemaker implanted during their first year of life, the battery of 6 of these devices failed at a mean of 4.1 years, necessitating implant revision [91]. In another study of sacral nerve stimulators used in patients with fecal incontinence, there were battery failures in 8 out of 87 device failures (6 were due to idiopathic causes and 2 due to accidental discharge caused by MRI) [92]. Attempts at making a rechargeable battery for these and other deeply implanted devices have failed due to ineffective wireless telemetry fields and external coils typically not tolerated by patients [93]. However, device battery life is improving rapidly and may eventually preclude the need for rechargeable batteries. Safety of electronic devices is essential and with development of the technology has become more challenging, as wireless transmission has to be protected from intentional and unintentional electrical interference, as well as possible access to private patient information.

Electrochemical changes within an implant may result in galvanic wear, even in implants without true electrical stimulation. Metal surfaces can serve as anodes, and given the high solute ionic environment of the body, many cathodes naturally exist and can "attack" the surface of the implant. When combined with mechanical wear from multiple loading cycles as discussed above, the results can be devastating. Some metals are more galvanically inert (titanium, chromium, and molybdenum), primarily due to their passivation layer formed during implant manufacturing. If this layer is destroyed, wear rates accelerate [84,94].

Wear damage can occur at any location in a mechanical construct, particularly where there is close proximity to metal-on-metal articulations. An example of this is the newly described phenomenon of trunnionosis in total hip arthroplasty implants. This occurs where there is galvanic corrosion between the metallic portion of the femoral head insert and the metal portion of the femoral component [95]. Unfortunately, in this situation, implant modularity has caused significant problems, resulting in similar maladies to those discovered with metal-on-metal hips. To combat this, ceramic heads on metal stems are being used at many locations.

## 7.7.3 EROSION AND MIGRATION

Regardless of the surgical approach or technique, it is essential that the device is placed in an optimal position and secured if necessary to avoid migration, dislodgement, or erosion. For instance, care must be taken to position subcutaneous devices deep enough within the adipose tissue to prevent skin tenting and potential pressure points. Skin erosions of deep brain stimulators [96], cardiac pacemakers [97], and implantable vascular access devices [98] have been described. Submucosal devices also have a significant risk of erosion, as illustrated in vaginal mesh used for prolapse repair with rates of mesh exposure of about 10% [99], with risk factors including smoking, amount of mesh used, and experience of the surgeon [100].

Device migration is a significant consideration in orthopedic and spine implants. Various methods have been utilized to try to secure these devices in place, including cement or screw-based fixation, as well as the more modern porous surfaces and coatings that promote biointegration. While most device migrations or erosions result in loss of the device and need for revision, some can result in even more devastating complications, such as cardiac or aortic wall erosion due to an atrial septal defect closure device [101] and pedicle screw penetration into either the abdominal aorta or the spinal cord [102].

## 7.8 REGULATORY CONSIDERATIONS: TACKLING THE FDA

Navigating regulations to obtain market clearance for a product is a challenging process with many potential pitfalls. Fortunately, the US Food and Drug Administration has provided comprehensive regulatory resources and guidance documents to assist manufacturers throughout almost every stage of the process [103]. This section will discuss the key steps in tackling the FDA and successfully obtaining market clearance for medical devices.

### 7.8.1 WHAT IS A DEVICE?

The Federal Food Drug and Cosmetic (FD&C) Act defines a medical device as:

> *"an instrument, apparatus, implement, machine, contrivance, implant, in vitro reagent, or other similar or related article, including a component part, or accessory which is:*

- *Recognized in the official National Formulary, or the United States Pharmacopoeia, or any supplement to them,*
- *Intended for use in the diagnosis of disease or other conditions, or in the cure, mitigation, treatment, or prevention of disease, in man or other animals, or*
- *Intended to affect the structure or any function of the body of man or other animals, and which does not achieve its primary intended purposes through chemical action within or on the body of man or other animals and which is not dependent upon being metabolized for the achievement of any of its primary intended purposes." [104]*

This definition provides clear distinction between what is considered a medical device and what is considered a drug. It is vital to confirm that the biomedical system being developed meets the regulatory definition of a device or the product may be regulated by a component of the FDA other than the Center for Devices and Radiological Health (CDRH) under different provisions of the FD&C Act. For example, if the primary intended use of the product is achieved through chemical action or by being metabolized by the body, the product is typically considered a drug. Human drugs are regulated by the Center for Drug Evaluation and Research of the FDA [105]. Clarification for product classification can be obtained through the Division of Small Manufacturers International and Consumer Assistance of the FDA to assist manufacturers in making the device determination.

## 7.8.2 CLASSIFY YOUR DEVICE: CLASS I, II, III

After establishing the product as a device, you must determine how the FDA will classify the device. The classification system establishes the level of regulatory control necessary to assure the safety and efficacy of the medical device. Additionally, the classification process dictates the marketing process the manufacturer must complete to obtain FDA clearance/approval. Classification is based on the intended use, indications for use, and the associated risks of the device [105]. The FD&C Act categorizes medical devices into three classes based on increasing complexity and risk:

- Class I (general controls)
- Class II (general controls and special controls)
- Class III (general controls and premarket approval)

Class I devices are the lowest-risk device category and require only general controls to demonstrate safety and effectiveness. General controls are authorized by the FD&C Act under sections 501 (adulteration), 502 (misbranding), 510 (registration), 516 (banned devices), 518 (notification and other remedies), 519 (records and reports), and 510 (general provisions). Most class I devices are not life-supporting or life-sustaining and do not present potential unreasonable risks of illness or injury. These devices are considered exempt from premarket notification in most cases; however, manufacturers must follow the general controls set out in the FD&C Act as in the US Code of Federal Regulations [106]. Class I device examples include manual surgical instruments for general use (chisel, clamp, contractor, cutter, file, forceps, hammer, pliers, scalpel, and measuring tape), elastic bandages, and examination gloves [107–109].

Class II devices are the next risk class category and require general controls and special controls to provide reasonable assurance of safety and effectiveness. Special controls may include performance standards, postmarket surveillance, labeling requirements, and development of guidance documents. Some class II devices are considered exempt from premarket notification; however, in most cases, class II devices require premarket review and clearance by FDA through the 510(k) process [106]. Class II device examples include powered wheelchairs, infusion pumps, surgical drapes, central venous catheters, tracheal tubes, and feeding tubes [110–115].

Class III devices are the highest-risk class and are categorized as such if insufficient information exists to determine that general controls and special controls are sufficient to provide reasonable assurance of safety and efficacy. Class III devices are typically life-supporting or life-sustaining or intended for a use that prevents the impairment of human health. Generally, class III devices are considered novel technologies with new intended uses or techniques that must be subjected to rigorous controls. Class III devices require premarket approval (PMA) by the FDA, and the manufacturer must establish reasonable assurance of the device's safety and efficacy, typically involving clinical data as the supporting material [106].

Class III device examples include heart valves, breast implants, peripheral stents, implantable infusion pumps, and cochlear implants. Bioimplantable systems that support life or sustain life will be classified into this device category and require rigorous scientific data to support safety and efficacy through the PMA process. A recent example of a novel bioimplantable system is the CardioMEMS Incorporated Champion Heart Failure Monitoring System [116]. This system provides pulmonary artery pressure data using a wireless sensor. The pulmonary artery pressure monitoring data are used by physicians in the management of heart failure by indicating changes in pressure and associated symptoms, enabling optimization of medications.

Manufacturers may request a classification determination from FDA in instances when a device is difficult to categorize via Section 513(g) of the FD&C Act. The FDA will respond within a 60-day period with classification information based on the description of the device provided [117].

### 7.8.3 PREMARKET NOTIFICATION AND APPROVAL

The FD&C Act was amended in 1976 to include premarket review of medical devices. Under the amendment, the FDA was authorized to set standards, with premarket clearance and approval for the device types listed in the previous section [118]. Devices posing little or no risk to users or patients were exempted from the premarket clearance process. A sponsor wishing to market a class I, II, or III device intended for human use in the United States, for which a PMA is not required, must submit a 510(k) exemption to the FDA unless the device is exempt from those requirements and does not exceed the limitations of exemptions in the Code of Federal Regulations (CFR) device classification [119].

When the FD&C Act was amended in 1976, all the devices currently on the market were established as preamendment devices. These preamendment devices, along with any devices that the FDA has down-classified from class III to class II, can serve as predicate devices in the 510(k) premarket clearance process [120]. The main objective of this clearance pathway is to establish that a device is as safe and effective as a legally marketed predicate device that is not subject to the PMA process. A device is considered substantially equivalent to a predicate device if, in comparison to the predicate, it has:

1. The same intended use *and*
2. The same technological characteristics as the predicate
   *or*
3. The same intended use as the predicate *and*

4. Different technological characteristics and the information submitted to FDA
   a. does not raise new questions of safety and effectiveness
   b. demonstrates that the device is at least as safe and effective as the legally marketed device [104,120].

A claim of substantial equivalence does not mean that the new device and the predicate device must be identical; rather, the sponsor must establish similarities with respect to intended use, design, energy used or delivered, material type, performance, safety, effectiveness, biocompatibility, standards, and other applicable characteristics. If FDA determines that a device is not substantially equivalent, the applicant may:

1. Resubmit another 510(k) with new data
2. Request class I or II designation through the de novo process
3. File a reclassification
4. Submit a PMA

The FD&C Act and 510(k) regulation do not specify who must apply for a 510(k); instead, they specify which actions, such as introduction of a device to the US market, require a 510(k) submission [104,120]. The following four parties must submit a 510(k) to the FDA prior to marketing a device in the United States:

1. Domestic manufacturers introducing a device to the US market
2. Specification developers introducing a device to the US market
3. Repackers or relabelers who make labeling changes or whose operations significantly affect the device
4. Foreign manufacturers/exporters or US representatives of foreign manufacturers/exporters introducing a device in the US market

Procedures for class III medical device PMA applications are provided in 21 CFR 814 of the US Code of Federal Regulations. PMAs require sufficient valid scientific evidence that assures that a device is safe and effective for its intended use(s) and is the most challenging type of device marketing application required by FDA [120]. Devices that are subject to PMA requirements include any class III medical device (unless exempt under 520(g) of the FD&C Act) and are typically novel device concepts without regulatory history associated with the medical device amendments. Categorization of device type is therefore normally not possible.

A PMA requires the highest level of scientific evidence and includes additional technical information in comparison to a 510k submission detailing nonclinical and clinical analyses of the device. Nonclinical sections typically include microbiology, toxicology, immunology, biocompatibility, stress, wear, shelf life, and other laboratory or animal tests performed following good laboratory practice regulation [121]. Clinical sections provide details on study protocols, safety and efficacy data, adverse reactions and complications, device failures and replacements, patient information, patient complaints, tabulations of data from all individual subjects, results of statistical analyses, and any other information from the clinical investigations that assists in providing scientific evidence supporting safety and effectiveness of the application device [120]. Due to the large amount of technical information, the FDA requires 180 days for review of PMA applications to make a determination of approval or disapproval. Typical review times are much longer and often include the use of FDA advisory committees, consisting of experts in the field of medicine applicable to the device type [121].

### 7.8.4 INVESTIGATIONAL DEVICE USE

Clinical studies are typically performed to support PMA applications and a small subset of class II 510k submissions. Manufacturers must obtain investigational device exemption (IDE) when distributing significant-risk devices that are not cleared to be marketed when performing clinical research to gather existing evidence to support the safety and effectiveness of the device [120]. The FDA provides guidance to assist manufacturers in determining the risk level of the device under evaluation and in the preparation of IDE applications. 21 CFR 812 of the US Code of Federal Regulations provides details on the procedures for conducting clinical investigations of devices with the intent of discovery and development of useful devices in an ethical manner [122].

A significant-risk device is defined as an investigational device that

1. is intended as an implant and presents a potential for serious risk to the health, safety, or welfare of a subject;
2. is purported or represented to be for use in supporting or sustaining human life and presents a potential for serious risk to the health, safety, or welfare of a subject;
3. is for use of substantial importance in diagnosing, curing, mitigating, or treating disease or otherwise preventing impairment of human health and presents a potential for serious risk to the health, safety, or welfare of a subject; or
4. otherwise presents a potential for serious risk to the health, safety, or welfare of a subject.

Nonsignificant devices do not pose a significant risk to human subjects participating in the clinical investigation [122].

Clinical research with investigational devices considered to be of significant risk, unless exempt, must have an approved IDE prior to study initiation. In addition to the approved IDE, clinical testing of devices that have not been cleared for marketing requires

- institutional review board (IRB) approval of the clinical investigation plan,
- informed consent documented for all participating subjects,
- labeling that states that the device is for investigational use only,
- that the study must be regularly monitored, and
- that documentation of records and reports must be maintained.

Nonsignificant-risk device studies only require IRB approval prior to study initiation; however, the study sponsor must still abide by the above-stated requirements to protect patient health and data integrity [122]. Abbreviated IDE requirements for nonsignificant-risk devices are listed in 21 CFR 812.2 (b) of the US Code of Federal Regulations.

The sponsor of the clinical investigation is the responsible party for submitting an IDE application to the FDA [122]. It is important for sponsors to communicate with the FDA early and often in the presubmission process to obtain guidance prior to the submission of an IDE application. IDE study approval must be obtained before study initiation and enrolling patients; therefore, sponsor interaction with the FDA can increase the speed of the regulatory process by minimizing delays associated with understanding FDA requirements, regulations, and guidance documents. In addition, pre-IDE meetings will allow FDA reviewers to clearly understand the device under evaluation and study design considerations.

### 7.8.5 QUALITY SYSTEMS REGULATION PRACTICES

The FDA established current good manufacturing practices (CGMPs) in 1978 to ensure that finished devices will be safe, effective, and in compliance with the FD&C Act [104]. Since their inception, CGMPs have been revised to add design controls, purchasing controls, validation requirements, and management requirements to be consistent with applicable international standards such as ISO 9001 and ISO 13485 [123]. The quality system regulations allow for much flexibility in how manufacturers produce a specific device by only providing a framework to follow in developing appropriate manufacturing procedures, rather than specific manufacturing steps/procedures. This allows manufacturers to tailor particular processes and innovative techniques for a broad spectrum of device types; however, each manufacturer shall establish and maintain a quality system that is appropriate for the specific medical device(s) designed or manufactured and that meets the regulation requirements [123].

21 CFR 820 lists the following quality system requirements and provisions:

- Management responsibility
- Quality audit procedures
- Personnel specifications
- Design controls
- Document controls
- Purchasing controls
- Identification and traceability
- Production and process controls
- Acceptance activities
- Nonconforming product
- Corrective and preventative action
- Labeling and packaging control
- Handling, storage, distribution, and installation
- Records
- Servicing

All of these areas of the quality system are subject to FDA facility inspection and auditing to confirm compliance with regulations. A robust quality system incorporating CGMPs shall be maintained by each manufacturer. The system shall be appropriate for the specific medical device(s) and meet the requirements discussed in this section [123].

### 7.8.6 DEVICE LABELING

Section 201 of the FD&C Act defines a label as a "display of written, printed, or graphic matter upon the immediate container of any article…" and defines labeling as "all labels and other written, printed, or graphic matter (1) upon any article or any of its containers or wrappers, or (2) accompanying such article" [104].

Medical device labeling regulations have been detailed for the following device types:

- General device labeling (21 CFR 801)
- *In vitro* diagnostic products (21 CFR 809)
- Investigational device exemptions (21 CFR 812)
- Good manufacturing practices (21 CFR 820)
- General electronic products (21 CFR 1010)

These regulations specify the minimum requirements for all device types in each specific category. The general labeling provisions indicate the following are requirements for device labeling:

- Name and place of business (21 CFR 801.1)
- Intended use (21 CFR 801.4)
- Adequate directions for use (21 CFR 801.5)
- Description of false or misleading statements (21 CFR 801.6)
- Prominence of statements (21 CFR 801.15)

Labeling and advertising are closely associated with one another. Any forms of publicity including posters, brochures, pamphlets, booklets, and claims associated with promotional materials are regulated by FDA and the Federal Trade Commission [124]. Two types of labeling are recognized by the FDA: FDA-approved labeling and promotional labeling [118]. FDA-approved medical labeling must accompany the device on or within the package, or else the device is considered misbranded since it does not contain adequate directions for use (21 CFR 801.109(c) and FD&C 502(f)). In addition to this requirement, a device is considered misbranded and in violation of the law if labeling "is false or misleading in any way" [104].

### 7.8.7 REPORTING OF ADVERSE EVENTS

Since 1984, medical device manufacturers have been required by the FDA to report all device-related deaths, serious injuries, and malfunctions. The Medical Device Reporting (MDR) regulations contain mandatory requirements for manufacturers, device importers, and device user facilities to report specific device-related incidents or adverse events to the FDA. This regulation calls out the type of incidents to report, whom to report the information, and how long the reporter has to complete this operation. The following table summarizes the information manufacturers, importers, and user facilities must report [125].

| Reporter | Information to Report | Whom to Report to | Timing of Report |
|----------|----------------------|-------------------|------------------|
| Manufacturer | Report of deaths, serious injury, and malfunctions | FDA | Within 30 calendar days of becoming aware of event |
| | Event designated by FDA or an event that requires remedial action to prevent an unreasonable risk of substantial harm to the public health | FDA | Within 5 work days of becoming aware of the event |
| Importers | Reports of deaths and serious injuries | FDA and manufacturer | Within 30 calendar days of becoming aware of the event |
| | Reports of malfunctions | Manufacturer | |
| User facility | Device-related death | FDA and manufacturer | Within 10 work days of becoming aware |
| | Device-related serious injury | Manufacturer | |
| | Annual summary of death and serious injury reports | FDA | January 1 for the preceding year |

Adverse event monitoring and reporting vigilance programs are important to safe medical device use. Tracking events can indicate trends in usage that may lead to improvements in user needs and patient safety. The fixed reporting windows provide an assurance that information is compiled and directed to the appropriate party in a timely manner. A judicious MDR program allows FDA the best opportunity to monitor device safety to protect patients from avoidable serious health hazards. In the case where a device may be considered defective, could be a risk to human health, or is in violation of the rules administered by the agency, a recall might be required [118]. Recalls are classified by the FDA to indicate the degree of health hazard potential the recalled product presents. Health hazards are evaluated based on the assessment of the disease or injury associated with the product, seriousness of the risk posed by the product, and likelihood of occurrence in the near-term and long-term consequences [118]. Definitions of the recall classification for FDA are provided below:

• Class I recall—a situation presenting a reasonable probability that the use of, or exposure to, a violative product will cause serious adverse health consequences or death

- Class II recall—a situation in which use of, or exposure to, a violative product may cause temporary or medically reversible adverse health consequences or where the probability of serious adverse health consequences is remote
- Class III recall—a situation in which use or exposure to a violative product is not likely to cause adverse health consequences

Supplementary materials and procedures concerning recalls, including guidance documents, effectiveness checks, model press releases, 510(k) requirements during firm-initiated recalls, and the FDA recall database, can be found on FDA's website under CDRH Device Advice [126]. Additionally, the FDA provides a database of recalls, market withdrawals, and safety alerts where press releases and public notices are posted for regulated products [126]. These press releases aid in alerting health professionals and the public when needed to ensure the health and safety of patients.

### 7.8.8 REGULATORY CONCLUSIONS

Communicating with the FDA early and often in the device development process is imperative to successfully obtaining market clearance for medical devices. Utilizing the proper regulatory strategy from the outset will remove many of the hurdles encountered by manufacturers during the process. The FDA provides a multitude of guidance documents and device advice to assist in the development and approval of products. The ultimate goal of the FDA is to protect and promote public health; therefore, a transparent process to bring innovative ideas to solving patient health problems is necessary for manufacturers and FDA to achieve their shared goals.

## 7.9 SUMMARY AND CONCLUSIONS

Implantable biomedical systems are advancing rapidly in today's technological environment, particularly in the areas of increased power, miniaturization, and wireless connectivity. Nonetheless, the clinical considerations for developing an implantable device have not changed. These include considerations for special populations, considerations of biocompatibility, considerations of the need for explantation, risk versus benefit analyses, and regulatory requirements to obtain clinical approval for implantation. This chapter summarizes these considerations and provides suggestions, examples, and references for further research for researchers developing implantable medical devices.

## REFERENCES

[1] US Census Bureau Public Information Office. U.S. Census Bureau projections show a slower growing, older, more diverse nation a half century from now—population—newsroom—U.S. Census Bureau, http://www.census.gov/newsroom/releases/archives/population/cb12-243.html; 2014 [accessed 27.05.14].

[2] US Department of Health and Human Services. Multiple chronic conditions: a strategic framework—optimum health and quality of life for individuals with multiple chronic conditions. Washington, DC: U.D.o.H.a.H. Services, Editor; 2010. Vogeli, et al. J Gen Intern Med 2007. PMID: 18026807.

[3] Epstein AE, et al. Implantable cardioverter-defibrillator prescription in the elderly. Heart Rhythm 2009;6(8):1136–43.

[4] Danielsson L, Lindberg H, Nilsson B. Prevalence of coxarthrosis. Clin Orthop Relat Res 1984;191:110–5; Merrill C (Thomson Healthcare), Elixhauser A (AHRQ). Hospital stays involving musculoskeletal procedures, 1997–2005. HCUP Statistical Brief #34. Rockville, MD: Agency for Healthcare Research and Quality; July 2007. http://www.hcup-us.ahrq.gov/reports/statbriefs/sb34.pdf.

[5] Kurtz SM, et al. Impact of the economic downturn on total joint replacement demand in the United States: updated projections to 2021. J Bone Joint Surg Am 2014;96(8):624–30.

[6] O'Connor RC, et al. Artificial urinary sphincter placement in elderly men. Urology 2007;69(1):126–8.

[7] Kosmadakis G, et al. End-stage renal disease due to hydronephrosis in a patient with artificial urinary sphincter. Int J Artif Organs 2011;34(11):1106–8.

[8] Centers for Disease Control, Prevention. Prevalence of doctor-diagnosed arthritis and arthritis-attributable activity limitation—United States, 2010–2012. MMWR Morb Mortal Wkly Rep 2013;62(44):869–73. Available from: http://www.ncbi.nlm.nih.gov/pubmed/24196662.

[9] Ganley JP, Roberts J. Eye conditions and related need for medical care. Vital Health Stat 11 1983;(228)1–69.

[10] Schiller JS, et al. Summary health statistics for U.S. adults: National Health Interview Survey 2010. Vital Health Stat 10 2012;(252)1–207.

[11] Plassman BL, et al. Prevalence of dementia in the United States: the aging, demographics, and memory study. Neuroepidemiology 2007;29(1–2):125–32.

[12] Gad BV, et al. Validity of patient-reported comorbidities before total knee and hip arthroplasty in patients older than 65 years. J Arthroplasty 2012;27(10):1750–6, e1.

[13] Stechmiller JK. Understanding the role of nutrition and wound healing. Nutr Clin Pract 2010;25(1):61–8.

[14] Lioupis C. Effects of diabetes mellitus on wound healing: an update. J Wound Care 2005;14(2):84–6.

[15] Treaba CA, et al. Cerebral lesions of multiple sclerosis: is gadolinium always irreplaceable in assessing lesion activity? Diagn Interv Radiol 2014;20(2):178–84.

[16] Kumar K, et al. Avoiding complications from spinal cord stimulation: practical recommendations from an international panel of experts. Neuromodulation 2007;10(1):24–33.

[17] Kumar R, Lerski RA, Gandy S, Clift BA, Abboud RJ. Safety of orthopedic implants in magnetic resonance imaging: an experimental verification. J Orthop Res 2006;24(9):1799–802.

[18] Shellock FG. Prosthetic heart valves and annuloplasty rings: assessment of magnetic field interactions, heating, and artifacts at 1.5 Tesla. J Cardiovasc Magn Reson 2001;3(4):317–24.

[19] Shellock FG. Metallic neurosurgical implants: evaluation of magnetic field interactions, heating, and artifacts at 1.5-Tesla. J Magn Reson Imaging 2001;14(3):295–9.

[20] Shellock FG. Biomedical implants and devices: assessment of magnetic field interactions with a 3.0-Tesla MR system. J Magn Reson Imaging 2002;16(6):721–32.

[21] Jung W, et al. Initial experience with magnetic resonance imaging-safe pacemakers: a review. J Interv Card Electrophysiol 2011;32(3):213–9; Forleo GB. Heart Rhythm 2010. PMID: 20167289.

[22] Woods TO. Standards for medical devices in MRI: present and future. J Magn Reson Imaging 2007;26(5):1186–9.

[23] Samuels-Reid JH, Blake ED. Pediatric medical devices: a look at significant US legislation to address unmet needs. Expert Rev Med Devices 2014;11(2):169–74.

[24] Dannawi Z, et al. Early results of a remotely-operated magnetic growth rod in early-onset scoliosis. Bone Joint J 2013;95-B(1):75–80.

[25] Hayden Gephart MG, et al. Using bioabsorbable fixation systems in the treatment of pediatric skull deformities leads to good outcomes and low morbidity. Childs Nerv Syst 2013;29(2):297–301.

[26] Fortescue EB, et al. Patient, procedural, and hardware factors associated with pacemaker lead failures in pediatrics and congenital heart disease. Heart Rhythm 2004;1(2):150–9.

[27] Tang L, Eaton JW. Natural responses to unnatural materials: a molecular mechanism for foreign body reactions. Mol Med 1999;5(6):351–8.

[28] Anderson JM, Rodriguez A, Chang DT. Foreign body reaction to biomaterials. Semin Immunol 2008;20(2):86–100.

[29] Haschek WM, Rousseaux CG, Wallig MA. Haschek and Rousseaux's handbook of toxicologic pathology. Amsterdam: Academic Press; 2013.

[30] Leto A, et al. Bioinertness and fracture toughness evaluation of the monoclinic zirconia surface film of Oxinium femoral head by Raman and cathodoluminescence spectroscopy. J Mech Behav Biomed Mater 2014;31:135–44.

[31] Schalock PC, Thyssen JP. Metal hypersensitivity reactions to implants: opinions and practices of patch testing dermatologists. Dermatitis 2013;24(6):313–20.

[32] Case CP, et al. Widespread dissemination of metal debris from implants. J Bone Joint Surg Br 1994;76(5):701–12.

[33] Urban RM, et al. Dissemination of wear particles to the liver, spleen, and abdominal lymph nodes of patients with hip or knee replacement. J Bone Joint Surg Am 2000;82(4):457–76.

[34] Elbetieha A, Al-Hamood MH. Long-term exposure of male and female mice to trivalent and hexavalent chromium compounds: effect on fertility. Toxicology 1997;116(1–3):39–47.

[35] Kumar S, et al. Semen quality of industrial workers occupationally exposed to chromium. J Occup Health 2005;47(5):424–30.

[36] Eastmond GC, Höcker H, Klee D, editors. Biomedical applications, polymer blends. Advances in polymer science. Berlin, New York: Springer; 1999.

[37] Bostman O, Pihlajamaki H. Clinical biocompatibility of biodegradable orthopaedic implants for internal fixation: a review. Biomaterials 2000;21(24):2615–21.

[38] Chaudhry ZA, et al. Detailed analysis of allergic cutaneous reactions to spinal cord stimulator devices. J Pain Res 2013;6:617–23.

[39] Agrawal CM, Athanasiou KA. Technique to control pH in vicinity of biodegrading PLA-PGA implants. J Biomed Mater Res 1997;38(2):105–14.

[40] Morais JM, Papadimitrakopoulos F, Burgess DJ. Biomaterials/tissue interactions: possible solutions to overcome foreign body response. AAPS J 2010;12(2):188–96.

[41] Korovessis P, Repanti M. Evolution of aggressive granulomatous periprosthetic lesions in cemented hip arthroplasties. Clin Orthop Relat Res 1994;300:155–61.

[42] Potter EH, Rohrich RJ, Bolden KM. The role of silicone granulomas in recurrent capsular contracture: a review of the literature and an approach to management. Plast Reconstr Surg 2013;131(6):888e–95e.

[43] Barnsley GP, Sigurdson LJ, Barnsley SE. Textured surface breast implants in the prevention of capsular contracture among breast augmentation patients: a meta-analysis of randomized controlled trials. Plast Reconstr Surg 2006;117(7):2182–90.

[44] Horenstein MS, et al. Chronic performance of steroid-eluting epicardial leads in a growing pediatric population: a 10-year comparison. Pacing Clin Electrophysiol 2003;26 (7 Pt 1):1467–71.

[45] Kurien T, Pearson RG, Scammell BE. Bone graft substitutes currently available in orthopaedic practice: the evidence for their use. Bone Joint J 2013;95-B(5):583–97.

[46] Shim IK, et al. Biofunctional porous anodized titanium implants for enhanced bone regeneration. J Biomed Mater Res A 2014;102(10):3639–48.

[47] Albrektsson T, Johansson C. Osteoinduction, osteoconduction and osseointegration. Eur Spine J 2001;10(Suppl. 2):S96–S101.

[48] Schurtz G, et al. Biodegradable polymer Biolimus-eluting stent (Nobori(R)) for the treatment of coronary artery lesions: review of concept and clinical results. Med Devices (Auckl) 2014;7:35–43.

[49] Funk S, et al. Safety and efficacy of Implanon, a single-rod implantable contraceptive containing etonogestrel. Contraception 2005;71(5):319–26.

[50] Iung B, et al. Decision-making in elderly patients with severe aortic stenosis: why are so many denied surgery? Eur Heart J 2005;26(24):2714–20.

[51] Varadarajan P, et al. Clinical profile and natural history of 453 nonsurgically managed patients with severe aortic stenosis. Ann Thorac Surg 2006;82(6):2111–5.

[52] American College of Surgeons. ACS NSQIP surgical risks calculator, http://riskcalculator.facs.org; 2014 [accessed 11.04.14].

[53] The Society of Thoracic Surgeons. Online STS risk calculator, http://riskcalc.sts.org/STSWebRiskCalc/; 2014 [accessed 11.04.14].

[54] Peltola M, Malmivaara A, Paavola M. Learning curve for new technology?: a nationwide register-based study of 46,363 total knee arthroplasties. J Bone Joint Surg Am 2013;95(23):2097–103.

[55] Alemozaffar M, et al. Benchmarks for operative outcomes of robotic and open radical prostatectomy: results from the health professionals follow-up study. Eur Urol 2014 Feb 11. http://dx.doi.org/10.1016/j.eururo.2014.01.039 pii: S0302-2838(14)00118-3. [Epub ahead of print].

[56] Linke A, et al. Treatment of aortic stenosis with a self-expanding transcatheter valve: the International Multi-centre ADVANCE Study. Eur Heart J 2014 Mar 28. [Epub ahead of print] PMID: 24682842.

[57] Majerus SJ, et al. Low-power wireless micromanometer system for acute and chronic bladder-pressure monitoring. IEEE Trans Biomed Eng 2011;58(3):763–7.

[58] Cribier A, et al. Percutaneous transcatheter implantation of an aortic valve prosthesis for calcific aortic stenosis: first human case description. Circulation 2002;106(24):3006–8.

[59] Rodes-Cabau J, et al. Feasibility and initial results of percutaneous aortic valve implantation including selection of the transfemoral or transapical approach in patients with severe aortic stenosis. Am J Cardiol 2008;102(9):1240–6.

[60] Wang JT, et al. Rates of revision and device failure in cochlear implant surgery: a 30-year experience. Laryngoscope 2014;124(10):2393–9.

[61] Brunner MP, et al. Outcomes of patients requiring emergent surgical or endovascular intervention for catastrophic complications during transvenous lead extraction. Heart Rhythm 2014;11(3):419–25.

[62] Buch E, Boyle NG, Belott PH. Pacemaker and defibrillator lead extraction. Circulation 2011;123(11):e378–80.

[63] Brunner MP, et al. Clinical predictors of adverse patient outcomes in an experience of more than 5000 chronic endovascular pacemaker and defibrillator lead extractions. Heart Rhythm 2014;11(5):799–805.

[64] Engesaeter LB, et al. Surgical procedures in the treatment of 784 infected THAs reported to the Norwegian Arthroplasty Register. Acta Orthop 2011;82(5):530–7.

[65] Silvestre A, et al. Revision of infected total knee arthroplasty: two-stage reimplantation using an antibiotic-impregnated static spacer. Clin Orthop Surg 2013;5(3):180–7.

[66] Sohail MR, et al. Management and outcome of permanent pacemaker and implantable cardioverter-defibrillator infections. J Am Coll Cardiol 2007;49(18):1851–9.

[67] Henry GD, et al. Revision washout decreases penile prosthesis infection in revision surgery: a multicenter study. J Urol 2005;173(1):89–92.

[68] Wilson SK, et al. Upsizing of inflatable penile implant cylinders in patients with corporal fibrosis. J Sex Med 2006;3(4):736–42.

[69] Woodcock J, et al. Do retained pediatric implants impact later total hip arthroplasty? J Pediatr Orthop 2013;33(3):339–44.

[70] Ceschi P, et al. Biodegradable polymeric coatings on cochlear implant surfaces and their influence on spiral ganglion cell survival. J Biomed Mater Res B Appl Biomater 2014;102(6):1255–67.

[71] Arnoldi J, Alves A, Procter P. Early tissue responses to zoledronate, locally delivered by bone screw, into a compromised cancellous bone site: a pilot study. BMC Musculoskelet Disord 2014;15:97.

[72] Matsukawa M, et al. Bacterial colonization on intraluminal surface of urethral catheter. Urology 2005;65(3):440–4.

[73] Costerton JW. Introduction to biofilm. Int J Antimicrob Agents 1999;11(3–4):217–21, discussion, p. 237–9.

[74] Silverstein AD, et al. Biofilm formation on clinically noninfected penile prostheses. J Urol 2006;176(3):1008–11.

[75] Nakamoto DA, et al. Use of fibrinolytic agents to coat wire implants to decrease infection. An animal model. Invest Radiol 1995;30(6):341–4.

[76] Aminoshariae A, Kulild J. Premedication of patients undergoing dental procedures causing bacteremia after total joint arthroplasty. J Endod 2010;36(6):974–7.

[77] del Rio A, et al. Surgical treatment of pacemaker and defibrillator lead endocarditis: the impact of electrode lead extraction on outcome. Chest 2003;124(4):1451–9.

[78] Leonard HA, et al. Single- or two-stage revision for infected total hip arthroplasty? A systematic review of the literature. Clin Orthop Relat Res 2014;472(3):1036–42.

[79] Parvizi J, Cavanaugh PK, Diaz-Ledezma C. Periprosthetic knee infection: ten strategies that work. Knee Surg Relat Res 2013;25(4):155–64.

[80] Simmons TD, Stern SH. Diagnosis and management of the infected total knee arthroplasty. Am J Knee Surg 1996;9(2):99–106.

[81] Lu YC, et al. Wear-pattern analysis in retrieved tibial inserts of mobile-bearing and fixed-bearing total knee prostheses. J Bone Joint Surg Br 2010;92(4):500–7.

[82] Scully WF, Teeny SM. Pseudotumor associated with metal-on-polyethylene total hip arthroplasty. Orthopedics 2013;36(5):e666–70.

[83] Agatstein EH, Farrer JH, Raz S. Fracture of semirigid penile prosthesis: a rare complication. J Urol 1986;135(2):376–7.

[84] Einhorn TA, et al. Orthopaedic basic science foundations of clinical practice. Rosemont, IL: American Academy of Orthopaedic Surgeons; 2007, p. 65–85.

[85] Rockwood CA, Green DP, Bucholz RW. Rockwood & Green's fractures in adults. Philadelphia, PA: Lippincott, Williams & Wilkins; 2010, p. 1 online resource (2 v. (xvii, 2174, 39 p.)).

[86] Gardner MJ, Silva MJ, Krieg JC. Biomechanical testing of fracture fixation constructs: variability, validity, and clinical applicability. J Am Acad Orthop Surg 2012;20(2):86–93.

[87] Smith TH, et al. Comparison of mechanical and nonmechanical failure rates associated with rotating hinged total knee arthroplasty in nontumor patients. J Arthroplasty 2013;28(1):62–7, e1.

[88] Einhorn K. Sesame subjects: reptiles. 1st ed. New York, NY: Random House; 2007.

[89] Deviri E, et al. Obstruction of mechanical heart valve prostheses: clinical aspects and surgical management. J Am Coll Cardiol 1991;17(3):646–50.

[90] Roudaut R, Serri K, Lafitte S. Thrombosis of prosthetic heart valves: diagnosis and therapeutic considerations. Heart 2007;93(1):137–42.

[91] Aellig NC, et al. Long-term follow-up after pacemaker implantation in neonates and infants. Ann Thorac Surg 2007;83(4):1420–3.

[92] Faucheron JL, Voirin D, Badic B. Sacral nerve stimulation for fecal incontinence: causes of surgical revision from a series of 87 consecutive patients operated on in a single institution. Dis Colon Rectum 2010;53(11):1501–7.

[93] Joung YH. Development of implantable medical devices: from an engineering perspective. Int Neurourol J 2013;17(3):98–106.

[94] Jacobs JJ, Gilbert JL, Urban RM. Corrosion of metal orthopaedic implants. J Bone Joint Surg Am 1998;80(2):268–82.

[95] Pastides PS, et al. Trunnionosis: a pain in the neck. World J Orthop 2013;4(4):161–6.

[96] Pena E, et al. Skin erosion over implants in deep brain stimulation patients. Stereotact Funct Neurosurg 2008;86(2):120–6.

[97] Griffith MJ, et al. Mechanical, but not infective, pacemaker erosion may be successfully managed by re-implantation of pacemakers. Br Heart J 1994;71(2):202–5.

[98] Bass J, Halton JM. Skin erosion over totally implanted vascular access devices in children. Semin Pediatr Surg 2009;18(2):84–6.

[99] Maher CM, et al. Surgical management of pelvic organ prolapse in women: the updated summary version Cochrane review. Int Urogynecol J 2011;22(11):1445–57.

[100] Withagen MI, et al. Risk factors for exposure, pain, and dyspareunia after tension-free vaginal mesh procedure. Obstet Gynecol 2011;118(3):629–36.

[101] Crawford GB, et al. Percutaneous atrial septal occluder devices and cardiac erosion: a review of the literature. Catheter Cardiovasc Interv 2012;80(2):157–67.

[102] Fukuda W, et al. Screw in the aorta: minimally invasive graft replacement for chronic aortic erosion by spinal instrument. Ann Thorac Cardiovasc Surg 2013;19(4):320–2.

[103] United States Food and Drug Administration. Device advice: overview of medical device regulation, http://www.fda.gov/MedicalDevices/DeviceRegulationandGuidance/Overview/default.htm; 2014 [accessed 22.04.14].

[104] Unites States 94th Congress. Medical device amendments: 90 Stat. 539. Public Law 94-295, 1976, http://www.gpo.gov/fdsys/pkg/STATUTE-90/pdf/STATUTE-90-Pg539.pdf; 2014 [accessed 22.04.14].

[105] United States Food and Drug Administration. Device advice: overview of medical device regulation—is the product a medical device, http://www.fda.gov/MedicalDevices/DeviceRegulationandGuidance/Overview/ClassifyYourDevice/ucm051512.htm; 2014 [accessed 22.04.14].

[106] Code of United States Regulations, 21CFR860.3. D.o.H.a.H. Services, Editor, National Archives and Records Administration, Office of the Federal Register; 2013.

[107] Code of United States Regulations, 21CFR878.4800. D.o.H.a.H. Services, Editor, National Archives and Records Administration, Office of the Federal Register; 2013.

[108] Code of Unites States Regulations, 21CFR880.5075. D.o.H.a.H. Services, Editor, National Archives and Records Administration, Office of the Federal Register; 2013.

[109] Code of United States Regulations, 21CFR880.6250. D.o.H.a.H. Services, Editor, National Archives and Records Administration, Office of the Federal Register; 2013.

[110] Code of United States Regulations, 21CFR890.3860. D.o.H.a.H. Services, Editor, National Archives and Records Administration, Office of the Federal Register; 2013.

[111] Code of United States Regulations, 21CFR880.5725. D.o.H.a.H. Services, Editor, National Archives and Records Administration, Office of the Federal Register; 2013.

[112] Code of United States Regulations, 21CFR878.4370. D.o.H.a.H. Services, Editor, National Archives and Records Administration, Office of the Federal Register; 2013.

[113] Code of United States Regulations, 21CFR870.1250. D.o.H.a.H. Services, Editor, National Archives and Records Administration, Office of the Federal Register; 2013.

[114] Code of United States Regulations, 21CFR868.5730. D.o.H.a.H. Services, Editor, National Archives and Records Administration, Office of the Federal Register; 2013.

[115] Code of United States Regulations, 21CFR876.5980. D.o.H.a.H. Services, Editor, National Archives and Records Administration, Office of the Federal Register; 2013.

[116] Loh JP, Barbash IM, Waksman R. Overview of the 2011 Food and Drug Administration Circulatory System Devices Panel of the medical devices advisory committee meeting on the CardioMEMS Champion Heart Failure Monitoring System. J Am Coll Cardiol 2013;61(15):1571–6.

[117] Center for Devices and Radiological Health. CDRHLearn—CDRH learn course list (English), http://www.fda.gov/MedicalDevices/DeviceRegulationandGuidance/default.htm; [accessed 03.10.14].

[118] Jones PA, Regulatory Affairs Professionals Society. Fundamentals of US regulatory affairs. 8th ed. Rockville, MD: Regulatory Affairs Professionals Society; 2013 viii, 540 p.

[119] Code of United States Regulations, 21 CFR 862-892. D.o.H.a.H. Services, Editor, National Archives and Records Administration, Office of the Federal Register; 2013.

[120] Code of United States Regulations, 21CFR814. D.o.H.a.H. Services, Editor, National Archives and Records Administration, Office fo the Federal Register; 2013.

[121] United States Food and Drug Administration. Device advice: premarket approval, http://www.fda.gov/MedicalDevices/DeviceRegulationandGuidance/HowtoMarketYourDevice/PremarketSubmissions/PremarketApprovalPMA/default.htm; 2014 [accessed 22.04.14].

[122] Code of United States Regulations, 21CFR812. D.o.H.a.H. Services, Editor, National Archives and Records Administration, Office of the Federal Register; 2013.

[123] Code of United States Regulations, 21CFR820. D.o.H.a.H. Services, Editor, National Archives and Records Administration, Office of the Federal Register; 2013.

[124] Loignon C, et al. What makes primary care effective for people in poverty living with multiple chronic conditions?: study protocol. BMC Health Serv Res 2010;10:320.

[125] Code of United States Regulations, 21CFR803. D.o.H.a.H. Services, Editor, National Archives and Records Administration, Office of the Federal Register; 2013.

[126] United States Food and Drug Administration. Device advice: recalls, corrections, and removals, http://www.fda.gov/medicaldevices/deviceregulationandguidance/postmarketrequirements/recallscorrectionsandremovals/default.htm; 2014 [accessed 22.04.14].

[127] Traboulsee AL. The role of MRI in the diagnosis of multiple sclerosis. Adv Neurol 2006. PMID: 16400831.

[128] Engh CA. J Bone Joint Surg Br 1987. PMID: 3818732.

[129] Evans JW, Ralph DJ. Removal of ureteric stents with a flexible cystoscope. Br J Urol 1991;109. PMID: 1993269.

[130] Bratzler DW, Dellinger EP, Olsen KM, Perl TM, Auwaerter PG, Bolon MK, et al. Clinical practice guidelines for antimicrobial prophylaxis in surgery. Am J Health Syst Pharm 2013;70(3):195–283.

# Reliability and security of implantable and wearable medical devices

**Younghyun Kim***, **Woosuk Lee***, **Anand Raghunathan***, **Vijay Raghunathan***,
**Niraj K. Jha**†

*Purdue University, West Lafayette, Indiana, USA*
†*Princeton University, Princeton, New Jersey, USA*

## CHAPTER CONTENTS

Bhunia et al. Implantable Biomedical Microsystems. http://dx.doi.org/10.1016/B978-0-323-26208-8.00008-X

## 8.1 INTRODUCTION

Over the past few decades, the emergence of implantable and wearable medical devices (IWMDs) has enabled fundamentally new options for monitoring, diagnosis, and treatment of a wide range of medical conditions. In the United States alone, about 300,000 new cardiac implantable medical devices (IMDs) are implanted every year [1]. A recent report forecasts that the US IMD market will grow at a compounded annual growth rate (CAGR) of 8%, reaching $73.9 billion by 2018 [2]. Another report predicts that the global wearable medical device (WMD) market will reach $5.8 billion by 2019 at a CAGR of 16.4% [3].

The rapid development of IWMDs has been largely driven by advances in electronics and computing technologies. For example, early pacemakers were the size of a table radio that could not be implanted and were powered by a 110 V AC line voltage or a heavy battery [4]. Highly improved functionality is now provided by much smaller devices, and the next generation of devices on the horizon is smaller than 1 cm$^3$ [5]. The expected lifetime of an IMD after implantation is now several years, sometimes even semipermanent when wireless recharging technology is used. In addition, as shown in Figure 8.1, IWMDs may be connected to various external devices, such as a programmer, personal health hub, or a healthcare professional's mobile device, for real-time monitoring and regular device updates. Thanks to such technologies, healthcare professionals have unprecedented access to the patients' physiological signs in real time from anywhere and can provide treatment as necessary.

Such advances, however, represent a double-edged sword: while they provide a fertile ground for sophisticated treatments, they also introduce the potential of adverse effects on patients. Various recent public announcements and academic research papers have revealed underlying causes of safety problems in IWMDs [6–12]. The common theme is a lack of reliability and security of the electronic systems

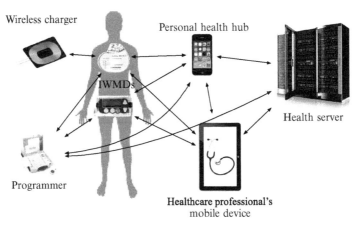

**FIGURE 8.1**

Integration of IWMDs into a personal healthcare system.

incorporated within the devices. Although manufacturers and government agencies have started to place emphasis on ensuring safety, a substantial number of warnings on malfunctions and security vulnerabilities continue to be reported regularly. Unfortunately, conventional reliability and security solutions employed in other types of embedded systems are often not applicable to IWMDs due to their severe resource constraints and unique usage model.

In this chapter, we identify the various challenges that undermine the safety of IWMDs and discuss potential countermeasures against them. The rest of this chapter is organized as follows. In Section 8.2, we discuss how IWMD safety is undermined and discuss the challenges associated with ensuring safety in IWMDs. In Sections 8.3 and 8.4, we describe various threats to safety using concrete examples. In Section 8.5, we present future directions of research and concluding remarks.

## 8.2 SAFETY OF IMPLANTABLE AND WEARABLE MEDICAL DEVICES

In this section, we look at what causes the safety of IWMDs to be undermined and discuss why it is more difficult to make IWMDs safe than other systems.

### 8.2.1 SAFETY CONCERNS FOR IWMDS

In the United States, the Food and Drug Administration (FDA) is responsible for premarket approval and postmarket tracking of IWMDs. After the FDA determines an IWMD to be safe and effective, the product is approved for release in the market. The FDA also requires manufacturers' reports on IWMD-related injuries or deaths and issues advisories or orders recalls of unsafe IWMDs. In this context, the definition of safety[1] of medical devices, as defined by the FDA, is given below [13]:

> There is reasonable assurance that the device is safe when it can be determined based on valid scientific evidence that the probable benefits to health from use of the device for its intended uses and conditions of use, when accompanied by adequate directions and warnings against unsafe use, outweigh the probable risks.

IWMDs are sophisticated devices that perform life-critical functions. Hence, any risks that may compromise safety should be thoroughly identified and addressed.

Unfortunately, advances in functionality of IWMDs are accompanied by an increase in device complexity, programmability, and connectivity [14], which often result in a decrease in reliability and security. The complexity of IWMDs, in terms

---

[1]In the dependable computing area, the term safety is an attribute of dependability and defined as a property of a system that ensures that it does not cause human injury or death nor damage the system's environment, irrespective of whether it is operating normally or abnormally.

of both hardware and software, is rapidly increasing to enable more sophisticated monitoring, diagnosis, and treatment. Consider, for example, a typical functional block diagram of an insulin pump, as shown in Figure 8.2. It contains various digital/analog integrated circuits (ICs), including the microcontroller for device control and memory card for data storage, sensors, and actuators. The RF link provides wireless connectivity to other devices, such as a glucose meter or a user remote. It is, essentially, a full-featured embedded system. Software in pacemakers and implantable cardioverter defibrillators (ICDs) has grown significantly in complexity. These systems now have 80k–100k lines of code [16]. Wireless connectivity, which is becoming common for remote monitoring, interdevice communication, and postimplant reprogramming, not only provides healthcare professionals with access to the device but also may leave the device open to malicious hacking. Not surprisingly, many FDA advisories and recalls issued recently are related to unreliable components or security loopholes. A few examples are given below:

- Between 2006 and 2011, 5294 recalls were reported to the FDA, and about 23% of the recalls were computer-related [8].
- Out of 23 class I—"a reasonable probability that use of these products will cause serious adverse health consequences or death"—recalls for defective medical devices issued by the FDA during the first half of 2010, at least six were likely caused by defects in software [9].
- In June 2013, the FDA warned medical device makers and hospitals about security threats that reconfigurable and interconnected medical devices may be vulnerable to malicious attacks when designed and operated without adequate security safeguards [6]. Possible breaches mentioned included malware, improperly managed passwords, failure to provide timely software updates, and insecure authorization.

More specific cases related to IWMD users show how real the concern about unreliable or unsecured IWMDs is:

- In 2005, a 21-year-old patient died from cardiac arrest. His ICD had short-circuited and failed to deliver an electric pulse shock when needed [10]. More than 70,000 implanted devices were recalled because of this malfunction.
- Due to worries about the risk of hacking, wireless connectivity in former US Vice President Dick Cheney's defibrillator was turned off when it was implanted in 2007 [17].
- In 2008, researchers demonstrated remote attacks on an ICD through its wireless interface [18].
- Academic researchers, followed by the hacker community, showed that one can take control of an insulin pump by reprogramming the device with widely available off-the-shelf components [11,12].

Since IWMDs are expected to become even more functionally complex and interconnected, reliability and security concerns need to be identified and addressed proactively.

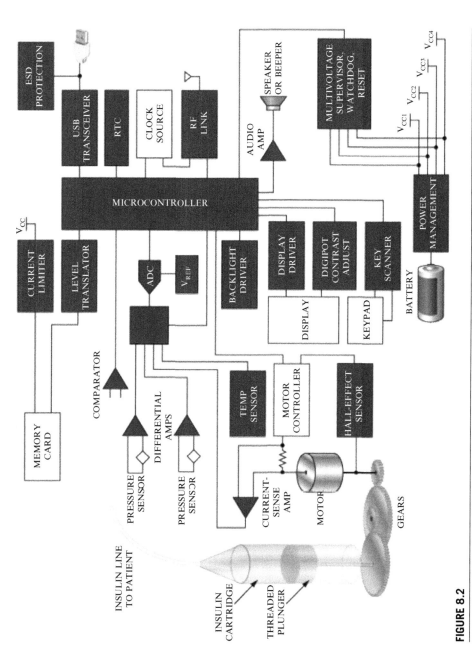

**FIGURE 8.2**

Functional block diagram of an insulin pump [15].

## 8.2.2 CHALLENGES IN RELIABLE AND SECURE IWMD DESIGN

For decades, extensive research efforts have been directed towards making computing systems more reliable and secure. This requires deployment of increased resources in terms of energy, time, computing power, physical dimension, or a combination. Depending on the required reliability and security levels and available resources, various hardware and software techniques have been developed and employed.

Figure 8.3 shows a plane with two axes: the required reliability and security levels on the x-axis and resource richness on the y-axis. Each computing platform is placed in one quadrant, requiring very different design approaches.

Mainframe computers are expected to be highly reliable and secure because they often store and process sensitive data and are expected to provide uninterrupted service for a very long time. On the other hand, they are relatively less constrained in area, energy, and computing power and can thus use various resource-demanding solutions.

In wireless sensor networks, in contrast, individual wireless sensor nodes (e.g., smart dust motes) are neither as reliable nor as secure as mainframe computers. Reliability and security are realized at the network level through the composition of unreliable and unsecured individual nodes. What dictate this approach are restrictions on the volume and cost of individual nodes, which in turn limit the amount of hardware resources that can be used.

For IWMDs, the level of reliability and security expected from individual devices is extremely high since they directly impact human health. Yet, this needs to be achieved with very limited resources. For example, for ease of implantation or wearability, the physical dimensions need to be highly restricted, thus also severely constraining the battery capacity and computation power. Unlike sensor networks, security and reliability must be achieved at the individual node level, not just at the network level.

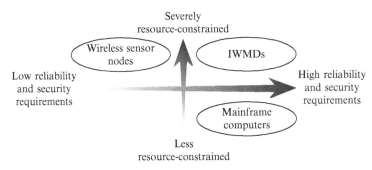

**FIGURE 8.3**

Resource richness versus single node-level reliability and security requirements of different computing platforms.

## 8.3 RELIABILITY CONCERNS AND SOLUTIONS

IWMDs do not always function as anticipated and may malfunction for various reasons. Since IWMDs are often life-critical, even a single malfunction may have catastrophic consequences. Even if potential malfunctions are identified before an adverse event occurs, deimplantation surgery to remove the IMD poses the risk of infection. Such a risk sometimes prompts manufacturers do not disclose the defect to the public [10]. Therefore, IWMDs should have a very high level of reliability by design.

There are various causes that undermine IWMD reliability. Manufacturers periodically publish performance reports for their products [19,20], which contain various possible causes for reported malfunctions and the number of affected devices, even if no related advisories or recalls were issued. A few example causes are the following:

- Electrical components failures, such as battery depletion, capacitor failures, and IC failures
- Software failures, such as firmware errors, parameter errors, and memory map errors
- Mechanical failures, such as magnetic switch failures, cracked solder joints, loose connectors, and setscrew failures

In this section, we discuss reliability issues associated with the electronic components of IWMDs. These include the power system, processor and memory, software, and radio communication. We also cover device packaging and operator errors. A taxonomy of reliability issues that may arise in IWMDs is shown in Figure 8.4. The issues can be categorized into hardware failure (Section 8.3.1), software and data reliability (Section 8.3.2), RF reliability (Section 8.3.3), and human reliability (Section 8.3.4).

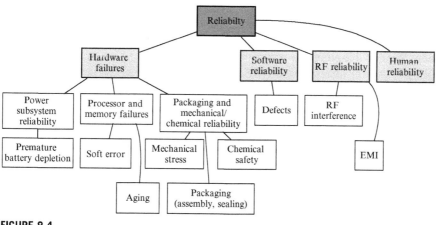

**FIGURE 8.4**

Reliability issues in IWMD components.

### 8.3.1 HARDWARE FAILURES

Hardware failures have been the major cause of IWMD malfunctions. An analysis of FDA reports [21] showed that about 80% of pacemaker and ICD malfunctions are caused by hardware failures. These include failures in the battery/capacitor, circuit, connector, and sealing. Recently, as IWMDs have become increasingly complex embedded systems, IC component failures have begun drawing attention. The rest of this subsection discusses component-specific reliability concerns and solutions.

#### 8.3.1.1 Power Subsystem Reliability

A reliable power supply is the key to stable operation of any electronic system. Since most IWMDs are battery-powered, prolonging battery life is an important design goal, particularly for IMDs where battery replacement is very costly and risky.

Despite its importance, and the fact that it is one of the major causes of IMD malfunctions [21], the reliability of batteries in IWMDs has not been addressed adequately. In [22], several concerns about batteries in medical devices are discussed. This report points out that it is important to develop an accurate battery capacity measurement technique to examine initial battery quality and suggests that IWMD designers should thoroughly understand battery characteristics, such as chemical and thermal behavior, in order to prevent errors in state-of-charge estimation or physical damage. Another article [23] suggests that variability in capacity and internal resistance among battery cells should be taken into account when integrating a battery pack in IWMDs.

In addition to energy capacity, special IWMD requirements should also be taken into account when designing a reliable power subsystem. IWMDs often run in a low-current standby mode for a long time and periodically draw a high current for actuation. Therefore, low-leakage and high-peak current supply capabilities are desirable [24]. The need for hermetic sealing of IMDs restricts the use of certain types of batteries. For example, zinc/air batteries have a very high energy density, but cannot be used for IMDs because they require oxygen to work.

Battery depletion, either normal or premature, is the most important concern in the design of the power supply subsystem. The risk of battery depletion would be significantly reduced if the battery can be recharged as needed without deimplantation. Wireless charging is a promising technique to prolong the longevity of IMDs [25–27]. As illustrated in Figure 8.5, a pair of aligned coils can be used to transfer energy by inductive coupling. In order for appropriate and efficient power transfer to take place, accurate coil design and alignment are important. An analytical model of

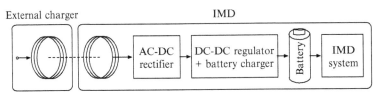

**FIGURE 8.5**

Wireless battery recharging.

power transfer efficiency, which depends on coil structure and lateral/angular misalignment, is presented in [28].

Energy harvesting for IMDs is a relatively less mature technique but can potentially make IMDs self-sustaining. Energy sources that an IWMD can tap into are very limited because it is either implanted in the body or covered by clothing. Practical sources for energy harvesting include inertial/direct-force kinetic energy and thermoelectric energy [29]. Kinetic energy harvesters convert kinetic energy from ambient human motion into electricity [30]. Thermoelectric generators (TEGs) convert temperature differentials into electricity and have the highest volumetric power density in practice [29]. To obtain a high temperature difference, TEGs are often attached to the skin or implanted near the skin surface [31,32]. Endocochlear potential, which is an electrochemical gradient that exists in the inner ear, can be a power source for inner-ear hearing aids or drug-delivery actuators [33]. Biofuel cells generate electrical power from biocatalytic activity of enzymes, microbial cells, or biological organelles. The power density of biofuel cells is not high enough for powering IMDs for long-term applications yet, but active research is ongoing on enhancing their efficiency [34,35].

Even if a power supply failure occurs, the device should avoid a complete loss of therapy, if possible, until it is detected and fixed. In [8], the use of fail-safe mechanisms to mitigate battery depletion is suggested. For example, when excessive current drain is detected and premature battery depletion is anticipated, unused system components can be turned off to maintain power for the critical parts. This can delay complete battery depletion and provide more time for detecting and replacing the battery before therapy is completely lost.

### 8.3.1.2 Processor and Memory Failures

IWMDs consist of various digital and analog ICs, as shown in Figure 8.2. As VLSI process technology scales, various challenges, such as process variations, single-event upsets (soft errors), and device wear-out, make ICs increasingly unreliable [36]. This trend will require a wide range of modifications to IC design, including in the process, design tools, and test methodologies [37,38].

Soft errors are single-event upsets that are caused by alpha particles (predominantly cosmic rays) hitting silicon chips. It does not physically damage the chip, but flips data bits in logic latches or memory cells and may cause a system failure. Low-voltage operation to ensure low power consumption increases IC susceptibility to soft errors. An early work [39] points out that soft errors encountered in ICDs from field tests can be attributed to cosmic radiation. A designer may attempt to mitigate the rate of soft errors, correct errors after detecting them, or both. Through judicious choice of package and substrate materials and careful device geometry, soft error resiliency can be improved at the device level [40]. For soft error detection and correction, fault-tolerant design can be used. Error-correcting codes (ECCs) or multistrobe prevent the propagation of soft error [41]. Triple modular redundancy (TMR) schemes employ three identical copies of a module, and a soft error in one of the three modules can be detected using majority voting [42]. However, improved resiliency is accompanied by area and power overheads because of the increased transistor geometries or redundant hardware, which pose a challenge in the context of resource-constrained IWMDs. Therefore, it is important to identify the most vulnerable nodes or circuits and apply these techniques selectively to them to minimize the overheads.

Aging also has a significant impact on transistor performance. Hot-carrier injection causes time-dependent shifts in the threshold voltage and saturation current. These adverse effects are expected to become worse with technology scaling [36]. In order for an IWMD to operate for a long time, aging effects need to be taken into account during reliability assessment.

### 8.3.1.3 Packaging and Mechanical/Chemical Reliability

Packaging is more challenging for IMDs than for WMDs because hermetic sealing is required to protect both the device and the human body from each other. Biocompatibility is essential for IMD packaging since it directly contacts human tissue. Packaging materials should not lead to any adverse effects on the human body. At the same time, they must be stable in the body environment. Biocompatible materials include titanium and its alloys, biograde stainless steels, tantalum, alumina, zirconia, and some biocompatible glass and polymers [43]. The reliability of hermetic packaging depends on the choice of material, thickness, final seal design, fabrication process, and so on. Designers should also be cognizant of other issues, such as outgassing of internal materials, eddy-current-induced heating, and thermal expansion compatibility. Repetitive mechanical stress damages weak components within IWMDs, such as connectors and solder joints. Damaged components may cause changes in measurement, loss of therapy, or loss of power. In order to meet the long-lifetime requirement, the reliability and long-term degradation of solder joints must be carefully studied [44].

### 8.3.2 SOFTWARE RELIABILITY

The software content of IWMDs is increasing at a rapid pace. Pacemakers and other ICDs have 80k–100k lines of code [16], which is likely to increase further in future versions of these devices. It is, therefore, not surprising that software error is one of the major causes of IWMD malfunctions. It appears that 33.3% of class I recalls,[2] 65.6% of class II recalls,[3] and 75.3% of class III recalls[4] on medical devices between 2006 and 2011 were caused by software errors [8]. The life-critical nature of IWMDs demands that the software be reliable and bug-free to ensure safe operation and meet all the requirements and standards.

---

[2]A situation in which there is a reasonable probability that the use of, or exposure to, a violative product will cause serious adverse health consequences or death [45].

[3]A situation in which use of, or exposure to, a violative product may cause temporary or medically reversible adverse health consequences or where the probability of serious adverse health consequences is remote.

[4]A situation in which use of, or exposure to, a violative product is not likely to cause adverse health consequences.

Formal verification-based techniques take a particular program implementation and some specification as inputs and analyze all possible program execution paths [46]. From the output traces, they detect any violation of the given specification (e.g., array bounds, pointer safety, or user-specified assertions). Formal verification consists of a rigorous deduction in a mathematical logic to ensure conformance with well-formed statements of specification. Therefore, program correctness for all possible inputs can be examined symbolically. The use of formal verification for medical device software has been proposed in [47–49]. In [46], two key challenges in leveraging formal verification in the context of medical devices are presented. First, medical device programs are highly hardware-dependent and interrupt-driven. Verification tools for high-level programming languages are not suitable for such programs. Second, domain-specific properties that deal with interaction of medical devices with the real world should be taken into account. Real-world device-specific safety properties should be reflected as assertions in the program. In order to address these two challenges, the authors proposed a domain-specific formal verification framework (shown in Figure 8.6). It involves the transformation of the original source code into verifiable C code and the transformation of medical device specifications into verifiable properties based on device-specific I/O mapping and memory maps. For rigorous verification, the specification must be not only formal but also meaningful for testing and analysis. The specification should encapsulate end-to-end system properties, rather than purely functional descriptions [50]. Any

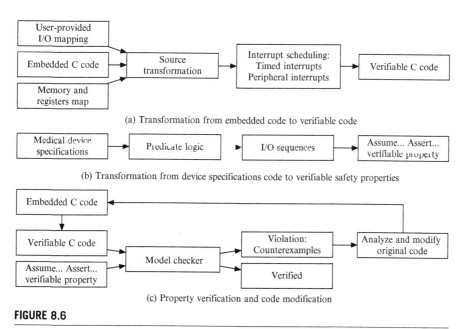

(a) Transformation from embedded code to verifiable code

(b) Transformation from device specifications code to verifiable safety properties

(c) Property verification and code modification

**FIGURE 8.6**

Formal verification workflow of medical device software [46].

shortfalls in the specification will result in a false sense of safety during subsequent design, implementation, and testing.

### 8.3.3 RF RELIABILITY

IWMDs communicate with other devices for various purposes. For example, they communicate with each other (e.g., continuous glucose meter and insulin pump) or are connected to external devices for delivering measured medical data or reprogramming of device settings. Communication is often done over a wireless network. A wireless network that involves IWMDs (often called a body area network or BAN) requires a more reliable and fail-safe setup and operation than general wireless sensor networks require. BAN requirements and their implementations are presented in [51]. The BAN should be as intuitive and easy to set up as wired communication: fast but fully controllable, flexible and scalable, and free of interference from other BANs. A healthcare professional may use a setup pen that emits his/her unique ID through infrared (IR) means. All the devices, including the patient's IWMDs and clinical medical devices, that receive the same ID form a BAN. This technique can be applied to WMDs that can accept ID from the setup pen, but IMDs would need a different mode of communication to deliver the ID (e.g., body-coupled communication [52]).

IWMDs may face interference not only from other IWMDs but also from other devices that generate electromagnetic interference (EMI). Modern IMDs incorporate hardware designs to reduce the pickup of EMI and software designs to detect and remove the interference [53]. However, the equipment that generates strong magnetic fields, such as magnetic medical navigation systems [54] or magnetic resonance imaging (MRI) systems, may cause significant interference. In an experiment discussed in [55], functional failures, such as electrical reset or device interrogation failure, were observed in the presence of interference. Surprisingly, personal devices, such as headphones [56], portable media players [57], or mobile phones [58], may also be a source of EMI, causing clinically significant interference to IWMDs. Another related issue is the temperature rise in IMDs when exposed to a strong EM field. For example, the IMD may cause a higher tissue temperature during MRI [59,60]. It is recommended that the temperature rise of the IMD surface should be no more than 2°C [43].

### 8.3.4 HUMAN RELIABILITY

Human reliability should not be overlooked while assessing the reliability of critical systems. Medical devices that are designed with poor understanding of human factors may result in injury or death. More than 50% of the technical medical equipment-related problems are due to operator errors [61], and at least 44,000 people (perhaps as many as 98,000 people) die each year as a result of medical errors that could have been prevented [62]. Although most IWMDs are autonomous and seldom require continuous interaction with users, potential sources of human error should not be neglected. For example, a patient with a drug pump died from an overdose because the injection interval was set to 20 minutes rather than 20 hours by a healthcare professional [63]. The delivered drug dosage was 60 times the desired amount and the patient passed out while driving, resulting in a motor vehicle accident. Systematic failure analysis techniques, such as failure mode and effects analysis (FMEA), can be utilized to assess human reliability. In [64], the human reliability factors of medical devices are modeled and evaluated by FMEA. Fuzzy linguistic theory is applied to convert the subjective cognition of experts regarding various risk factors into numerical values.

## 8.4 SECURITY CONCERNS AND SOLUTIONS

As mentioned in Section 8.2, patients may be concerned about the risk that their IWMD may be attacked by a malicious entity. Several recent works demonstrate security attacks on IWMDs [11,12,18]. The security mechanisms in the breached devices were easily broken in these studies. However, the risk of hacking represents only a small portion of security vulnerabilities, as observed in [65].

Security is a relatively new concern for regulatory agencies, such as the FDA. Manufacturers, being concerned about potential problems or delays associated with regulatory approval, have traditionally had little incentive to add security mechanisms to their devices [66]. However, as the regulatory agencies now realize the criticality of security for IWMDs, manufacturers are increasing the amount of attention given to implementing appropriate security mechanisms in their products. Addressing security risks is essential, not only for wide acceptance of IWMDs but also for the success of home healthcare [67].

The three fundamental building blocks of security are confidentiality, integrity, and availability. The implications of these properties for IWMDs are as follows:

- Confidentiality—Only intended parties should be able to access the medical information. Unauthorized parties should not be able to interpret the information.
- Integrity—Accuracy and consistency of medical information must be maintained. The medical information should not be altered by unauthorized parties.
- Availability—The IWMD and its medical information should be available when needed.

We highlight the security vulnerabilities by describing the attacks that target these key properties in IWMDs. We also describe defensive measures against those vulnerabilities.

In the rest of this section, we categorize security vulnerabilities into radio attacks (Section 8.4.1), side-channel attacks (Section 8.4.2), hardware security (Section 8.4.3), software security (Section 8.4.4), and human errors (Section 8.4.5). The taxonomy is illustrated in Figure 8.7.

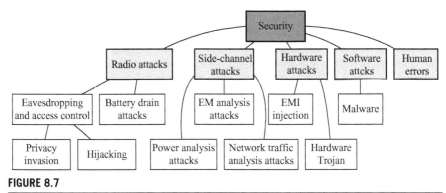

**FIGURE 8.7**

Security concerns in IWMD components.

## 8.4.1 **RADIO ATTACKS**

As IWMDs increasingly support wireless connectivity for remote monitoring, treatment, and software reprogramming, radio attacks have become one of the major threats to IWMD security. Successful attacks on an insulin pump [11,12] and an ICD [18] have been demonstrated recently. In these demonstrations, the communication protocol was reverse engineered using off-the-shelf radio devices (e.g., software radios), and the potential for various kinds of attacks, including extracting private data, altering device settings, and manipulating treatment, was discussed.

### 8.4.1.1 Eavesdropping and Access Control

Eavesdropping by a passive adversary refers to the process of overhearing the radio communication from/to an IWMD, without actually interfering with the communication, in order to extract sensitive medical data. An active adversary, on the other hand, has more capabilities than a passive adversary. The active adversary can generate radio transmissions to trigger unintended behavior, disable the IWMD, or even take full control of it.

For transmitting sensitive data, especially over a wireless channel, cryptography is a fundamental technique for securing the communication. Though there are numerous studies on cryptography, due to the resource constraints and usage models of IWMDs, selection of an appropriate cryptographic technique is not a straightforward process [68]. The additional energy required for encryption is one of the major concerns in adopting cryptography in resource-constrained IWMDs. Rostami *et al.* [69] observed that low power consumption will be a key requirement for cryptography. Generally, symmetrical encryption algorithms consume less power than asymmetrical encryption algorithms and are, hence, more practical for IWMDs [70]. Lightweight hardware-oriented block ciphers, such as PRESENT [71] and KATAN/KTANTAN [72], have been developed for resource-constrained platforms. Their hardware implementations are as small as 1000–1500 gate equivalent for a 64-bit block size encryption.

Beck *et al.* [73] proposed a low-power block cipher-based security protocol in IWMDs. This protocol has two modes: a stream mode and a session mode. The stream mode is for short message transmissions; it utilizes output feedback (OFB) to obtain a scalable stream cipher that enables strict duty cycling for energy efficiency. The session mode utilizes cipher-block chaining (CBC) and a challenge-response scheme to provide high security for longer messages.

Even if relatively low-power symmetrical encryption algorithms are used, the energy consumed in encryption will shorten battery life. Encompression [74] combines compressive sensing, which is a power-saving approach that exploits the sparsity of sensor data, with encryption and integrity checking, as shown in Figure 8.8. It is shown that the total energy consumption with encompression can be even lower than that of a baseline system that does not include security.

One major concern in symmetrical encryption algorithms is the distribution of the shared key between devices. IWMDs are often required to communicate with previously unknown devices. Sometimes, the key distribution needs to be done without the patient's intervention. If the patient is unconscious and unable to provide the shared key to healthcare professionals, encryption may impede emergency access to the IWMD for timely treatment. Healthcare professionals may have a universal key to access any IWMD. However, this is inherently unsafe since once the key is obtained by an adversary, it can be used to attack all other devices. A practical solution is physically carrying the secret key so that the healthcare professionals can easily access it, but malicious attackers cannot without patient's recognition. The key can be imprinted on a card or on the wearable device itself if it can be hidden securely [75] or printed directly onto the patient's skin using ultraviolet-ink micropigmentation, which only becomes visible under ultraviolet light [76].

Even after the key is securely shared, periodically changing the shared key is important to maintain security, but it is not practical to rely on the user to do this. Using symmetrical properties of the wireless channel between devices (e.g., received signal strength indicator (RSSI)), two devices can generate the symmetrical key from the communication [77–80]. An eavesdropper located at a third location will measure a different, uncorrelated radio channel and cannot generate the same key. These techniques can be used not only for key sharing during initial setup but also for renewing the key subsequently.

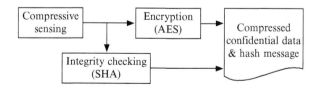

**FIGURE 8.8**

Encompression architecture. The Advanced Encryption Standard (AES) is used for the symmetrical cipher and Secure Hash Algorithm (SHA) for integrity checking [74].

When both the communicating devices are worn on the same body, a biometric signal can be used for key generation [81–88]. For example, blood glucose, blood pressure, and temperature have enough randomness that they can be exploited for key generation [86]. Electrocardiography (ECG) signals can be utilized too for this purpose [87,88].

An auxiliary channel or out-of-band (OOB) communication uses audio, visual, or haptic channels for authentication [75,76,89,90]. However, Halevi *et al.* [91] demonstrated that authentications through auxiliary channels can be breached by close-range eavesdropping.

Another approach for preventing unauthorized access is the use of close-range communication that limits the range of the communication. Since patients would be aware if somebody approached abnormally close to them, close-range communication reduces the chances of attacks. In this context, an RFID-based channel and near-field communication (NFC) can be utilized since they are designed for communications within a relatively short range.

Another close-range communication mechanism is body-coupled communication (BCC) that uses the human body as a signal propagation medium [92]. There are two BCC mechanisms, as shown in Figure 8.9. The transmission line approach uses the human body as a transmission line (Figure 8.9(a)). The capacitive approach uses the human body as a floating conductor whose electric potential relative to the environment (reference) is modulated (Figure 8.9(b)). However, the use of BCC is not ideal because physiological signals can be read during normal physical contact, such as a handshake [93].

Instead of using inherently short-range channels, IWMDs can enforce distance-bounding of communication by measuring the distance to an external device. After measuring the distance, access is granted only when an external device is within a safe range. The distance between devices can be measured using various techniques.

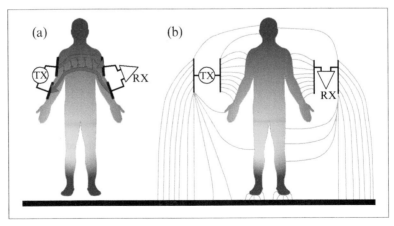

**FIGURE 8.9**

Coupling mechanisms for BCC: (a) transmission line approach, (b) capacitive approach [92].

A typical way is to limit the response time to a verification request [94]. For devices on the same body, Rasmussen *et al.* [95] used ultrasonic waves to directly measure the distance, and Shi *et al.* [96] used the stability of RSSI as the distance indicator. Distance bounding of communication relies on the assumption that an attacker cannot approach too close to the patient. However, Denning *et al.* [97] pointed out that this assumption is flawed. Adversaries are often adept at extending communication ranges and may make physical contact with the patient in a crowded space without raising suspicion. For example, a successful RFID eavesdropping attack was demonstrated at a distance of a few meters with off-the-shelf antenna kits [98]. Cremers *et al.* [94] demonstrated that an out-of-bound attacker can exploit other legitimate devices to gain access.

### 8.4.1.2 *Battery Drain Attacks*

Battery drain attacks involve sending a large number of service requests in order to exhaust the battery prematurely to deteriorate device availability [99]. Even if the attacker is not able to breach the encryption to deliver valid-looking requests to the application, a simple jamming-based denial-of-sleep attack that prevents the device from entering a power-down mode can effectively exhaust the battery [100]. These attacks can be identified by detecting anomalies in the access pattern [101,102].

Defense mechanisms against the battery drain attacks should be particularly power-efficient because a power-hungry defense mechanism may consume more energy than the attack itself. Wireless charging or kinetic/thermoelectric energy harvesting discussed in Section 8.3.1.1 can mitigate the risk of battery exhaustion by battery drain attacks, but is not an ideal solution. With RF power harvesting, the defense mechanisms can even be powered by energy harvested from the attack. In [18], zero-power notification and zero-power authentication with RF power harvesting are proposed. Zero-power notification provides an audible alert to the user when a battery drain attack is detected. In zero-power authentication, the IMD is powered by the RF carrier signal, and acoustic communication is used for key exchange. Close-range communications and distance-bounding protocols discussed in the previous section may also be utilized to prevent battery drain attacks.

### 8.4.1.3 *External Security Devices to Defend Against Radio Attacks*

Enhancing the security of existing IWMDs requires significant modifications to their hardware and software, which may cause unintended changes in behavior. To address this challenge, some recent studies propose the use of an external device to enhance radio security without any modification to the IWMD itself.

When a new external programmer attempts to connect to the IWMD, it is hard for the IWMD to tell whether it is a legitimate healthcare professional or an adversary. If the IWMD blocks all attempts from the new external programmer, legitimate access to the device will be delayed in an emergency situation. Therefore, a communication cloaker is proposed in [97] to implement fail-open access to all external programmers. In the normal mode, it allows only preauthorized programmers to access the IWMD; in the emergency mode, the cloaker is turned off and any programmer can access the IWMD.

Shield, shown in Figure 8.10, is a personal base station placed in between the IWMD and external programmer [103]. Only a legitimate programmer can communicate with the Shield using encryption, and all other attempts to directly communicate with the IWMD are jammed by Shield. Shield decrypts the commands from the programmer and delivers them to the IWMD. It also receives transmissions from the IWMD, encrypts them, and relays them to the programmer. While receiving a transmission, a jamming signal is generated by Shield so that other devices cannot receive the transmission directly.

IMDGuard is an external security system designed for ICDs [88]. This system includes Guardian, an external wearable device, to coordinate interactions between the ICD and other external programmers. It exploits the patient's ECG signals to extract keys exclusively shared between the ICD and Guardian. Any other external programmers need to be authenticated by Guardian before they can communicate with the ICD. The ICD needs no special hardware modification since it utilizes its existing capability to measure the ECG signals.

MedMon, shown in Figure 8.11, is an external security device that detects abnormal communication from/to the IWMDs [102]. Based on the physical characteristics

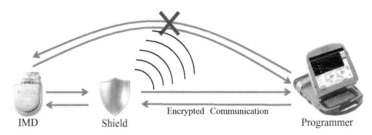

**FIGURE 8.10**

Shield jams all direct communications between the IWMD and external programmers. The programmer communicates with the IWMD only through the Shield [103].

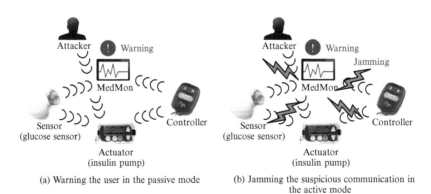

**FIGURE 8.11**

MedMon detects anomalies in communications and warns the user or jams the communication [102].

of the transmitted signal, such as RSSI, time of arrival (TOA), differential time of arrival (DTOA), and angle of arrival (AOA), it detects anomalies in the transmission. Upon identifying an attack, it provides a warning to the user and/or generates jamming signals. MedMon also captures behavioral anomalies, such as vital signs or commands, that lie outside the historical records of the patient. For example, if a command to inject a too large a dose of a drug is received, it prevents it from being executed and prompts the user for authorization.

Many external security devices provide jamming-based protection to IWMDs. However, Tippenhauer *et al.* [104] demonstrated that jamming-based device protection is not always effective. In their experiments, using two antennas, a multiple-input and multiple-output (MIMO)-based attack was able to recover the data protected by jamming.

### 8.4.2 SIDE-CHANNEL ATTACKS

Side-channel attacks are attempts to uncover secret information based on physical property (e.g., power consumption or EM radiation) of a cryptosystem, rather than exploiting the theoretical weaknesses in the implemented cryptographic algorithm.

Power analysis attacks begin with precisely measuring the power consumption of the target device many times. Depending on the secret key used in the algorithm, the power consumption of an unprotected implementation shows a unique power consumption profile. By matching the profile against the power profiles predicted with every possible key, the secret key can be deduced without accessing the data in the system [105].

Since power analysis attacks are based on an accurate power consumption measurement of the target device, it is likely that such attacks on IMDs would be attempted before implantation. An attacker, upon acquiring a target IWMD, can disassemble the device, hook it up to a power measurement equipment, and recover its secret key, which can be used to take control of the device or potentially other devices that use the same key. Two principles of countermeasures against this attack are hiding and masking [106,107]. Hiding removes the data dependency of power consumption, while the key is not changed in a cryptographic algorithm. It can be realized using special hardware. Circuit-level techniques make the power consumption of 0-to-1 and 1-to-0 transitions undistinguishable [108,109]. Special hardware, such as a current-flattening circuit [110] or a band-pass filter [111], can also be used to suppress information leakage. Masking randomizes the key in each execution of the algorithm [112–114]. Key masking by multiplication, random key rotation, and random insertion of redundant symbols in the key can be used to mask the secret key. Most of the abovementioned techniques, however, achieve resiliency against power analysis attacks at the cost of large area or power overheads.

EMI is a source of chip failures, as discussed in Section 8.3.1.2. It can also be exploited for security attacks. EM analysis attacks are a variation of power analysis attacks. Here, power consumption is measured by capturing EM radiation that is induced by changes in current [115]. Although EM radiation is detectable wirelessly,

EM analysis attacks are not an *in vivo* threat because it requires accurate sensing at a close distance. Defense mechanisms against power analysis attacks can also be adopted for defense against EM analysis attacks since they reduce the data dependency of power consumption. Additional countermeasures against EM analysis attacks introduced in [115] include the use of a Faraday cage (to block EM radiation) and asynchronous circuit design to reduce the EM radiation signature.

Network traffic analysis attacks are attempts at monitoring communication packets in order to deduce some information, even if an attacker is not able to decrypt them. For example, an adversary can uncover the existence and location of IWMDs, which can be an invasion of the patient's privacy. The volume or rate of traffic may expose the type of IWMDs or the health condition of patients. For example, a regular traffic pattern may imply that multiple IWMDs are running in a closed-loop control. A sudden increase in the frequency or length of communication between IWMDs and remote hub may imply a significant change in health condition [116]. Several defense mechanisms, such as generating random dummy traffic, pretending to be a fake device, and adjusting traffic delay, have been introduced for traffic obfuscation [117]. However, these techniques do not completely hide the existence of the IWMD and consume more power to generate the dummy traffic.

### 8.4.3 HARDWARE ATTACKS

#### 8.4.3.1 Hardware Trojan

These days, most ICs are manufactured in outsourced fabrication facilities and often include intellectual property (IP) cores supplied by third-party vendors. Some IP cores and manufacturing steps are not verifiable by IC or system designers and so cannot be fully trusted. Hardware Trojan is a malicious function added to an IC in order to destroy or disable the target system at some future time or to leak confidential information from it. When a hardware Trojan is inserted in an IWMD, an attacker may threaten the device manufacturer or individual patients to disable the device. It can also leak sensitive medical information.

Many research papers have proposed hardware Trojan detection techniques [118,119]. They exploit various fingerprints of ICs, such as transient or static power consumption [120], path delay [121], and thermal characterization [122]. A hardware Trojan can also be detected by adding a circuit to identify each IC, such as a physical unclonable function (PUF) or an on-die temperature monitor [123], but the additional power and area overheads may not be acceptable due to the low power requirements of IWMDs.

#### 8.4.3.2 EMI Injection Attacks

EM radiation can be exploited to recover internal information from IWMDs, as described in Section 8.4.2. In addition, it can also be exploited to inject signals into them. IWMDs have various analog sensors to measure patients' physiological signals. These sensors need to be extremely accurate. Injecting EMI signals can alter the sensor readings and trick the IWMDs. Kune *et al.* [124] demonstrated that ICDs can be stopped from delivering the pacing signal by employing EMI injection into their sensing leads. While the distance for launching an effective attack is significantly shortened when implanted in the body, this kind of attack is possible on similar IWMDs with sensing leads. Analog defense mechanisms against EMI injection attacks include the use of a differential comparator to eliminate common-mode voltage, use of more tight signal filters, and shielding [124]. Digital approaches, such as signal processing to detect and filter out EMI injection, are also applicable.

### 8.4.4 SOFTWARE ATTACKS

Malware (malicious software) continuously adapts to various new types of computing platforms. Handheld mobile device platforms, such as smartphones, have already been breached by malware [125]. As more IWMDs are designed to support wireless reprogrammability for parameter changes and software updates, malware-infected IWMDs may emerge soon.

Various software vulnerabilities in reprogrammable medical devices, starting with external devices, have been studied recently. For example, an automated external defibrillator (AED) is shown in [126] that accepts counterfeit firmware updates. Various security vulnerabilities, such as improper handling of integer/buffer overflow, recklessly stored passwords and credentials, and improper use of cyclic redundancy check (CRC) of firmware as a digital signature, are identified and exploited. Although this is an example of an external medical device, IWMD manufacturers also should address these security vulnerabilities as well.

Secure execution environments provide a trusted, isolated execution environment for security-critical applications. By running security-critical medical applications in a secure execution environment, they can be protected from a compromised operating system (OS) or other applications. Secure virtual machines (VMs) provide a secure network interface, secure storage, and secure execution environment under an untrusted management OS [127]. In a secure VM, only security-critical medical applications and supporting programs are executed, isolated from other applications that may be malware. Another approach is physical separation that executes security-critical applications on an isolated hardware to provide a safe environment that is free of observation and interference through direct physical access [128]. Amulet [129] is an external wearable device, which plays a role of a trustworthy hub between the sensors on the same body (Figure 8.12). All health-related programs run on Amulet, isolated from other programs on other untrustworthy computing platforms (e.g., smartphones and PCs).

Trusted computing that uses a trusted platform module (TPM), which is a device that stores cryptographic keys of a specific host computer, can be employed in IWMDs. For resource-constrained platforms, where the size and cost overheads of a separate TPM chip are not acceptable, a software-based TPM that has low energy and computation overheads can be used [130]. Plug-n-Trust [131] is a microSD card that provides a trusted computing environment to medical applications on a smartphone (Figure 8.13). In this system, all sensitive data are encrypted before they

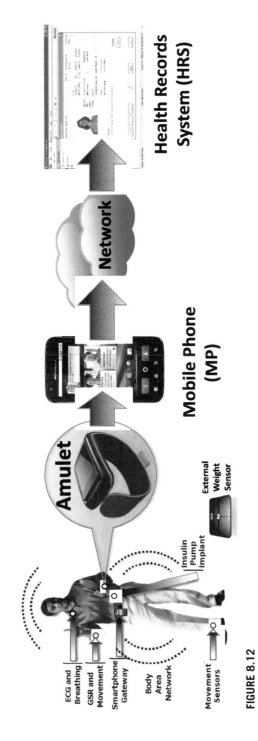

**FIGURE 8.12**

Amulet is a wrist-worn device that provides a trustworthy execution environment for health-related programs [129].

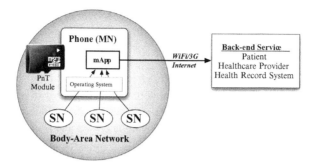

**FIGURE 8.13**

Plug-n-Trust is a trusted computing environment based on a microSD card plugged into a smartphone [131].

leave the sensors and remain encrypted while stored on the smartphone. They are decrypted for processing only in the card through provided application programming interfaces (APIs).

While secure execution environments protect medical applications from a compromised OS or other compromised applications, they cannot protect the system if the medical application itself is the malware. Static analysis techniques characterize the execution behavior of a program and form an important step in detecting program flaws. They can be utilized to identify security vulnerabilities at design time. A static analysis tool called CodeSonar is able to automatically detect various errors, such as buffer overrun, uninitialized variables, and null pointer dereference, in commercial medical devices [132]. Based on the properties of programs extracted through static program analysis, run-time monitoring detects unintended behavior at run time. In [133], a dynamic binary instrumentation-based malware detection framework is introduced. An untrusted program is traced during execution in a virtualized testing environment with a large number of input values and checked with extensive security policies to detect malicious behavior. If the program passes the test, a behavioral model is constructed, which is used as a proxy for the security policies. During execution, a run-time monitor compares system behavior with the model, and any deviation is detected as an anomaly. Due to the rigorous security policy checking in the test environment, run-time performance penalty is reduced. To reduce the run-time monitoring overhead further, a hardware-assisted monitoring technique can be utilized [134].

Power consumption analysis can be used as a security attack, as described in Section 8.4.2. However, it can also be utilized as a defense mechanism to detect malware in embedded systems [135,136]. This technique measures the power consumption signatures of normal programs and known malware and compares run-time power consumption with the power signatures. Any anomaly in the power consumption is flagged as malicious behavior. In [137], a similar technique is demonstrated for medical devices. Instead of relying on a preconstructed power consumption model, this technique utilizes supervised learning to classify normal and abnormal activities. The key to malware detection based on power consumption characteristics is a high sampling rate and high-accuracy power measurement. However, accurate power measurement requires a sophisticated power measurement circuit, which may be not feasible on resource-constrained IWMDs.

Besides the abovementioned technical approaches, a change in regulatory policy may also help detect and eliminate software-based security vulnerabilities. Due to patent issues, a vast majority of IWMD software is closed and proprietary. However, "security through obscurity" should be strongly discouraged, because publicly available source code may improve its security through continuous and broad peer review to identify and eliminate the vulnerabilities that might otherwise go unrecognized by the original developers [138]. In this context, Sandler *et al.* [9] discussed the security benefits of open-source software compared with closed-source, proprietary programs and urged the FDA to require manufacturers to publish their source code to the public to be examined and evaluated.

### 8.4.5 HUMAN ERRORS

Human errors are often overlooked but are a significant cause of many security problems. Password management is a common human risk factor [139]. Improperly generated or managed passwords, such as weak, common, and visible passwords, potentially weaken system security. For example, wireless access points left unencrypted are easily detectable. They can expose private data or can be exploited for malicious anonymous access. Similarly, the users of IWMDs may be unaware of the importance of security or may not have the ability to change device settings properly. It is quite likely since they are often elderly or weak. Therefore, security mechanisms in IWMDs should not rely on patients' active intervention. Surprisingly, these mistakes are even made by IWMD manufacturers. In June 2013, the Industrial Control Systems Cyber Emergency Response Team (ICS-CERT) warned that a hard-coded password in medical devices can be exploited for changing critical settings or modifying firmware [7]. This affected roughly 300 medical devices across approximately 40 vendors.

### 8.5 CONCLUSIONS

IWMDs tremendously improve patients' quality of life by enabling pervasive healthcare. Various functionalities implemented by IWMDs, such as insulin injection, cardioverter defibrillation, and neurostimulation, are often life-critical. Hence, a very high level of reliability and security needs to be ensured. While advancements in embedded system technologies have enabled sophisticated therapy using miniaturized devices, increasing complexity, programmability, and connectivity pose challenges to the reliability and security of the devices.

In this chapter, we discussed various reliability and security issues related to IWMDs. We saw that every component of an IWMD, such as hardware, software, communication channel, and even the human operator, can be affected by these issues, possibly compromising patient safety. Through a comprehensive survey, we examined and discussed the benefits and limitations of various potential security solutions. However, despite active research effort and progress in developing reliable and secure IWMDs, there are still several limitations that need to be overcome before they can be adopted in IWMDs.

A general set of principles for reliability and security solutions can be summarized as follows:

- IWMDs are an extremely resource-constrained computing platform. The solutions should be very energy-efficient and not computation-intensive. Cross layer design to minimize overhead and fully exploit the given resources may be beneficial.
- Incorporating safety solutions often requires a deep understanding of the application. Medical domain knowledge would be helpful for generating a better specification of the requirements of IWMDs.
- The tension between utility and safety should be managed well. IWMDs may be used in an unanticipated environment, for example, emergency room, and the safety safeguards should not delay the necessary treatment.
- Ease of use is highly desirable. Patients are not professionals at manipulating IWMDs, and even healthcare professionals can make mistakes. User intervention should be minimized and intuitive to prevent human errors.

For successful deployment of IWMDs and improved healthcare quality, it would be helpful to adequately incorporate the above design considerations and principles in the IWMD development process.

## ACKNOWLEDGMENT

This work was supported in part by the National Science Foundation (NSF) under grants CNS-1219570, CNS-0953468 and CCF-1018358.

## REFERENCES

[1] Zhan C, Baine WB, Sedrakyan A, Steiner C. Cardiac device implantation in the United States from 1997 through 2004: a population-based analysis. J Gen Intern Med 2008;23(1):13–9.
[2] Transparency Market Research, Implantable medical devices market (reconstructive joint replacement, spinal implants, cardiovascular implants, dental implants, intraocular lens and breast implants) - U.S. industry analysis, size, share, trends, growth and forecast 2012–2018; 2013.
[3] Transparency Market Research, Wearable medical devices market (heart rate monitors, activity monitors, ECG, pulse oximeters, EEG, EMG, glucose/insulin management, pain management, wearable respiratory therapy) - global industry analysis, size, share, growth, trends and forecast, 2013–2019; 2013.
[4] Greatbatch W, Holmes CF. History of implantable devices. IEEE Eng Med Biol Mag 1991;10(3):38–41.

[5] Cheng A, Tereshchenko LG. Evolutionary innovations in cardiac pacing. J Electrocardiol 2011;44(6):611–5.

[6] U.S. Food and Drug Administration. FDA safety communication: cybersecurity for medical devices and hospital networks, http://www.fda.gov/medicaldevices/safety/alertsandnotices/ucm356423.htm; 2013.

[7] Industrial Control Systems Cyber Emergency Response Team. Medical devices hard-coded passwords, http://www.ics-cert.us-cert.gov/alerts/ICS-ALERT-13–164–01; 2013.

[8] Alemzadeh H, Iyer RK, Kalbarczyk Z, Raman J. Analysis of safety-critical computer failures in medical devices. IEEE Security Privacy 2013;11(4):14–26.

[9] Sandler K, Ohrstrom L, Moy L, McVay R. Killed by code: software transparency in implantable medical devices, Software Freedom Law Center. http://www.softwarefreedom.org/resources/2010/transparent-medical-devices.html; 2010 1–12.

[10] Hauser RG, Maron BJ. Lessons from the failure and recall of an implantable cardioverter-defibrillator. Circulation 2005;112(13):2040–2.

[11] Li C, Raghunathan A, Jha NK. Hijacking an insulin pump: Security attacks and defenses for a diabetes therapy system. In: IEEE international conference on e-health networking applications and services; 2011. p. 150–6.

[12] Radcliffe J. Hacking medical devices for fun and insulin: breaking the human SCADA system. In: Black hat conference; 2011.

[13] U.S. Food and Drug Administration. Title 21–Food and Drugs Chapter I–Food and Drug Administration Department of Health and Human Services Subchapter H–Medical Devices, http://www.accessdata.fda.gov/scripts/cdrh/cfdocs/cfCFR/CFRSearch.cfm?FR=860.7; 2013.

[14] Zhang M, Raghunathan A, Jha NK. Trustworthiness of medical devices and body area networks. Proc IEEE 2014;102(8):1174–88.

[15] Mossman J. Important considerations for insulin pump and portable medical designs. http://www.maximintegrated.com/app-notes/index.mvp/id/4675; 2010.

[16] Jiang Z, Pajic M, Mangharam R. Cyber-physical modeling of implantable cardiac medical devices. Proc IEEE 2012;100(1):122–37.

[17] Ford D. Cheney's defibrillator was modified to prevent hacking, CNN News, http://www.cnn.com/2013/10/20/us/dick-cheney-gupta-interview/; 2013.

[18] Halperin D, Heydt-Benjamin T, Ransford B, Clark SS, Defend B, Morgan W, et al. Pacemakers and implantable cardiac defibrillators: software radio attacks and zero-power defenses. In: IEEE symposium on security and privacy; 2008. p. 129–42.

[19] Boston Scientific. CRM product performance report, http://www.bostonscientific.com/templatedata/imports/HTML/CRM/Product_Performance_Resource_Center/report_archives/q1_14_ppr.pdf; 2014.

[20] Medtronic. Cardiac rhythm disease management: product performance report. http://www.medtronic.com/productperformance-files/Issue%2069%20MDT%20CRDM%20PPR%202013%202nd%20Edition.pdf; 2013.

[21] Maisel WH, Moynahan M, Zuckerman BD, Gross TP, Tovar OH, Tillman D-B, et al. Pacemaker and ICD generator malfunctions: Analysis of Food and Drug Administration annual reports. JAMA 2006;295(16):1901–6.

[22] Buchmann I. How to make batteries in medical devices more reliable, Tech. rep., Cadex Electronics Inc. http://www.cadex.com/_content/Device_Manufacturer_Medical.pdf; 2013.

[23] Shin D, Poncino M, Macii E, Chang N. A statistical model of cell-to-cell variation in Li-ion batteries for system-level design. In: IEEE international symposium on low power electronics and design; 2013. p. 94–9.

[24] Mallela VS, Ilankumaran V, Rao NS. Trends in cardiac pacemaker batteries. Indian Pacing Electrophysiol J 2004;4(4):201–12.

[25] Li P, Bashirullah R. A wireless power interface for rechargeable battery operated medical implants. IEEE Trans Circuits Syst II: Exp Briefs 2007;54(10):912–6.

[26] Dissanayake T, Budgett D, Hu A, Malpas S, Bennet L. Transcutaneous energy transfer system for powering implantable biomedical devices. In: Lim C, Goh J, editors. International conference on biomedical engineering. IFMBE proceedings, vol. 23. Berlin-Heidelberg: Springer; 2009. p. 235–9.

[27] Wang P, Liang B, Ye X, Ko W, Cong P. A simple novel wireless integrated power management unit (PMU) for rechargeable battery-operated implantable biomedical telemetry systems. In: International conference on bioinformatics and biomedical engineering; 2010. p. 1–4.

[28] Fotopoulou K, Flynn BW. Optimum antenna coil structure for inductive powering of passive RFID tags. In: IEEE international conference on RFID; 2007. p. 71–7.

[29] Mitcheson PD. Energy harvesting for human wearable and implantable biosensors. In: International conference of the IEEE Engineering in Medicine and Biology Society; 2010. p. 3432–6.

[30] Mitcheson PD, Yeatman EM, Rao GK, Holmes AS, Green TC. Energy harvesting from human and machine motion for wireless electronic devices. Proc IEEE 2008;96(9):1457–86.

[31] Yang Y, Wei X-J, Liu J. Suitability of a thermoelectric power generator for implantable medical electronic devices. J Phys D Appl Phys 2007;40(18):5790.

[32] Ramadass YK, Chandrakasan AP. A battery-less thermoelectric energy harvesting interface circuit with 35 mV startup voltage. IEEE J Solid State Circuits 2011;46(1):333–41.

[33] Mercier PP, Lysaght AC, Bandyopadhyay S, Chandrakasan AP, Stankovic KM. Energy extraction from the biologic battery in the inner ear. Nat Biotechnol 2012;1240–3.

[34] Southcott M, MacVittie K, Halamek J, Halamkova L, Jemison WD, Lobel R, et al. A pacemaker powered by an implantable biofuel cell operating under conditions mimicking the human blood circulatory system – battery not included. Phys Chem Chem Phys 2013;15:6278–83.

[35] Dong K, Jia B, Yu C, Dong W, Du F, Liu H. Microbial fuel cell as power supply for implantable medical devices: a novel configuration design for simulating colonic environment. Biosens Bioelectron 2013;41:916–9.

[36] Borkar S. Designing reliable systems from unreliable components: the challenges of transistor variability and degradation. Micro, IEEE 2005;25(6):10–6.

[37] Gerrish P, Herrmann E, Tyler L, Walsh K. Challenges and constraints in designing implantable medical ICs. IEEE Trans Device Mat Rel 2005;5(3):435–44.

[38] Porter M, Gerrish P, Tyler L, Murray S, Mauriello R, Soto F, et al. Reliability considerations for implantable medical ICs. In: IEEE international reliability physics symposium; 2008. p. 516–23.

[39] Bradley PD, Normand E. Single event upsets in implantable cardioverter defibrillators. IEEE Trans Nucl Sci 1998;45(6):2929–40.

[40] Dodd PE, Massengill LW. Basic mechanisms and modeling of single-event upset in digital microelectronics. IEEE Trans Nucl Sci 2003;50(3):583–602.

[41] Mitra S, Seifert N, Zhang M, Shi Q, Kim KS. Robust system design with built-in soft-error resilience. Computer 2005;38(2):43–52.

[42] Lyons RE, Vanderkulk W. The use of triple-modular redundancy to improve computer reliability. IBM J Res Dev 1962;6:200–9.

[43] Jiang G, Zhou DD. Technology advances and challenges in hermetic packaging for implantable medical devices. In: Zhou D, Greenbaum E, editors. Implantable neural prostheses 2. Biological and medical physics, biomedical engineering. New York: Springer; 2010. p. 27–61.

[44] Borgesen P, Cotts E. Implantable medical electronics assembly quality and reliability considerations. SMTA News J Surf Mt Technol 2004;17:25–33.

[45] U.S. Food and Drug Administration. Introduction to medical device recalls: industry responsibilities. http://www.fda.gov/downloads/Training/CDRHLearn/UCM209281.pdf.

[46] Li C, Raghunathan A, Jha NK. Improving the trustworthiness of medical device software with formal verification methods. IEEE Embed Syst Lett 2013;5(3):50–3.

[47] Jetley R, Iyer SP, Jones PL. A formal methods approach to medical device review. Computer 2006;39(4):61–7.

[48] Arney D, Jetley R, Jones P, Lee I, Sokolsky O. Formal methods based development of a PCA infusion pump reference model: generic infusion pump (GIP) project. In: Joint workshop on high confidence medical devices, software, and systems and medical device plug-and-play interoperability; 2007. p. 23–33.

[49] Cordeiro L, Fischer B, Chen H, Marques-Silva J, Cordeiro L, Fischer B, et al. Semiformal verification of embedded software in medical devices considering stringent hardware constraints. In: International conference on embedded software and systems; 2009. p. 396–403.

[50] Trustworthy medical device software. In: Public health effectiveness of the FDA 510(k) clearance process: measuring postmarket performance and other select topics, workshop report; 2011.

[51] Baldus H, Klabunde K, Musch G. Reliable set-up of medical body-sensor networks. Wirel Sensor Netw 2004;2920:353–63.

[52] Falck T, Baldus H, Espina J, Klabunde K. Plug 'n play simplicity for wireless medical body sensors. Mobile Netw Appl 2007;12(2–3):143–53.

[53] Wessels D. Implantable pacemakers and defibrillators: device overview & EMI considerations. In: IEEE international symposium on electromagnetic compatibility, 2; 2002. p. 911–5.

[54] Jilek C, Tzeis S, Reents T, Estner H-L, Fichtner S, Ammar S, et al. Safety of implantable pacemakers and cardioverter defibrillators in the magnetic field of a novel remote magnetic navigation system. J Cardiovasc Electrophysiol 2010;21(10):1136–41.

[55] Roguin A, Zviman MM, Meininger GR, Rodrigues ER, Dickfeld TM, Bluemke DA, et al. Modern pacemaker and implantable cardioverter/defibrillator systems can be magnetic resonance imaging safe: in vitro and in vivo assessment of safety and function at 1.5 T. Circulation 2004;110(5):475–82.

[56] Lee S, Fu K, Kohno T, Ransford B, Maisel WH. Clinically significant magnetic interference of implanted cardiac devices by portable headphones. Heart Rhythm 2009;6(10):1432–6.

[57] Thaker JP, Patel MB, Jongnarangsin K, Liepa VV, Thakur RK. Electromagnetic interference with pacemakers caused by portable media players. Heart Rhythm 2008;5(4):538–44.

[58] Hayes DL, Wang PJ, Reynolds DW, Estes NAM, Griffith JL, Steffens RA, et al. Interference with cardiac pacemakers by cellular telephones. N Engl J Med 1997;336(21):1473–9.

[59] Nyenhuis JA, Kildishev AV, Bourland JD, Foster KS, Graber G, Athey TW. Heating near implanted medical devices by the MRI RF-magnetic field. IEEE Trans Magn 1999;35(5):4133–5.

[60] Nyenhuis JA, Park S-M, Kamondetdacha R, Amjad A, Shellock FG, Rezai AR. MRI and implanted medical devices: Basic interactions with an emphasis on heating. IEEE Trans Device Mat Rel 2005;5(3):467–80.

[61] Dhillon BS. Medical device reliability and associated areas. Boca Raton, FL; CRC Press; 2000.

[62] Kohn LT, Corrigan JM, Donaldson MS, et al. To err is human: building a safer health system, 627. Washington, DC: National Academies Press; 2000.

[63] MAUDE adverse event report: Neuro N'Vision programmer, http://www.accessdata.fda.gov/scripts/cdrh/cfdocs/cfmaude/Detail.CFM?MDRFOI__ID=527622; 2004.

[64] Lin Q-L, Wang D-J, Lin W-G, Liu H-C. Human reliability assessment for medical devices based on failure mode and effects analysis and fuzzy linguistic theory. Saf Sci 2014;62:248–56.

[65] Fu K, Blum J. Controlling for cybersecurity risks of medical device software. Commun ACM 2013;56(10):35–7.

[66] Burleson W, Clark SS, Ransford B, Fu K. Design challenges for secure implantable medical devices. In: Design automation conference; 2012. p. 12–7.

[67] Hall JL, McGraw D. For telehealth to succeed, privacy and security risks must be identified and addressed. Health Aff 2014;33(2):216–21.

[68] Hosseini-Khayat S. A lightweight security protocol for ultra-low power ASIC implementation for wireless implantable medical devices. In: International symposium on medical information communication technology; 2011. p. 6–9.

[69] Rostami M, Burleson W, Koushanfar F, Juels A. Balancing security and utility in medical devices? In: Design automation conference; 2013. p. 13:1–6.

[70] Potlapally NR, Ravi S, Raghunathan A, Jha NK. Analyzing the energy consumption of security protocols. In: International symposium on low power electronics and design; 2003. p. 30–5.

[71] Bogdanov A, Knudsen LR, Leander G, Paar C, Poschmann A, Robshaw MJB, et al. Present: an ultra-lightweight block cipher. In: Paillier P, Verbauwhede I, editors. Cryptographic hardware and embedded systems. Lecture Notes in Computer Science, 4727. Verlag, Berlin-Heidelberg: Springer; 2007. p. 450–66.

[72] Canniere C, Dunkelman O, Knezevic M. KATAN and KTANTAN – a family of small and efficient hardware-oriented block ciphers. In: Clavier C, Gaj K, editors. Cryptographic hardware and embedded systems. Lecture Notes in Computer Science, 5747. Verlag, Berlin-Heidelberg: Springer; 2009. p. 272–88.

[73] Beck C, Masny D, Geiselmann W, Bretthauer G. Block cipher based security for severely resource-constrained implantable medical devices. In: International symposium on applied sciences in biomedical and communication technologies; 2011. p. 62:1–5.

[74] Zhang M, Kermani MM, Raghunathan A, Jha NK. Energy-efficient and secure sensor data transmission using encompression. In: International conference on VLSI design and international conference on embedded systems; 2013. p. 31–6.

[75] Denning T, Borning A, Friedman B, Gill BT, Kohno T, Maisel WH. Patients, pacemakers, and implantable defibrillators: human values and security for wireless implantable medical devices. In: SIGCHI conference on human factors in computing systems; 2010. p. 917–26.

[76] S. Schechter, Security that is meant to be skin deep: using ultraviolet micropigmentation to store emergency-access keys for implantable medical devices, Tech. rep., Microsoft Research; 2010.

[77] Mathur S, Trappe W, Mandayam N, Ye C, Reznik A. Radio-telepathy: extracting a secret key from an unauthenticated wireless channel. In: ACM international conference on mobile computing and networking; 2008. p. 128–39.

[78] Jana S, Premnath SN, Clark M, Kasera SK, Patwari N, Krishnamurthy SV. On the effectiveness of secret key extraction from wireless signal strength in real environments. In: International conference on mobile computing and networking; 2009. p. 321–32.

[79] Ali ST, Sivaraman V, Ostry D. Zero reconciliation secret key generation for body-worn health monitoring devices. In: ACM conference on security and privacy in wireless and mobile, networks; 2012. p. 39–50.

[80] Shi L, Yuan J, Yu S, Li M. ASK-BAN: authenticated secret key extraction utilizing channel characteristics for body area networks. In: ACM conference on security and privacy in wireless and mobile, networks; 2013. p. 155–66.

[81] Venkatasubramanian KK, Banerjee A, Gupta SKS. PSKA: usable and secure key agreement scheme for body area networks. IEEE Trans Inf Technol Biomed 2010;14(1):60–8.

[82] Chang S-Y, Hu Y-C, Anderson H, Fu T, Huang EYL. Body area network security: robust key establishment using human body channel. In: USENIX conference on health security and privacy; 2012. p. 5.

[83] Rostami M, Juels A, Koushanfar F. Heart-to-heart (H2H): authentication for implanted medical devices. In: ACM SIGSAC conference on computer & communications security; 2013. p. 1099–112.

[84] Hu C, Cheng X, Zhang F, Wu D, Liao X, Chen D. OPFKA: secure and efficient ordered-physiological-feature-based key agreement for wireless body area networks. In: IEEE international conference on computer communications; 2013. p. 2274–82.

[85] Jurik AD, Weaver AC. Securing mobile devices with biotelemetry. In: International conference on computer communications and, networks; 2011. p. 1–6.

[86] Cherukuri S, Venkatasubramanian KK, Gupta SKS. Biosec: a biometric based approach for securing communication in wireless networks of biosensors implanted in the human body. In: International conference on parallel processing workshops; 2003. p. 432–9.

[87] Venkatasubramanian KK, Banerjee A, Gupta SKS. EKG-based key agreement in body sensor networks. In: IEEE international conference on computer communications; 2008. p. 1–6.

[88] Xu F, Qin Z, Tan CC, Wang B, Li Q. IMDGuard: securing implantable medical devices with the external wearable guardian. In: IEEE international conference on computer communications; 2011. p. 1862–70.

[89] Goodrich MT, Sirivianos M, Solis J, Tsudik G, Uzun E. Loud and clear: human-verifiable authentication based on audio. In: IEEE international conference on distributed computing systems; 2006. p. 10.

[90] Li M, Yu S, Guttman JD, Lou W, Ren K. Secure ad hoc trust initialization and key management in wireless body area networks. ACM Trans Sensor Netw 2013;9(2):18:1–35.

[91] Halevi T, Saxena N. On pairing constrained wireless devices based on secrecy of auxiliary channels: the case of acoustic eavesdropping. In: ACM conference on computer and communications security; 2010. p. 97–108.

[92] Baldus H, Corroy S, Fazzi A, Klabunde K, Schenk T. Human-centric connectivity enabled by body-coupled communications. IEEE Commun Mag 2009;47(6):172–8.

[93] Bagade P, Banerjee A, Milazzo J, Gupta SKS, Bagade P, Banerjee A, Milazzo J, Gupta SKS. Protect your BSN: no handshakes, just Namaste! In: IEEE international conference on body sensor, networks; 2013. p. 1–6.

[94] Cremers C, Rasmussen KB, Schmidt B, Capkun S. Distance hijacking attacks on distance bounding protocols. In: IEEE symposium on security and privacy; 2012. p. 113–27.

[95] Rasmussen KB, Castelluccia C, Heydt-Benjamin TS, Capkun S. Proximity-based access control for implantable medical devices. In: ACM conference on computer and communications security; 2009. p. 410–9.

[96] Shi L, Li M, Yu S, Yuan J. BANA: body area network authentication exploiting channel characteristics. IEEE J Sel Areas Commun 2013;31(9):1803–16.

[97] Denning T, Fu K, Kohno T. Absence makes the heart grow fonder: new directions for implantable medical device security. In: Conference on hot topics in security; 2008. p. 5:1–7.

[98] Hancke GP, Centre SC. Eavesdropping attacks on high-frequency RFID tokens. In: Workshop on RFID security; 2008. p. 100–13.

[99] Martin T, Hsiao M, Ha D-S, Krishnaswami J. Denial-of-service attacks on battery-powered mobile computers. In: IEEE conference on pervasive computing and, communications; 2004. p. 309–18.

[100] Raymond DR, Marchany R, Brownfield MI, Midkiff SF. Effects of denial-of-sleep attacks on wireless sensor network MAC protocols. IEEE Trans Veh Technol 2009;58(1):367–80.

[101] Hei X, Du X, Wu J, Hu F. Defending resource depletion attacks on implantable medical devices. In: IEEE global telecommunications conference; 2010. p. 1–5.

[102] Zhang M, Raghunathan A, Jha NK. MedMon: securing medical devices through wireless monitoring and anomaly detection. IEEE Trans Biomed Circuits Syst 2013;7(6):871–81.

[103] Gollakota S, Hassanieh H, Ransford B, Katabi D, Fu K. They can hear your heartbeats: non-invasive security for implantable medical devices. In: ACM SIGCOMM conference; 2011. p. 2–13.

[104] Tippenhauer NO, Malisa L, Ranganathan A, Capkun S. On limitations of friendly jamming for confidentiality. In: IEEE symposium on security and privacy; 2013. p. 160–73.

[105] Kocher P, Jaffe J, Jun B. Differential power analysis. In: Wiener M, editor. Advances in cryptology. Lecture Notes in Computer Science, 1666. Berlin-Heidelberg: Springer; 1999. p. 388–97.

[106] Mangard S, Oswald E, Popp T. Power analysis attacks: revealing the secrets of smart cards. Heidelberg: Springer; 2010.

[107] Tillich S, Herbst C. Attacking state-of-the-art software countermeasures–a case study for AES. In: International workshop on cryptographic hardware and embedded systems; 2008. p. 228–43.

[108] Tiri K, Akmal M, Verbauwhede I. A dynamic and differential CMOS logic with signal independent power consumption to withstand differential power analysis on smart cards. In: European solid-state circuits conference; 2002. p. 403–6.

[109] Tiri K, Verbauwhede I. Charge recycling sense amplifier based logic: securing low power security ICs against DPA. In: European solid-state circuits conference; 2004. p. 179–82.

[110] Muresan R, Gregori S. Protection circuit against differential power analysis attacks for smart cards. IEEE Trans Comput 2008;57:1540–9.

[111] Ratanpal GB, Williams RD, Blalock TN. An on-chip signal suppression countermeasure to power analysis attacks. IEEE Trans Dependable Secure Comput 2004;1:179–89.

[112] Coron J-S, Goubin L. On Boolean and arithmetic masking against differential power analysis. In: International workshop on cryptographic hardware and embedded systems; 2000. p. 231–7.

[113] Golić JD, Tymen C. Multiplicative masking and power analysis of AES. In: International workshop on cryptographic hardware and embedded systems; 2003. p. 198–212.

[114] Hasan MA. Power analysis attacks and algorithmic approaches to their countermeasures for Koblitz curve cryptosystems. IEEE Trans Comput 2001;50:1071–83.

[115] Quisquater J-J, Samyde D. ElectroMagnetic analysis (EMA): measures and counter-measures for smart cards. In: International conference on research in smart cards: smart card programming and security; 2001. p. 200–10.

[116] Arney D, Venkatasubramanian KK, Sokolsky O, Lee I. Biomedical devices and systems security. In: International conference of the IEEE engineering in medicine and biology society; 2011. p. 2376–9.

[117] Buttyan L, Holczer T. Traffic analysis attacks and countermeasures in wireless body area sensor networks. In: IEEE international symposium on a world of wireless, mobile and multimedia networks; 2012. p. 1–6.

[118] Chakraborty RS, Narasimhan S, Bhunia S. Hardware Trojan: threats and emerging solutions. In: IEEE international high level design validation and test workshop; 2009. p. 166–71.

[119] Tehranipoor M, Koushanfar F. A survey of hardware Trojan taxonomy and detection. IEEE Design Test 2010;27(1):10–25.

[120] Agrawal D, Baktir S, Karakoyunlu D, Rohatgi P, Sunar B. Trojan detection using IC fingerprinting. In: IEEE symposium on security and privacy; 2007. p. 296–310.

[121] Jin Y, Makris Y. Hardware Trojan detection using path delay fingerprint. In: IEEE symposium on hardware-oriented security and trust; 2008. p. 51–7.

[122] Hu K, Nowroz AN, Reda S, Koushanfar F. High-sensitivity hardware Trojan detection using multimodal characterization. In: Design, automation & test in Europe conference; 2013. p. 1271–6.

[123] Li J, Lach J. At-speed delay characterization for IC authentication and Trojan horse detection. In: IEEE international workshop on hardware-oriented security and trust; 2008. p. 8–14.

[124] Kune DF, Backes J, Clark SS, Kramer D, Reynolds M, Fu K, et al. Ghost talk: mitigating EMI signal injection attacks against analog sensors. In: IEEE symposium on security and privacy; 2013. p. 145–59.

[125] Peng S, Yu S, Yang A. Smartphone malware and its propagation modeling: a survey. IEEE Commun Surveys Tutorials 2014;16(2):925–41.

[126] Hanna S, Rolles R, Molina-Markham A, Poosankam P, Fu K, Song D. Take two software updates and see me in the morning: the case for software security evaluations of medical devices. In: Workshop on health security and privacy; 2011. p. 1–5.

[127] Li C, Raghunathan A, Jha NK. Secure virtual machine execution under an untrusted management OS. In: IEEE international conference on cloud, computing; 2010. p. 172–9.

[128] Smith SW, Weingart S. Building a high-performance, programmable secure coprocessor. Comput Netw 1999;31(8):831–60.

[129] Sorber J, Shin M, Peterson R, Cornelius C, Mare S, Prasad A, et al. An amulet for trustworthy wearable mHealth. In: Workshop on mobile computing systems and applications; 2012. p. 7:1–6.

[130] Aaraj N, Raghunathan A, Jha NK. Analysis and design of a hardware/software trusted platform module for embedded systems. ACM Trans Embedded Comput 2009;8(1):8:1–31.

[131] Sorber JM, Shin M, Peterson R, Kotz D. Plug-n-trust: practical trusted sensing for mHealth. In: International conference on mobile systems, applications, and services; 2012. p. 309–22.

[132] Jetley RP, Jones PL, Anderson P. Static analysis of medical device software using CodeSonar. In: Workshop on static analysis; 2008. p. 22–9.

[133] Aaraj N, Raghunathan A, Jha NK. Dynamic binary instrumentation-based framework for malware defense. In: International conference on detection of intrusions and malware, and vulnerability, assessment; 2008. p. 64–87.

[134] Arora D, Ravi S, Raghunathan A, Jha NK. Secure embedded processing through hardware-assisted run-time monitoring. In: Design, automation & test in Europe conference; 2005. p. 178–83.

[135] Kim H, Smith J, Shin KG. Detecting energy-greedy anomalies and mobile malware variants. In: International conference on mobile systems, applications, and services; 2008. p. 239–52.

[136] Liu L, Yan G, Zhang X, Chen S. Virusmeter: preventing your cellphone from spies. In: Kirda E, Jha S, Jha S, Balzarotti D, editors. Recent advances in intrusion detection. Lecture Notes in Computer Science, 5758. Berlin-Heidelberg: Springer; 2009. p. 244–64.

[137] Clark SS, Ransford B, Rahmati A, Guineau S, Sorber J, Xu W, et al. WattsUpDoc: power side channels to nonintrusively discover untargeted malware on embedded medical devices. In: USENIX workshop on health information technologies; 2013.

[138] U.S. Department of Defense, DoD open source software (OSS) FAQ. http://dodcio.defense.gov/OpenSourceSoftwareFAQ.aspx.

[139] Carstens DS, McCauley-Bell PR, Malone LC, DeMara RF. Evaluation of the human impact of password authentication practices on information security. Informing Sci J 2004;67–85.

# Applications of Bioimplantable Systems

# Biochips

## Electrical Biosensors: Peripheral Nerve Sensors

# 9

**Brian Wodlinger\*, Yazan Dweiri†, Dominique M. Durand†**

*Electrical and Computer Engineering Department, University of Utah, Salt Lake City, Utah, USA*

*†Urology Institute, University Hospitals Case Medical Center, Cleveland, Ohio, USA*

## CHAPTER CONTENTS

## 9.1 INTRODUCTION

While the central nervous system (CNS) provides many data storage and processing functions of interest to neuroscientists, creating a robust artificial interface to record these signals has proven difficult. In contrast, the peripheral nervous system (PNS) is responsible mainly for signal transmission. These signals, however, contain a large amount of information that could be very useful to medical science. For example, lower motor neurons carry muscle activation signals that do not require decoding—a large advantage compared with their complex cortical equivalents. Similarly, nerves provide a connection between various biological sensors and their associated control systems; a suitable peripheral neural interface would allow us to robustly monitor a large variety of biological signals such as blood sugar and blood pressure, as well as a variety of sensory modalities, and augment failing or insufficient biological controllers. A wide variety of designs have been demonstrated to provide biocompatible, safe electrodes capable of recovering these signals.

Bhunia et al. Implantable Biomedical Microsystems. http://dx.doi.org/10.1016/B978-0-323-26208-8.00009-1

Peripheral nerve electrodes may be divided based on their level of invasiveness. A convenient and physiologically meaningful distinction is based on the penetration of neural membranes. Cuff-style electrodes including oval or spiral cuffs [1] and flat-interface nerve electrodes (FINEs) [2,3] wrap around the outside of the nerve and do not penetrate it. Other designs such as the Utah Slanted Electrode Array (USEA) [4] and longitudinal intrafascicular electrode (LIFE) [5] penetrate both the loose outer epineurium and the perineurium whose tight junctions maintain positive pressure inside the fascicles. Penetrating electrodes are able to achieve far greater signal-to-noise ratios (SNRs) due to their tight spatial proximity to the nerve axons within the fascicles; however, the long-term stability of these recordings has yet to be demonstrated.

The small signal level, particularly in cuff-style sensors, has motivated the design of several amplification and processing systems designed to reduce losses by providing amplification very close to the sensor itself (e.g., the RHA series of amplifiers from Intan Technologies). Similar on- or pericuff devices may also be used to transmit telemetry through the body to other internal or external systems [6].

This article begins with an overview of the relevant anatomy of the PNS in general and the common structure of the peripheral nerves, followed by sections on various sensor technologies and their design, fabrication, and academic or commercial use. A review of recording circuitry is then provided, followed by novel techniques that have recently been proposed and a summary on the current state of the art.

## 9.2 PERIPHERAL NERVE ANATOMY

The PNS is composed of the all neural tissues in the body outside of the brain and spinal cord. This includes 11 pairs of cranial nerves, 31 pairs of spinal nerves, and a multitude of derivative nerves, plexuses, and ganglia.

Peripheral nerves contain the axons of neurons whose cell bodies are located in either the spinal cord, ganglia, or adjacent to (or part of) a sensor (e.g., a vestibular hair cell). These axons may be either afferent (sensory) carrying signals towards the CNS or efferent (motor) carrying commands from the CNS to the muscles and other effectors throughout the body.

Axons within a nerve are bundled in tight groups known as fascicles, which are surrounded by the perineurium membrane. This membrane is analogous to the dura in the CNS in that it is composed of tight junctions and maintains positive pressure within the fascicles. Fascicles are subsequently bound together with loose epineural tissue to form a single nerve. Recent work suggests that axons within the same fascicle, as well as spatially nearby fascicles, tend to have similar anatomical end points [7].

## 9.3 CUFF-STYLE ELECTRODES

Cuff-style electrodes wrap around the outside of the nerve, puncturing neither the epineurium nor the perineurium. Because of this, their electrical contacts sit farther from the neural source and so tend to be larger (aggregating the activity of many sources). These larger contacts result in lower input impedance that provides opportunities for amplifier design. A low-noise amplifier design is particularly important because of the very small signal levels (0.5–2.0 μV RMS).

Cuff electrodes are often designed with a tripolar structure [8,9] where reference electrodes at the proximal and distal openings of the cuff are used to both amplify the apparent nerve signals (due to the shape of the action potential as it moves down the nerve) and reduce interference from sources outside the nerve.

### 9.3.1 TYPES

Cuff electrodes are generally grouped based on shape. Cylindrical cuffs [1] have a generally oval cross section that attempts to match the resting shape of the nerve. In contrast, helical cuffs [10] do not fully insulate the nerve, but rather coil around it. By applying light pressure, they ensure solid contact at all points. Further along the same lines is the FINE [2,3]. This design has a more rectangular cross section, reshaping the soft epineurium so that the fascicles are more spread out. This results in the contacts being closer to the neural sources, improving SNR and also improving fascicle selectivity (recording signals from each fascicle independently) [3,11,12].

### 9.3.2 FABRICATION

Cuff electrodes are generally fabricated from a silicone elastomer, with the outer shell either molded or constructed from preformed silicone pieces. Individual contacts may be platinum (particularly if the electrode is going to be used for stimulation), iridium oxide, gold, stainless steel, or other biocompatible metal with reasonable charge-density properties. Contacts are often welded to braided stainless steel wires to form a cable that connects to the amplifier. A procedure for the fabrication of high-contact-density FINEs is shown in Figure 9.1. The finished contacts are then glued into a premade cuff using silicone adhesive, along with platinum reference electrodes to form the tripole (see finished electrode in Figure 9.2).

### 9.3.3 USES

Cuff electrodes of various types have been used to record both efferent motor commands and afferent sensation signals in preclinical (animal) [1,3,13–15] and clinical (human) [16–18] settings with promising results. Some important clinical uses are recording on the branches of the sciatic nerve for patients with foot drop to permit more normal walking gait [18] and recently control of prosthetic limbs in amputees including restoration of sensation using tactile sensors embedded in the prosthesis. These electrodes may be used to record both spontaneous neural activity that occurs as a result of natural sensation and motor function and the much larger evoked activity generated by reflex twitches or direct neural stimulation, as may be done to determine nerve health during surgery [19].

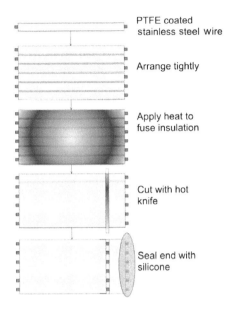

PTFE coated
stainless steel wire

Arrange tightly

Apply heat to
fuse insulation

Cut with hot
knife

Seal end with
silicone

**FIGURE 9.1**

High-contact-density electrode fabrication technique. From top to bottom: PTFE-coated
stainless steel medical wires are first arranged tightly next to each other as a ribbon cable.
A heat gun or the flat side of a hot knife is then used to fuse the insulation, locking the
wires in place. A hot knife or wire with a guide is then used to cut a slit in the insulation,
exposing the metal underneath. Finally, the unwanted end is sealed in silicone to prevent
current leakage.

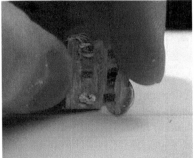

**FIGURE 9.2**

Left: Array of 16 contacts produced by the technique in Figure 9.1. The array is 4 mm
wide. Right: Completed electrode with FINE structure made from silicone sheeting and
tubing and platinum tripolar references.

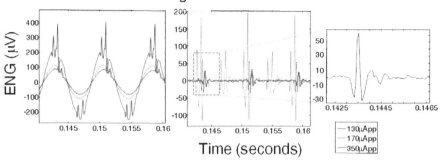

**FIGURE 9.3**

130 Hz sinusoidal stimulation produces CAP-like neural responses, which change amplitude, delay, and complexity as the stimulation intensity is raised. Left: Recording high-pass filtered at 100 Hz, center: the same signal high-pass filtered at 800 Hz. Note that the filter has effectively removed all traces of the stimulus artifact. Right: A single neural response during 170 μApp stimulation.

Sinusoidal stimulation may be used to approximate both types of activity, as shown in Figures 9.3 and 9.4. A FINE cuff was placed on a nerve while a distal branch of the nerve was stimulated using a low-frequency (Figure 9.3) or high-frequency (Figure 9.4) sinusoid. The evoked neural activity was then recorded, and the stimulation artifact filtered out.

## 9.4 PENETRATING AND SIEVE ELECTRODES

Penetrating electrodes pass through at least one of the two membranes surrounding the axons in the nerve. They vary significantly in shape, size, and material but are generally unified in having "microelectrode"-style contacts, which are generally much smaller than 1 mm in size. These higher-impedance devices tend to have higher SNRs than cuff electrodes since their contacts are much closer to the axons, which also makes them more selective at the axonal level. However, this should not be confused with functional selectivity as information encoding in the PNS often relies on the statistical properties of a large number of fibers. While these electrodes are more susceptible to micromotion and encapsulation-related signal changes due to their proximity to the neural source, it is unclear how nerve health is affected by chronic placement.

Sieve or regeneration-style electrodes use the opposite approach: the nerve is severed and the electrode sutured between the two halves. The axons are then forced to regenerate through holes that are surrounded by electrical contacts

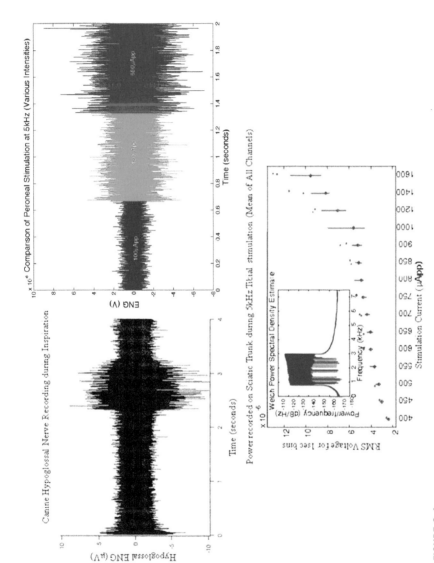

**FIGURE 9.4**

Left: ENG recorded during passive breathing from a cuff electrode implanted chronically on the hypoglossal nerve of a canine. Note the small amplitude, low SNR, and chaotic nature of the signal [32]. Right: ENG signals during 5 kHz stimulation of the peroneal branch, filtered 800–3000 Hz at three different intensities. Each intensity was recorded during a separate but sequential trial and a segment from each is shown side by side here for comparison. The power of the signal increases as the stimulation intensity (given in white at the center of each segment) increases. Below: Box plot of the RMS amplitude of 1 s windows for each stimulation intensity. Note the increase in signal with increasing stimulation power and the rejection of 5 kHz stimulation artifact in the inset power spectral density (from the highest stimulation level).

[10,20,21]. Clearly, this approach causes significant trauma to the nerve and regeneration is not assured; however, animal studies have demonstrated that, with correctly sized holes and noncytotoxic materials, recordings can be stable over the long term [21].

### 9.4.1 TYPES

Figure 9.5 provides examples of some of the more popular types of penetrating and sieve electrodes, including the USEA whose silicon spikes are metallized at their tips with iridium oxide or titanium to allow them to record from within fascicles. These electrodes have a very high contact density, with up to 100 channels on a single 4×4 mm array. The LIFE and transverse intrafascicular multichannel electrode (TIME), on the other hand, are threadlike electrodes that are meant to be drawn through the epineurium without piercing the perineurium in order to record from just outside the fascicles. Sieve electrodes vary principally by hole size and material. Smaller-holed versions provide more selectivity but may impede axonal regeneration.

### 9.4.2 FABRICATION

Fabrication techniques for these electrodes generally require photolithography-based techniques due to the small sizes involved, very similar to techniques used for penetrating cortical array fabrication. For example, the fabrication of USEAs involves backside dicing, glassing, backside metallization, front-side dicing, wet etching, tip metallization, Parylene-C deposition, tip deisolation, and wire

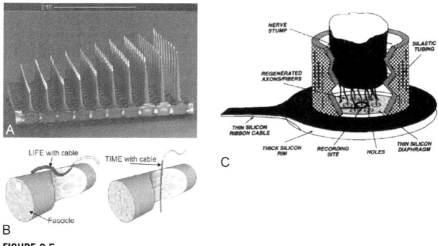

**FIGURE 9.5**

Penetrating and sieve electrode arrays. (a) Scanning electron microscope image of a USEA [25]. (b) LIFE and TIME electrode diagrams [23]. (c) Schematic of sieve electrode [20].

bonding [22]. However, a simpler form of penetrating electrode can be made from partially deinsulated stainless steel or platinum–iridium wire that is drawn through the nerve using a suture. Advances on this general design (which forms the basis for both LIFE and TIME styles) include thin-film multicontact versions and corrugation of the electrode to increase contact density to improve isolation between channels [23,24].

### 9.4.3 USES

Penetrating nerve electrodes have seen similar experimental use to cuff electrodes [4,25,26]; however, while they have been used in human clinical trials [27], they have not yet become clinically available for peripheral nerve use, though similar electrodes are clinically available for temporary implantation during epilepsy monitoring. Sieve electrodes have to date only seen use in animal experiments.

## 9.5 NEURAL RECORDING AMPLIFIERS

The primary concern with all peripheral nerve electrodes is eliminating long cables over which noise may contaminate the small signals. For this reason, several groups have developed preamplifiers that sit on or very close to the electrodes. In many cases, these amplifiers are designed specifically for a particular electrode to ensure the proper trade-off between voltage and current noise and a controlled bandwidth.

The main drawback of cuff electrode recording is the low SNR. Input-referred noise of less than $1\,\mu V_{RMS}$ is desired for a useful SNR. Consequently, customized ultralow-noise circuitry is essential in order to minimize the noise induced by the electronics, which can easily cover the targeted neural activity.

For an acquisition device with a given input voltage and current noise ($e_n$ and $i_n$, respectively), the best noise performance with the highest SNR and lowest noise figure occurs when the source resistance is equal to an optimum value ($R_{s(optimum)}$) which equals to $e_n/i_n$. $R_{s(optimum)}$ is usually larger than the actual source resistance associated with cuff electrode's contacts, which can be interpreted as needing to introduce an additional resistor at the input in order to reach $R_{s(optimum)}$. This scenario is misleading since that would add another noise source and degrade the total SNR.

Two techniques can be used to match the effective source resistance seen by the input device. The first method involves power matching with the use of a transformer. This has been used for whole nerve recording [28–31]. The use of the transformer increases the effective source resistance by the turn ratio $n$ leading to

$$R_{s(optimum)} = \frac{e_n}{i_n \cdot n^2}$$

As such, the source impedance can be matched to the optimum source impedance by varying the transformer turn ratio leading to the optimum SNR. The limitations of this technique are the narrow range of sources for which the transformer turn ratio is optimized and the large size of the transformers restricting the number of channels that can be implemented in an amplifier of a given size.

The second technique is hardware averaging. Each electrode is connected to multiple identical devices in parallel and the outputs are then averaged. This technique improves the noise performance in two ways: Initially, the device-induced voltage noise is reduced by the square root of the number of parallel devices ($N$). Secondly, the optimum source resistance is reduced such that

$$R_{s(optimum)} = \frac{e_n}{i_n \cdot N}$$

Since the contribution of the intrinsic voltage noise of the device is reduced independently of the source resistance, the total noise is still reduced even before reaching the optimum source resistance; hence, the noise improvement from this technique is not limited to a set of sources with a particular resistance. Additionally, hardware averaging can be implemented in high-channel count devices thanks to the high-channel count of many amplifier-integrated circuits. The limitations of this technique are the cumulative increase of input current noise and power consumption from adding more devices in parallel. As a result, low current noise devices ($<100\,fA/\sqrt{Hz}$) should be considered, and the trade-off between noise improvement and power consumption should be accounted for in the design of such a system.

Figure 9.6 presents a schematic example of circuitry designed to be placed directly on a FINE cuff, incorporating an ultralow-noise amplifier design ($0.78\,\mu V_{RMS}$ for 700 Hz–5 kHz) with the use of hardware averaging ($N=4$). The amplifier IC selected (RHD 2164) contains a built-in multiplexer and analog-to-digital converter, which reduces the number of wires connected to the nerve, reducing tethering forces that could cause damage over time. Furthermore, triboelectric noise from the contact leads is reduced by digitizing the data at the electrode site.

Neural recordings with cuff electrodes in the PNS represent a significant challenge since very small signals ($0.5$–$2\,\mu V_{RMS}$) must be obtained in the presence of much larger EMG signals (1–2 mV) without damaging the nerve. However, very low-noise amplifier designs can meet this challenge. With the aid of these amplifiers, both cuff and penetrating electrodes may be used to chronically record from a multitude of channels on a single peripheral nerve, providing the signals necessary to advance our understanding of the nervous system, restore voluntary control of paralyzed or prosthetic limbs, and monitor our intrinsic biochemical and perceptual sensors.

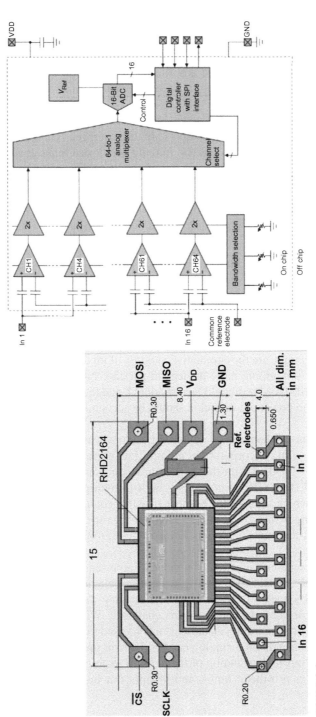

**FIGURE 9.6**

Left: Circuit layout for on-cuff electronics. This system is capable of recording 16 channels of ENG at 15 kHz each with <1 $\mu V_{RMS}$ input-referred noise and will fit on the back of a typical FINE. One of these circuits may be placed on each side for 32 channel recordings. A multiplexer and analog-to-digital converter are included in the package, reducing the number of wires and hence the force on the cuff, from 20 to 6. Right: Modified schematic of the Intan RHD2164 chip with the external connection of four parallel devices per electrode contact to implement hardware averaging for noise reduction.

# REFERENCES

[1] Riso RR, Mosallaie FK, Jensen W, Sinkjaer T. Nerve cuff recordings of muscle afferent activity from tibial and peroneal nerves in rabbit during passive ankle motion. IEEE Trans Rehabil Eng 2000;8:244–58.

[2] Tesfayesus W, Durand DM. Blind source separation of peripheral nerve recordings. J Neural Eng 2007;4:S157–67.

[3] Wodlinger B, Durand DM. Selective recovery of fascicular activity in peripheral nerves. J Neural Eng 2011;8:056005.

[4] Branner A, Stein RB, Fernandez E, Aoyagi Y, Normann RA. Long-term stimulation and recording with a penetrating microelectrode array in cat sciatic nerve. IEEE Trans Biomed Eng 2004;51:146–57.

[5] Micera S, Navarro X, Carpaneto J, Citi L, Tonet O, Rossini PM, Carrozza MC, Hoffmann KP, Vivó M, Yoshida K, Dario P. On the use of longitudinal intrafascicular peripheral interfaces for the control of cybernetic hand prostheses in amputees. IEEE Trans Neural Syst Rehabil Eng 2008;16:453–72.

[6] Smith B, Crish TJ, Buckett JR, Kilgore KL, Peckham PH. Development of an implantable networked neuroprosthesis. In: 10th annual conference of the international FES society, July 2005, Montreal, Canada; 2005. p. 454–7.

[7] Gustafson KJ, Pinault GCJ, Neville JJ, Syed I, Davis JA, Jean-Claude J, et al. Fascicular anatomy of human femoral nerve: implications for neural prostheses using nerve cuff electrodes. J Rehabil Res Dev 2009;46:973–84.

[8] Struijk JJ, Thomsen M. Tripolar nerve cuff recording: stimulus artifact, EMG and the recorded nerve signal. In: Proceedings of 17th international conference of the engineering in medicine and biology society, vol. 2. Montreal, Canada: IEEE; 1995. p. 1105–6.

[9] Rahal M, Taylor J, Donaldson N. The effect of nerve cuff geometry on interference reduction: a study by computer modeling. IEEE Trans Biomed Eng 2000;47:136–8.

[10] Navarro X, Krueger TB, Lago N, Micera S, Stieglitz T, Dario P. A critical review of interfaces with the peripheral nervous system for the control of neuroprostheses and hybrid bionic systems. J Peripher Nerv Syst 2005;10:229–58.

[11] Tyler DJ, Durand DM. Chronic response of the rat sciatic nerve to the flat interface nerve electrode. Ann Biomed Eng 2003;31:633–42.

[12] Perez-Orive J, Durand DM. Modeling study of peripheral nerve recording selectivity. IEEE Trans Rehabil Eng 2000;8:320–9.

[13] Yoo PB, Sahin M, Durand DM. Selective stimulation of the canine hypoglossal nerve using a multi-contact cuff electrode. Ann Biomed Eng 2004;32:511–9.

[14] Zariffa J, Nagai MK, Daskalakis ZJ, Popovic MR. Influence of the number and location of recording contacts on the selectivity of a nerve cuff electrode. IEEE Trans Neural Syst Rehabil Eng 2009;17:420–7.

[15] Cavallaro E, Micera S, Dario P, Jensen W, Sinkjaer T. On the intersubject generalization ability in extracting kinematic information from afferent nervous signals. IEEE Trans Biomed Eng 2003;50:1063–73.

[16] Fisher LE, Tyler DJ, Anderson JS, Triolo RJ. Chronic stability and selectivity of four-contact spiral nerve-cuff electrodes in stimulating the human femoral nerve. J Neural Eng 2009;6:046010.

[17] Polasek KH, Hoyen HA, Keith MW, Tyler DJ. Human nerve stimulation thresholds and selectivity using a multi-contact nerve cuff electrode. IEEE Trans Neural Syst Rehabil Eng 2007;15:76–82.

[18] Haugland MK, Sinkjar T. Cutaneous whole nerve recordings used for correction of foot-drop in hemiplegic man. IEEE Trans Rehabil Eng 1995;3:307–17.

[19] Brunnett WC, Hacker DC, Pagotto CA, Grant D. Nerve electrode USTA Patent 8515520; 2013.

[20] Akin T, Najafi K, Smoke RH, Bradley RM. A micromachined silicon sieve electrode for nerve regeneration applications. IEEE Trans Biomed Eng 1994;41:305–13.

[21] Bradley RM, Cao X, Akin T, Najafi K. Long term chronic recordings from peripheral sensory fibers using a sieve electrode array. J Neurosci Methods 1997;73:177–86.

[22] Nordhausen CT, Maynard EM, Normann RA. Single unit recording capabilities of a 100 microelectrode array. Brain Res 1996;726:129–40.

[23] Boretius T, Badia J, Pascual-Font A, Schuettler M, Navarro X, Yoshida K, et al. A transverse intrafascicular multichannel electrode (TIME) to interface with the peripheral nerve. Biosens Bioelectron 2010;26:62–9.

[24] Farina D, Yoshida K, Stieglitz T, Koch KP. Multichannel thin-film electrode for intramuscular electromyographic recordings. J Appl Physiol 2008;104:821–7.

[25] Branner A, Stein RB, Normann RA. Selective stimulation of cat sciatic nerve using an array of varying-length microelectrodes. J Neurophysiol 2001;85:1585–94.

[26] Bruns TM, Wagenaar JB, Bauman MJ, Gaunt RA, Weber DJ. Real-time control of hind limb functional electrical stimulation using feedback from dorsal root ganglia recordings. J Neural Eng 2013;10:026020.

[27] Micera S, Citi L, Rigosa J, Carpaneto J, Raspopovic S, Di Pino G, et al. Decoding information from neural signals recorded using intraneural electrodes: toward the development of a neurocontrolled hand prosthesis. Proc IEEE 2010;98:407–17.

[28] Sahin M, Durand D. An interface for nerve recording and stimulation with cuff electrodes. In: Proceedings of the 19th annual international conference of the IEEE engineering in medicine and biology society, vol. 5. Chicago: IEEE; 1997. p. 2004–5.

[29] Sahin M, Durand DM, Haxhiu MA. Chronic recordings of hypoglossal nerve activity in a dog model of upper airway obstruction. J Appl Physiol 1999;87:2197–206.

[30] Loeb GE, Peck RA. Cuff electrodes for chronic stimulation and recording of peripheral nerve activity. J Neurosci Methods 1996;64:95–103.

[31] Nikolić ZM, Popović DB, Stein RB, Kenwell Z. Instrumentation for ENG and EMG recordings in FES systems. IEEE Trans Biomed Eng 1994;41:703–6.

[32] Sahin M. Chronic recordings of hypoglossal nerve in a dog model of Upper Airway Obstruction [Doctoral dissertation]. Cleveland: Case Western Reserve University; 1998.

# Electrodes for electrical conduction block of the peripheral nerve

# 10

**Niloy Bhadra, Tina Vrabec**

*Department of Biomedical Engineering, Case Western Reserve University, Cleveland, Ohio, USA*

## CHAPTER CONTENTS

## 10.1 INTRODUCTION

Many neurological disorders are caused by pathological neuronal firing leading to adverse effects mediated through the innervated end organs. Examples are chronic pain, motor impairments, abnormal postures of the limb and torso, and spasticity due to lesions in the brain or the spinal cord. These pathologies are manifested in stroke, traumatic brain injury, spinal cord injury, cerebral palsy, multiple sclerosis, and amputation neuromas. Medical treatment is based on a hierarchy of increasingly invasive modalities. Pharmacological agents directly targeting the impairments are used in addition to other interventions including physiotherapy, splinting, botulinum toxin injections and implanted baclofen pumps, and in worst case scenarios irreversible surgical procedures. However, there are many side effects with pharmacotherapy. There is still a need for a reliable, fast-acting, nontoxic, and reversible method of nerve block for many of these neurological conditions.

Bhunia et al. Implantable Biomedical Microsystems. http://dx.doi.org/10.1016/B978-0-323-26208-8.00010-8

There are numerous problems with current therapeutic measures. The therapy is usually unable to target specific nerves and therefore has generalized side effects. The time needed to reach a therapeutic effect is very long and also not quickly reversible. Electrical nerve block has features that can fulfill many of these desired specifications. It targets specific peripheral nerves, which are blocked rapidly and unblocked soon after being turned off. Electrical nerve block can be obtained in three distinct ways: collision block, kilohertz frequency alternating current (KHFAC) block, and direct current (DC) nerve block. Collision block stimulates the nerve at a rapid rate and uses antidromic annihilation of action potentials of interest. The clinical use of this technique is limited and will not be discussed further. KHFAC block produces a rapid and reversible nerve block and is potentially safe for long duration applications (Section 10.2). DC blocks action potentials by two different mechanisms—depolarization block and hyperpolarization block—but typically damages the nerve within a few seconds (Section 10.3). Current research has focused on nondamaging, short-duration DC block.

Electrical nerve block is mediated through axonal ion channels. DC block affects the sodium channel gates [1]. In depolarization DC block, the sodium h-gate channel (inactivation parameter) assumes a value of almost zero, which means that probabilistically most of the gates in the block region are inactive. In hyperpolarization DC block, the sodium m-gate channel (activation parameter) approaches a zero value. This indicates a channel-closed state where the probability of the activation gate being open is very low in the blocked region. KHFAC block is mediated through depolarization mechanisms as has been discussed previously [2–4].

KHFAC nerve block is already being evaluated in humans by three companies. Intermittent block of the vagal nerve is being studied for obesity control. KHFAC is being used for spinal cord stimulation for pain control and KHFAC nerve block in the lower limb is being investigated for the amelioration of post-amputation neuroma pain. DC nerve block has not yet been implemented clinically.

## 10.2 KILOHERTZ FREQUENCY ALTERNATING CURRENT (KHFAC) NERVE BLOCK

Innovative research, over the last decade, has examined the possibility of using charge-balanced KHFAC, in the range of 5–40 kHz, as a means for quickly and reversibly blocking peripheral nerve conduction. This has been based on the preliminary work of others, particularly by Bowman and McNeal [5], who performed single-fiber recordings in the cat to examine the behavior of single neurons in high frequencies of stimulation (100 Hz–10 kHz), demonstrating nerve block for frequencies greater than 4 kHz. Using similar *in vivo* experimentation, it has been conclusively demonstrated that high-frequency alternating currents can be used to quickly and reversibly block conduction in the peripheral nerves [6,7]. KHFAC nerve block has been demonstrated in multiple mammalian *in vivo* models, including rat, cat, dog, and monkey [8]. Similar results have now been reported by others [9–12].

The typical experimental setup for evaluating electrical nerve block is shown in Figure 10.1. An example of the typical experimental results using KHFAC nerve block is shown in Figure 10.2. KHFAC neural conduction block can occur as quickly as 50 ms [13]. KHFAC block can be reversed simply by turning the high frequency

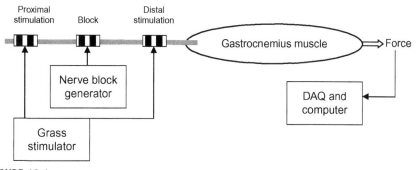

**FIGURE 10.1**

Typical *in vivo* experimental setup for testing nerve block. This example shows the sciatic nerve and gastrocnemius muscle model. Typically, three electrodes are placed on the nerve. The proximal electrode delivers test pulse that result in muscle twitches. The block electrode impedes nerve conduction and therefore eliminates the twitches evoked by PS. DS is used to confirm that the block is under the block electrode and not at the neuromuscular junction.

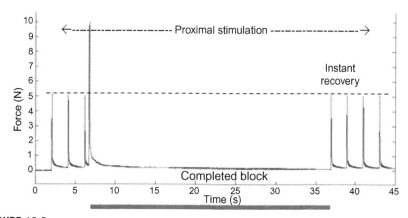

**FIGURE 10.2**

KHFAC block of the rat sciatic nerve at 30 kHz. A test stimulus at 0.5 Hz is delivered proximally to generate muscle twitches (2–43 s). KHFAC block is delivered from 7 s (red bar), blocking conduction. Initially, the KHFAC produces increased activity (the onset response) and then is fully blocked. After 30 s of KHFAC, the block is turned off and twitches return immediately. Twitch height before and after blocks is identical, showing the instantaneous reversibility of this method.

off, and in most cases, the nerve returns to normal conduction in less than a second [6]. It is also possible to obtain a graded/partial KHFAC neural conduction block by using amplitudes that are less than the block threshold (minimum current required to completely block a nerve) [6]. We have determined that the most ideal block characteristics are found at frequencies in the range of 20–40 kHz delivered through electrodes that surround the nerve (nerve cuff electrodes) [14]. Computer simulation studies suggest that the mechanism of the KHFAC-induced conduction block is a depolarization-induced sodium channel inactivation [7,12,15]. There is also some experimental evidence of this depolarization mechanism [2–4,16].

KHFAC produces a transient neural activity when turned on, an effect we have termed the "onset response" [6]. The onset response can take seconds to diminish and cease. If KHFAC block were applied to a mixed nerve in a sensate individual, the onset response would produce a brief painful sensation coupled with a brief muscle spasm. The onset response can be eliminated by combining a low duty-cycle DC block [1] with the KHFAC block [2–4] and will be discussed in the DC block section.

## 10.2.1 ELECTRODES FOR KHFAC NERVE BLOCK

Historically, many different electrode materials, shapes, sizes, and configurations have been used for KHFAC nerve block (Table 1 in [17,18]). Our research has focused on identifying optimal electrodes through predominantly acute *in vivo* experiments. We have designed a new electrode (J-cuff electrode) that has been very successful in KHFAC nerve block. It includes three major features. First, the cross section of the electrode allows for simple placement and removal from a nerve (similar to the placement of a hook electrode). This is particularly useful when placing an electrode in an anatomical region with difficult surgical access. Second, the electrode cuff diameter can easily be modified to ensure a close-fitting electrode–nerve interface at the time of placement on the nerve by reforming the electrode around a mandrel with the desired diameter. Third, the electrode fabrication process allows for the design and fabrication of custom electrode configurations with an arbitrary diameter, number of longitudinal contacts (as many as six in some studies), contact configuration (monopolar or multipolar), and longitudinal dimensions with very low failure rates during fabrication.

This electrode design has been successfully used in multiple whole-nerve studies to date and has been shown to produce robust and repeatable recruitment of nerve fibers during stimulation [17–20]. KHFAC nerve block can be obtained by monopolar, bipolar, or tripolar electrodes. Monopolar electrodes tend to have higher-block threshold and more onset responses. Tripolar electrodes have the least onset response. Bipolar electrodes have been methodically studied [17,18,21] and shown to be very effective. A 0.5 mm interpolar spacing was found to have the lowest block thresholds.

However, KHFAC nerve block can be achieved with many other electrode designs. A recent study examined four electrode types (case spiral cuff electrode, J-cuff electrode (Figure 10.3), FINE (flat-interface nerve electrode), and the Huntington spiral nerve cuff electrode) [22]. All electrodes were successful in producing KHFAC nerve block.

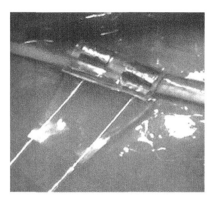

**FIGURE 10.3**

Bipolar J-cuff platinum electrode placed on the rat sciatic nerve.

## 10.3 **DIRECT CURRENT NERVE BLOCK**

Historically, the use of DC nerve block was for the purpose of blocking larger axons so that the function of small axons could be studied independently [23]. This was tested both in the periphery of the sciatic [24–26] and sural nerves [27–31] and in the autonomic system of the vagus [32–35] and phrenic [36] nerves. This property of DC block was also exploited to demonstrate micturition in the paralyzed bladder by preferentially blocking the larger fibers controlling the sphincter to allow micturition to occur [29,37].

A wide variety of electrode types, sizes, and configurations have been explored as shown in Table 1 of Bhadra and Kilgore [1]. Conventional electrode materials such as platinum and Ag/AgCl were used and saline-based electrodes [33,34,38–41]. Saline-based electrodes have more recently been explored as a way of delivering DC to mitigate the onset caused by KHFAC nerve block [2–4]. Predominately, bipolar electrodes have been used, but some tripolar [29,37] and monopolar electrodes [33,34] have been used as well. The electrode geometry varied widely as well with sizes from 0.5 to 6 mm and interelectrode spacing of 1.5 to 20 mm for bipolar/tripolar electrodes.

In most cases, it was presumed that the block occurred under the anode due to a positive potential, but the underlying mechanism was inconsistently reported as either hyperpolarization or depolarization. These contradictions make it difficult to tell what mechanism is actually causing the block in each experiment. In order to clarify these discrepancies, simulation and experiments performed by Bhadra and Kilgore [1] attempted to understand the mechanisms behind DC block. It was demonstrated that a depolarization block can be achieved at a lower current than hyperpolarization, which implies that in most cases depolarization was the underlying mechanism.

In addition to different electrode materials, a large range of waveform parameters were used depending on the requirements of the experimental setup. Pulse durations ranged from 140 ms [25] to look at unmyelinated versus myelinated fiber in the rat sciatic nerve to 10 min [39] in the cat vagus nerve. Current amplitudes needed to block ranged from 0.06 mA [36] in the rabbit phrenic nerve to 0.8 mA [32] in the cat vagus nerve.

## 10.3.1 ELECTRODE CHEMISTRY AND NERVE DAMAGE

Typically, parameters for electrical stimulation use a biphasic charge-balanced waveform with a pulse width below 1 ms. However, to provide a functional DC block, the pulse width would need to be several orders of magnitude larger. Shannon presented a model that specifies safe stimulation parameters based on charge density of the electrode and the charge per phase [42], which shows that longer pulse widths can be implemented provided that the electrode surface area is increased. However, for effective DC block, the electrode would have to be prohibitively large to remain in the safe region.

In order to determine how to develop safe electrodes for DC block, it is necessary to look at the mechanisms at the electrode interface that lead to neural damage. In a metal electrode and the instrumentation that controls it, the charge is carried by electrons. In the body, the charge is carried by ions such as sodium, potassium, and chloride. Different electrode materials utilize a different mechanism to convert the electrodes from the metal into the ions in the body. These mechanisms are furthermore controlled by the magnitude and shape of the stimulus waveform. Both the electrode material and the waveform shape must be designed carefully to avoid a reduction in the conduction of the nerve due to the DC block.

Charge transfer at the electrode/electrolyte interface is caused by one of two mechanisms. In a nonfaradic reaction, no chemical reactions occur and no electrons are transferred into the solution. The transfer of charge occurs by the movement of electrolyte ions, reorientation of solvent dipoles, or adsorption/desorption at the interface. This process is completely reversible. A faradic reaction involves the transfer of electrons from the metal into the electrolyte resulting in the reduction or oxidation of chemical species in the electrolyte. A faradic reaction can be reversible depending on the kinetics of the reaction. Faradic reactions occur at higher currents than nonfaradic reactions.

Neural damage due to stimulation is a result of reactive species that are generated during the stimulation cycle due to nonreversible faradic reactions that cause changes in pH. Examples of this for platinum are the oxide formation and reduction:

Oxidation/reduction: $Pt + H_2O \leftrightarrow PtO + 2H^+ + 2e-$

Reactions that are reversible or "pseudocapacitive" occur when the product of the reaction is bound to the electrode surface allowing it to be recovered when the current is reversed. For platinum, this is the adsorption of hydrogen on the platinum surface:

Hydrogen adsorption: $Pt + H^+ + e^- \leftrightarrow Pt\text{-}H$

The pseudocapacitance of an electrode depends on the effective surface area of the electrode with a "rougher" surface having a higher effective surface area. The pseudocapacitance for bare platinum with a roughness factor of 1.4 is 294 $\mu C/cm^2$ [43].

If the current is applied for a long enough period of time or at a high enough current, the total charge will exceed the pseudocapacitance of the electrode and the oxidation/reduction reaction will occur. Also, note that the total pseudocapacitance for a particular electrode depends on the geometric area of the electrode, so increasing the geometric area increases the acceptable charge/area as described in the Shannon model.

In order to reverse the reaction and restore the electrode potential to its resting value, a biphasic waveform must be used. To completely reverse the reaction, the recharge phase of the waveform must have a total charge that is equal but opposite to the charge of the initial phase. The current level of the recharge does not need to be the same as the initial pulse as long as the total charge is equal. This may be an important design decision depending on the application.

## 10.3.2 TYPES OF HIGH-CAPACITANCE ELECTRODES

In order to generate suitable electrodes for DC block, it is necessary to fabricate electrodes that have a high enough capacitive/pseudocapacitive region to be able to safely apply prolonged DC for block. For platinum, increasing the effective surface area of the electrode is one way to increase this region by providing more sites for hydrogen adsorption to occur. For small increases in capacitance, smooth electrodes have been abraded to create a rougher electrode with more effective surface area. However, this is insufficient for the large increases in charge needed for DC nerve block. Platinum black is a material that is fabricated by plating a bare platinum electrode with a coating of platinum powder thereby increasing the effective surface area. The mechanical properties of platinum black are not appropriate for a chronic implementation, but have been used in acute *in vivo* experiments to demonstrate that increasing the capacitive region can make DC nerve block feasible [44].

While platinum black uses a surface area phenomenon to increase the capacitance of the electrode, other materials increase the capacitance in different ways. In iridium oxide, the bulk iridium oxide material is converted between the Ir(III) and Ir(IV) oxidation states in addition to the charge transfer at the surface [45]. So the reactants from the faradic reaction are contained within the bulk of the material and cannot diffuse into the tissue. Iridium oxide has the potential to provide a more mechanically stable surface than platinum black, and since the reactions are contained in the bulk of the material, the capacitance cannot be effected by the buildup of proteins on the surface of the electrode.

In addition to high-capacitance electrodes, another way to prevent exposure of the tissue to reactive species is to provide a buffer that isolates the reactive species from the tissue. The separated interface neural electrode [2–4] consists of a platinum electrode contained in a saline-filled syringe that then interfaces to the nerve to a polymer nerve cuff via a 15 cm silicone tube. The saline buffers the reactive species preventing them from interfacing with the tissue. This technique was able to increase the cumulative DC delivery time by 6 times without any apparent nerve damage.

Although high-capacitance electrodes have been implemented in microarrays [46–49], few studies have attempted to deliver DC for the length of time needed for useful DC nerve block. Most studies have used high-capacitance materials to fabricate electrodes with very small geometric surface areas that still have reasonable stimulation/recording characteristics. New fabrication techniques were developed to produce the charge capacity needed to deliver multiple seconds of DC delivery.

### 10.3.3 FABRICATION AND TESTING OF HIGH-CAPACITANCE ELECTRODES FOR DC NERVE BLOCK

Monopolar nerve cuff electrodes were fabricated using platinum foil with a platinum lead wire welded to the contact surface. The platinum wire was then welded to a junction that was attached to a stainless steel lead crimped to a gold pin. The foil and lead wire were encapsulated in silicone elastomer and a window was cut on the contact surface to create an interface with the nerve [14]. Rectangular windows were cut to one of three sizes, $3 \times 1$, $3 \times 2$, or $3 \times 3$ mm. Electrodes were either left as bare platinum or electrochemically treated to produce high-capacitance electrodes.

The electrodes were electrochemically cleaned before creating the high-capacitance surfaces using a square wave of $+3$ mA for 10 s and $-3$ mA for 10 s, 25 cycles in 0.1 M $H_2SO_4$. The electrode was then electrochemically treated to create either a platinum black surface or an iridium oxide surface. The platinum black was created using a chloroplatinic acid solution (5 g $H_2PtCl_6$ in 500 ml $H_2O$, with NaCl (2.9 g) and lead acetate (0.3 g)) and applying a galvanic square wave (0 mA "on" current for 5 s, followed by 5 s at open circuit $= 10$ s/cycle) to platinize the surface. The total number of cycles determined the total charge capacity of the electrode.

The iridium oxide surface was created by depositing iridium oxide on the platinum foil via an oxidative process at a constant potential of 0.8 V versus a Ag/AgCl reference electrode (model MF-2052, Bioanalytical Systems Inc.) using an iridium–oxalate solution [50]. The deposition current was roughly constant between 0.8 and 1.0 mA/cm². The deposition process was carried out until a sufficiently large capacitance was obtained (typically 2–3 h).

To determine the capacitive region of the electrode, a cyclic voltammogram for each of the electrodes was generated using a Model 1280B potentiostat, Solartron Inc. (accurate to within 10 mV). The cyclic voltammogram shows the currents that occur when the potential is cycled over a given range. The shape of the curve is unique for different materials and reflects reactions that are occurring at each potential. For platinum and platinum black, the amount of charge that could be safely delivered by these electrodes (the "$Q$ value") was estimated by calculating the charge associated with hydrogen adsorption by integrating between approximately $-0.25$ and $+0.05$ V. For iridium oxide, the $Q$ value was determined by integrating the oxide reduction peaks between 0.125 and 1.2 V (Figure 10.4).

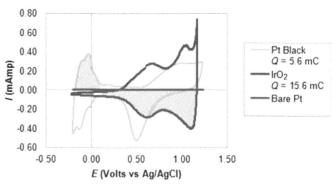

**FIGURE 10.4**

Cyclic voltammograms comparing Pt black, $IrO_2$, and bare platinum.

## 10.3.4 ELECTRODE POTENTIAL MEASUREMENTS

In order to determine if the electrode potential remained in the water window during the *in vivo* experiment, the potential at the DC electrode was measured versus a Ag/AgCl electrode using a saline-filled catheter placed intraperitoneally in the abdomen. A Model 1280B potentiostat, Solartron Inc. (accurate to within 10 mV), with a sampling rate of 1 Hz was used to record the potential. The impedance of the electrode due to the return path through the needle was determined by measuring the change of potential when the recharge step was applied in the middle of the charge-balanced DC (CBDC) waveform. The potential drop due to this impedance was removed from the potential measurements to determine if the water window was breached. The shape of the potential curve shows a larger capacitive effect for the Pt black and $IrO_2$ as compared with the bare Pt electrode (Figure 10.5). The bare Pt shows resistive behavior during the DC plateau, indicating hydrogen evolution.

## 10.3.5 DC WAVEFORM PARAMETERS

For all DC block experiments, a CBDC waveform was used. This waveform consists of a cathodic block phase and an anodic recharge phase as shown in Figure 10.6. The cathodic blocking part of the waveform consisted of a ramp-down phase, plateau phase, and ramp-up phase. The ramp-down and ramp-up phases ensured that no onset firing occurred as a result of the application of the DC [25,27]. The recharge consisted of a square charge-balanced recharge phase. For each trial, waveform parameters were limited to make sure that the total charge delivered was less than the $Q$ value for the particular Pt black electrode. The duration of the recharge phase was selected to balance 100% of the charge at a current level of 10% of the cathodic DC plateau level.

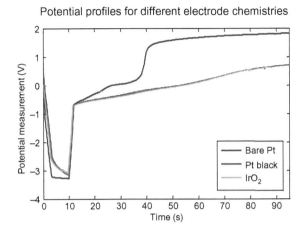

**FIGURE 10.5**

Potential readings for Pt black and $IrO_2$ show that the electrodes remain in the capacitive region.

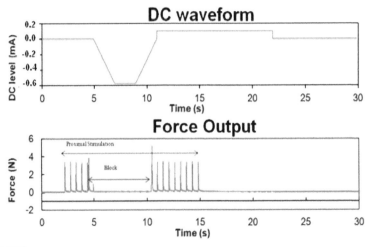

**FIGURE 10.6**

DC waveform delivered and resulting force output. Complete block occurs where the force output goes to zero.

### 10.3.6 DC BLOCK SUCCESS

A complete block was demonstrated by measuring the force produced by stimulation through the proximal electrode. When the force was reduced to a level indistinguishable from baseline noise, the motor part of the sciatic nerve to the gastrocnemius muscle was considered to be "completely blocked" (Figure 10.7). The lowest cathodic amplitude that causes block to occur is referred to as the "block threshold." The block threshold is an important metric to determine if DC nerve block is feasible with high-capacitance electrodes.

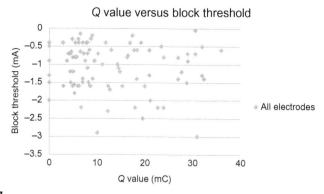

**FIGURE 10.7**

Comparison of block threshold for different $Q$ values.

DC block threshold values were compared with the $Q$ value of the electrodes over all electrode chemistries for 21 different electrodes (Figure 10.7). The correlation coefficient of $Q$ value versus block threshold is −0.065, indicating no correlation between $Q$ value and block threshold. This allows large $Q$ values to be used without a detrimental effect on the DC block threshold.

## 10.3.7 NERVE INTEGRITY TESTING

The proximal stimulation (PS) and distal stimulation (DS) electrodes were used in conjunction to monitor the integrity of nerve conduction over the duration of the experiment. At the start of the experiment, the force resulting from stimulation of each of these electrodes was determined and the ratio of PS to DS was recorded (PS/DS ratio). For a healthy nerve, the PS and DS should produce muscle twitches that were equal in magnitude, making the PS/DS ratio close to one. If the conductivity of the nerve was compromised under the CBDC electrode, the PS/DS ratio would drop below the ratio recorded at the start of the experiment (Figure 10.8). This ratio was recorded periodically during the experiment to monitor nerve conduction.

## 10.3.8 CUMULATIVE DC DELIVERY TESTING

CBDC waveforms were applied in cycles of 10–20 times and the PS/DS ratio was recorded after each multicycle trial. This procedure was repeated until one of three things occurred: The PS/DS ratio dropped below 0.5, 200 CBDC cycles were applied, or the animal expired. As shown in Figure 10.8, the high-capacitive electrodes were able to deliver over 250 mC of charge without a reduction of the nerve conduction while the bare Pt demonstrated a 20% conduction reduction after 125 mC of delivery. This indicates that DC nerve block is feasible provided that electrodes can be developed that have sufficient capacitance to deliver the charge required for a particular application.

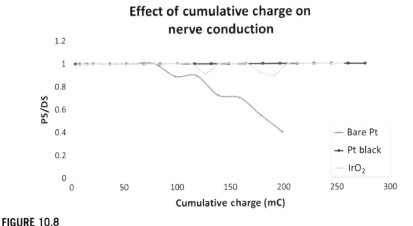

**FIGURE 10.8**

Change in PS/DS ratio for different electrode chemistries.

## 10.4 CONCLUSION

KHFAC block is achievable using multiple electrode designs. DC block requires high charge capacity electrodes to ensure that the blocking currents do not cause the electrode to operate outside its safe water window. While KHFAC nerve block is already being implemented in multiple human studies, DC block has yet to be evaluated in humans. Chronic safety studies and additional electrode chemistries will be explored in the near future and will potentially lead to a short duration electrical block that does not have an onset response.

## REFERENCES

[1] Bhadra N, Kilgore KL. Direct current electrical conduction block of peripheral nerve. IEEE Trans Neural Syst Rehabil Eng 2004;12(3):313–24.

[2] Ackermann DM, Bhadra N, Gerges M, Thomas PJ. Dynamics and sensitivity analysis of high-frequency conduction block. J Neural Eng 2011;8(6):065007.

[3] Ackermann Jr. DM, Bhadra N, Foldes EL, Kilgore KL. Conduction block of whole nerve without onset firing using combined high frequency and direct current. Med Biol Eng Comput 2011;49(2):241–51.

[4] Ackermann Jr. DM, Bhadra N, Foldes EL, Kilgore KL. Separated interface nerve electrode prevents direct current induced nerve damage. J Neurosci Methods 2011;201(1):173–6.

[5] Bowman BR, McNeal DR. Response of single alpha motoneurons to high-frequency pulse trains. Firing behavior and conduction block phenomenon. Appl Neurophysiol 1986;49(3):121–38.

[6] Bhadra N, Kilgore KL. High-frequency electrical conduction block of mammalian peripheral motor nerve. Muscle Nerve 2005;32(6):782–90.

[7] Kilgore K, Bhadra N. Nerve conduction block utilising high-frequency alternating current. Med Biol Eng Comput 2004;42(3):394–406.

[8] Ackermann Jr. DM, Ethier C, Foldes EL, Oby ER, Tyler D, Bauman M, Bhadra N, et al. Electrical conduction block in large nerves: high-frequency current delivery in the nonhuman primate. Muscle Nerve 2011;43(6):897–9.

[9] Gaunt RA, Prochazka A. Transcutaneously coupled, high-frequency electrical stimulation of the pudendal nerve blocks external urethral sphincter contractions. Neurorehabil Neural Repair 2009;23(6):615–26.

[10] Joseph L, Haeffele BD, Butera RJ. Conduction block induced by high frequency AC stimulation in unmyelinated nerves. Conf Proc IEEE Eng Med Biol Soc 2007;2007:1719–22.

[11] Tai C, Roppolo JR, de Groat WC. Block of external urethral sphincter contraction by high frequency electrical stimulation of pudendal nerve. J Urol 2004;172(5 Pt 1):2069–72.

[12] Williamson RP, Andrews BJ. Localized electrical nerve blocking. IEEE Trans Biomed Eng 2005;52(3):362–70.

[13] Foldes EL, Ackermann D, Bhadra N, Kilgore KL. Counted cycles method to quantify the onset response in high-frequency peripheral nerve block. Conf Proc IEEE Eng Med Biol Soc 2009;2009:614–7.

[14] Foldes EL, Ackermann DM, Bhadra N, Kilgore KL, Bhadra N. Design, fabrication and evaluation of a conforming circumpolar peripheral nerve cuff electrode for acute experimental use. J Neurosci Methods 2011;196(1):31–7.

[15] Elbasiouny SM, Mushahwar VK. Modulation of motoneuronal firing behavior after spinal cord injury using intraspinal microstimulation current pulses: a modeling study. J Appl Physiol 2007;103(1):276–86.

[16] Bromm B. Spike frequency of the nodal membrane generated by high-frequency alternating current. Pflügers Archiv Eur J Physiol 1975;353(1):1–19.

[17] Ackermann D, Foldes EL, Bhadra N, Kilgore KL. Electrode design for high frequency block: effect of bipolar separation on block thresholds and the onset response. Conf Proc IEEE Eng Med Biol Soc 2009;2009:654–7.

[18] Ackermann Jr. DM, Foldes EL, Bhadra N, Kilgore KL. Effect of bipolar cuff electrode design on block thresholds in high-frequency electrical neural conduction block. IEEE Trans Neural Syst Rehabil Eng 2009;17(5):469–77.

[19] Bhadra N, Kilgore K, Gustafson KJ. High frequency electrical conduction block of the pudendal nerve. J Neural Eng 2006;3(2):180–7.

[20] Miles JD, Kilgore KL, Bhadra N, Lahowetz EA. Effects of ramped amplitude waveforms on the onset response of high-frequency mammalian nerve block. J Neural Eng 2007;4(4):390–8.

[21] Ackermann Jr. DM, Bhadra N, Foldes EL, Wang XF, Kilgore KL. Effect of nerve cuff electrode geometry on onset response firing in high-frequency nerve conduction block. IEEE Trans Neural Syst Rehabil Eng 2010;18(6):658–65.

[22] Boger A, Bhadra N, Gustafson KJ. Different clinical electrodes achieve similar electrical nerve conduction block. J Neural Eng 2013;10(5):056016.

[23] Whitman JG, Kidd C. The use of direct current to cause selective block of large fibres in peripheral nerves. Br J Anaesth 1975;47(11):1123–33. http://www.ncbi.nlm.nih.gov/pubmed/1218139.

[24] Cangiano A, Lutzemberger L. The action of selectively activated group II muscle afferent fibers on extensor motoneurons. Brain Res 1972;41(2):475–8.

[25] Petruska JC, Hubscher CH, Johnson RD. Anodally focused polarization of peripheral nerve allows discrimination of myelinated and unmyelinated fiber input to brainstem nuclei. Exp Brain Res 1998;121(4):379–90.

[26] Sweeney JD, Mortimer JT. An asymmetric two electrode cuff for generation of unidirectionally propagated action potentials. IEEE Trans Biomed Eng 1986;33(6):541–9.

[27] Accornero N, Bini G, Lenzi GL, Manfredi M. Selective activation of peripheral nerve fibre groups of different diameter by triangular shaped stimulus pulses. J Physiol 1977;273(3):539–60.

[28] Casey KL, Blick M. Observations on anodal polarization of cutaneous nerve. Brain Res 1969;13(1):155–67.

[29] Rijkhoff NJ, Hendrikx LB, van Kerrebroeck PE, Debruyne FM, Wijkstra H. Selective detrusor activation by electrical stimulation of the human sacral nerve roots. Artif Organs 1997;21(3):223–6.

[30] Sassen M, Zimmermann M. Differential blocking of myelinated nerve fibres by transient depolarization. Pflugers Arch 1973;341(3):179–95.

[31] Zimmermann M. Selective activation of C-fibers. Pflugers Arch Gesamte Physiol Menschen Tiere 1968;301(4):329–33.

[32] Coleridge HM, Coleridge JC, Dangel A, Kidd C, Luck JC, Sleight P. Impulses in slowly conducting vagal fibers from afferent endings in the veins, atria, and arteries of dogs and cats. Circ Res 1973;33(1):87–97.

[33] Hopp FA, Seagard JL. Respiratory responses to selective blockade of carotid sinus baroreceptors in the dog. Am J Physiol 1998;275(1 Pt 2):R10–8.

[34] Hopp FA, Zuperku EJ, Coon RL, Kampine JP. Effect of anodal blockade of myelinated fibers on vagal C-fiber afferents. Am J Physiol 1980;239(5):R454–62.

[35] Thoren P, Shepherd JT, Donald DE. Anodal block of medullated cardiopulmonary vagal afferents in cats. J Appl Physiol Respir Environ Exerc Physiol 1977;42(3):461–5.

[36] Sant'Ambrogio G, Dccandia M, Provini L. Diaphragmatic contribution to respiration in the rabbit. J Appl Physiol 1966;21(3):843–7.

[37] Bhadra N, Grunewald V, Creasey G, Mortimer JT. Selective suppression of sphincter activation during sacral anterior nerve root stimulation. Neurourol Urodyn 2002;21(1):55–64.

[38] Fukushima K, Yahara O, Kato M. Differential blocking of motor fibers by direct current. Pflugers Arch 1975;358(3):235–42.

[39] Guz A, Trenchard DW. Pulmonary stretch receptor activity in man: a comparison with dog and cat. J Physiol 1971;213(2):329–43.

[40] Kato M, Fukushima K. Effect of differential blocking of motor axons on antidromic activation of Renshaw cells in the cat. Exp Brain Res 1974;20(2):135–43.

[41] McCloskey DI, Mitchell JH. The use of differential nerve blocking techniques to show that the cardiovascular and respirator reflexes originating in exercising muscle are not mediated by large myelinated afferents. J Anat 1972;111(Pt 2):331–2.

[42] Shannon RV. A model of safe levels for electrical stimulation. IEEE Trans Biomed Eng 1992;39(4):424–6.

[43] Merrill DR, Bikson M, Jefferys JG. Electrical stimulation of excitable tissue: design of efficacious and safe protocols. J Neurosci Methods 2005;141(2):171–98.

[44] Vrabec T, Wainright J, Bhadra N, Bhadra N, Kilgore K. Use of high surface area electrodes for safe delivery of direct current for nerve conduction block. Meeting abstracts. The Electrochemical Society; 2012.

[45] Wang K, Liu CC, Durand DM. Flexible nerve stimulation electrode with iridium oxide sputtered on liquid crystal polymer. IEEE Trans Biomed Eng 2009;56(1):6–14.

[46] Cogan SF, Ehrlich J, Plante TD, Smirnov A, Shire DB, Gingerich M, et al. Sputtered iridium oxide films for neural stimulation electrodes. J Biomed Mater Res B Appl Biomater 2009;89(2):353–61.

[47] Desai SA, Rolston JD, Guo L, Potter SM. Improving impedance of implantable microwire multi-electrode arrays by ultrasonic electroplating of durable platinum black. Front Neuroeng 2010;3:5.

[48] Meyer RD, Cogan SF, Nguyen TH, Rauh RD. Electrodeposited iridium oxide for neural stimulation and recording electrodes. IEEE Trans Neural Syst Rehabil Eng 2001;9(1):2–11.

[49] Negi S, Bhandari R, Rieth L, Solzbacher F. In vitro comparison of sputtered iridium oxide and platinum-coated neural implantable microelectrode arrays. Biomed Mater 2010;5(1):15007.

[50] Elsen HA, Monson CF, Majda M. Effects of electrodeposition conditions and protocol on the properties of iridium oxide pH sensor electrodes. J Electrochem Soc 2009;156(1):F1–6.

# Implantable bladder pressure sensor for chronic application: a case study

# 11

## Steve J.A. Majerus*,†, Paul C. Fletter*, Hui Zhu*,‡,§, Margot S. Damaser*,§,¶

*Advanced Platform Technology Center of Excellence, Louis Stokes Cleveland Department of Veterans Affairs Medical Center, Cleveland, Ohio, USA

†Department of Electrical Engineering and Computer Science, Case Western Reserve University, Cleveland, Ohio, USA

‡Urology Service, Louis Stokes Cleveland Department of Veterans Affairs Medical Center, Cleveland, Ohio, USA

§Glickman Urological and Kidney Institute, Cleveland Clinic, Cleveland, Ohio, USA

¶Department of Biomedical Engineering, Cleveland Clinic, Cleveland, Ohio, USA

## CHAPTER CONTENTS

## 11.1 INTRODUCTION

Control of urine leakage is of major importance to spinal cord injured (SCI) patients and the elderly, with untreated urinary incontinence leading to contact dermatitis and pressure ulcers due to prolonged exposure to excess moisture [1]. In one study, 81%

Bhunia et al. Implantable Biomedical Microsystems. http://dx.doi.org/10.1016/B978-0-323-26208-8.00011-X

of all patients with pressure ulcers had urinary incontinence [2]. In addition, pressure ulcers exposed to urine are more susceptible to infection since the urine alters the pH balance of the skin and reduces resistance to bacterial invasion [1]. Both SCI patients and the elderly are highly susceptible to incontinence [3,4], which is classically divided into two major types: overactive bladder (OAB) and stress urinary incontinence, both of which can lead to uncontrolled urine leakage.

Urinary incontinence is one of the leading causes of morbidity in persons with SCI and is associated with decreased quality of life [5,6] and high financial costs [7–9]. Bladder dysfunction can interfere with relationships and socialization, work, school, and family life and lead to depression, anger, poor self-image, embarrassment, and frustration. This is in part because, in addition to the complications above, SCI also results in bladder hyperreflexia, in which the bladder contracts spontaneously at small fluid volumes, and detrusor sphincter dyssynergia, in which the external urethral sphincter contracts, rather than relaxes, during voiding [10]. Hyperreflexia and dyssynergia cause high-pressure voiding with little urination, large postvoid residual volumes, and low bladder compliance. These factors cause frequent urinary tract infections, bladder trabeculation, ureteric reflux and obstruction, renal infections, long-term renal damage, and episodes of autonomic dysreflexia with dangerous rises in blood pressure, in addition to infected pressure ulcers and septicemia [11]. Therefore, control of urine leakage is of major importance for SCI patients.

Diagnosis of urinary incontinence is typically performed with urodynamics, in which saline is pumped into the bladder to simulate bladder filling, while detrusor pressure is monitored to observe bladder activity. Urologists rely on catheters for bladder pressure measurements, and while catheters have been improved by MEMS sensors and packaging, the shortcomings of catheterization remain. While catheters can successfully be used in acute applications, in chronic applications, catheterization presents risks to the patient (infection and stone formation), limits patient mobility, and reduces quality of life. The use of catheters in urodynamics is also limited because symptomatic bladder leakage is often irreproducible in a static clinical setting [12]. Some bladder symptoms might be induced through ambulation or activities of daily living, and clinicians must rely on patient reporting of these symptoms [13]. Even if a diagnosis is made, long-term confirmation of the correctness and efficacy of treatment is difficult due to the lack of chronic, tether-free bladder pressure sensors.

Neuromodulation has been shown to effectively arrest reflex bladder contractions and OAB activity in patients [14,15]. Existing neuromodulation systems, including the Medtronic InterStim® system, operate in an open-loop fashion by constantly stimulating a set of nerves. The main drawback of such devices is patient habituation to constant stimulation. As the effectiveness wanes, frequent doctor visits are required to adjust stimulation parameters. Closed-loop, or conditional, stimulation is only activated when the bladder state demands it and is more effective than continuous stimulation, leading to increased bladder capacity [16] while using less overall electrical power [17]. Conditional stimulation is often used for acute research purposes using catheter-based systems since an implantable chronic bladder pressure sensor is unavailable [7,18,19].

Both urodynamics and neuromodulation could be enhanced by a wireless, catheter-free, implanted bladder pressure sensor that could send real-time bladder pressure telemetry to an external receiver. Such a device would enable clinicians to properly diagnose and monitor bladder disease in real-world environments. Moreover, the telemetry could be sent directly to another implanted device, such as a modern neurostimulator. The devices could be chronically implanted, permitting closed-loop control of bladder dysfunction.

## 11.2 CHALLENGES AND CONSTRAINTS FOR A CHRONICALLY IMPLANTED BLADDER PRESSURE SENSOR

Development of an implantable pressure sensor for chronic application in the human bladder requires adherence to strict constraints on size, implantation depth, packaging, and overall system efficacy. The requirements for the implanted device in this work were determined through literature review [18] and consultation with Dr. Hui Zhu, Chief of Urology at the Louis Stokes Cleveland VA Medical Center. A summary of the important constraints is listed in Table 11.1.

To minimize patient risk and discomfort, the implanted device must be delivered in a minimally invasive manner using existing urological instruments. Furthermore, the implanted device must be thin enough to fit beneath the mucosa layer, where it is shielded from the urine stream, to prevent mineral encrustation and stone formation. The device powering and telemetry methods must accommodate an implantation depth of up to 20 cm when implanted in obese patients. Finally, the device packaging must be biocompatible and the overall system should not limit patient mobility during ambulatory operation.

## 11.3 WIRELESS IMPLANTABLE MICROMANOMETER CONCEPT

After a thorough review of the clinical requirements for a chronically implantable bladder pressure sensor, a literature search of previously published wireless pressure

**Table 11.1** Summary of Implantable Pressure Sensor Constraints

| Driving Constraint | System Implication |
|---|---|
| Implantation via 24 French[a] cystoscope | Device maximum dimension = 8 mm |
| Submucosal implantation site | Device thickness = 4 mm |
| Implantation in obese patients | Implantation depth = 20 mm |
| Suitable for urodynamics/neuromodulation | Pressure range = 200 cmH$_2$O |
| Reject aliasing of high-frequency motion artifacts | Max. sampling frequency = 100 Hz |
| Chronic implantation (>14 days) | Biocompatible packaging |
| Maximize patient mobility | Wireless operation |

[a]*The French scale is used to define catheter sizes, with the equivalent diameter in mm obtained by dividing the French size by 3. Cystoscopes can be ovoid, so the French size corresponds to the length of the major diameter.*

sensors was performed to further develop the concept. The existing published work mostly focused on measuring vascular, cranial, or ocular pressure using devices that are passive or powered by radio-frequency (RF) electromagnetic fields. Devices intended to measure pressure in the bladder, gastrointestinal tract, or other deep organs are typically designed for acute use, meaning they operate for a period of several days before being forcibly or naturally explanted. Generally, RF-powered devices are less suitable for chronic, real-time monitoring applications requiring an implantation depth (minimum distance from outer surface of skin to implant) greater than about 10 cm, because patient ambulation would be hampered by the bulky RF-powering electronics.

Since the existing body of work did not specifically address the issue of chronic bladder sensing, a concept for the wireless implantable micro-manometer (WIMM) was formed, as illustrated in Figure 11.1.

The specifications for the WIMM were chosen based on the desired functionality, implant location, and implant method and as a compromise between low-power consumption and sensing accuracy, with basic system requirements summarized in Table 11.1 and compared to other relevant sensors. To maximize patient mobility, the WIMM is powered from a chemical battery (as used in acute devices), but RF power transfer is used (as in RF-powered devices) for intermittent periods of RF recharge. Due to the expected implant location within the dorsal wall of the bladder, RF recharge would be most easily accomplished by placing the recharge coil on the patient's lower back, perhaps permitting daily recharge during periods of sleep. The microsystem electronics were designed specifically to meet the size limitations imposed by the application and limited by existing commercially available battery technology. A summary of the proposed specifications for the WIMM relative to existing implantable pressure sensors is shown in Table 11.2.

Compared to existing work, the WIMM was designed for chronic application. By integrating wireless recharge circuitry and designing for a low average current draw, the use of a rechargeable secondary cell is enabled. The inclusion of a shutdown

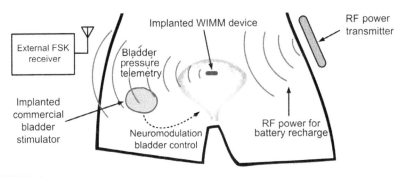

**FIGURE 11.1**

Illustration of WIMM system concept. The WIMM is implanted within the bladder wall and transmits telemetry to an external receiver for control of an implanted bladder stimulator. Intermittent RF recharge is used to maintain the battery charge.

*Modified after Majerus et al. [19].*

**Table 11.2** Proposed Specifications Relative to Existing Implantable Pressure Sensors

| Ref. | Dimensions (mm) | Current Draw | Shutdown Mode | Power Source | Lifespan/ Charge | Wireless Telemetry | Transmit Distance | Sample Rate | Resolution (cmH$_2$O) | Range (cmH$_2$O) |
|---|---|---|---|---|---|---|---|---|---|---|
| [23] | 220×20×4 | 150 µA | N/A | RF | N/A | 133 kHz LSK | 10 cm | 30 Hz | 1.4 | 367 |
| [24] | 29×7×6 | 610 µA | N/A | RF | N/A | 132 kHz LSK | 8.5 cm | 10 Hz | 0.4 | 306 |
| [25] | 25 mm sphere | 417 µA | No | Primary cell | 14 days | 434 MHz FSK | 100 cm | 3 mHz | 3.1 | 351 |
| [26] | 40×5 mm cylinder | 77 µA | No | Primary cell | 7 days | None | NA | 2 Hz | 9.2 | 295 |
| [27] | 27×19×19 | 535 µA | Yes | Primary cell | 2.3 days | 434 MHz ASK | 5 m | 25 Hz | 1.0 | 1022 |
| [28] | 30×4×0.3 | 4.5 nA | No | Capacitor | 1 days | 2.4 GHz OOK | 10 cm | 3 mHz | 1.7 | 82 |
| [29] | 25×3.15 cylinder | 33 µA | No | Secondary cell | 4.2 days | 400–460 MHz QPSK | 10 cm | 90 Hz | 13.6 | 816 |
| WIMM | 17×7×4 | 15 µA | Yes | Secondary cell | 8.3 days | 27.12 MHz FSK/4 MHz ASK | 30 cm | 100 Hz | 0.8 | 2200 |

mode limits battery discharge during packaging, implantation, and inactive periods. The transmission distance of 30 cm was slightly below the average among existing work but is sufficient for transmission through the tissue of obese patients. A pressure-sensing resolution of less than 1 cmH$_2$O is in line with existing implantable systems, but a much larger total dynamic range of 2200 cmH$_2$O ensures accurate operation in a chronic application, where baseline pressure drift may change due to scarification or migration of the implant.

The ultimate vision for the WIMM, as depicted in Figure 11.2, is a wireless pressure sensor that is small enough to be chronically implanted within the bladder wall. Onboard power storage permits greater patient mobility, and the implanted battery can be wirelessly recharged during patient rest periods. An external wireless receiver can detect transmitted pressure telemetry, and an RF-powering source can provide coupled energy for battery recharge. A block diagram of the wireless system is shown in Figure 11.2a.

The WIMM consists primarily of an application-specific integrated circuit (ASIC), pressure sensor, and lithium-ion secondary battery. The ASIC integrates the circuitry for pressure measurement, power management and wireless telemetry, and interfaces directly with the piezoresistive MEMS pressure transducer. Wireless telemetry is transmitted using a discrete inductor; battery recharge is accomplished using another custom-wound coil tuned to a different frequency. The only other microsystem components are a few passive capacitors, used to tune the antennas and to reduce voltage ripple from the battery supply.

The schematic for the microsystem shown in Figure 11.2b depicts the relative sizes of the system components. The battery and associated wireless recharge antenna are the size-limiting factors for the system since they consume over half of the WIMM volume. The implant includes two ferrite rods, which improve the wireless recharge efficiency [20] by shielding the inductive recharge coil from the metal battery casing. All circuit elements and the MEMS pressure transducer are affixed to a system PCB. A rectangular solenoidal coil used for wireless battery recharge is wound around the implant perimeter.

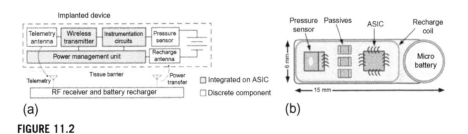

(a)                                                              (b)

**FIGURE 11.2**

Schematic views of the WIMM: (a) system block diagram consisting of implanted device communicating with external RF receiver and battery recharger. A physical layout for the implanted device is illustrated in (b).

## 11.4 IMPLANTABLE MICROSYSTEM DESIGN

The size constraints of the implantable bladder pressure sensor require that the active circuitry be highly integrated while consuming very little power. High integration minimizes off-chip passive components and wire bonds, which require a surprising amount of area within an implantable device (unless flip chip or newer bonding techniques are used). Low power consumption enables the use of a microbattery as the power source for the system. Standard instrumentation and digital circuitry can meet ultralow-power-consumption requirements, but peripheral components such as bias circuitry, regulators, and clock generators are sometimes omitted from these specifications. Furthermore, the power consumption of wireless transmission can eclipse that of the rest of the electronics. The WIMM ASIC was designed for ultralow-power operation at the system level and achieved that goal mainly through precise use of low-duty-cycle operation.

The WIMM ASIC block diagram is presented in Figure 11.3. The ASIC circuitry includes pressure-sensing and telemetry circuits, power control circuits, and RF battery recharge circuits.

The pressure-sensing and telemetry circuits form the instrumentation aspect of the bladder pressure sensor. The circuitry interfaces with a MEMS piezoresistive absolute pressure transducer (EPCOS C29 [21]). The piezoresistive sensor type was selected over capacitive due to greater process maturity and commercial availability. The sensor is excited with an intermittent stimulus to avoid large static power dissipation. A programmable-gain instrumentation amplifier (PG INA) and successive-approximation analog-to-digital converter (SAR ADC) amplify and

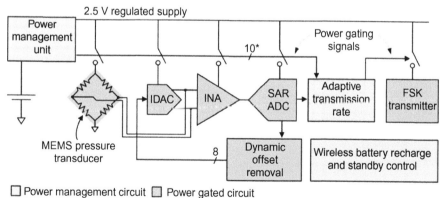

☐ Power management circuit ☐ Power gated circuit

**FIGURE 11.3**

Block diagram of the WIMM ASIC. Circuits are categorized as power management (blue) of power gated (yellow).

*Reproduced with permission from Majerus and Garverick [30].*

convert the transducer output to an 8-bit binary sample. The sample is sent to an auto-offset removal processor, which continuously removes pressure baseline drift to maximize the resolution of the instrumentation system. Finally, each sample is transmitted as a data packet using a frequency shift-keyed (FSK) transmitter and a 3 mm inductive antenna. An adaptive rate transmitter with parameters customized for bladder pressure signals determines the optimum transmission rate of the 27.12 MHz FSK transmitter.

### 11.4.1 POWER MANAGEMENT UNIT

The key power-saving feature of the WIMM ASIC is the power management unit (PMU). The PMU is a suite of very low-power circuits that are always running in the background but sequentially turn on and off the vital instrumentation and telemetry circuits such that power is not consumed when it is not needed. At the minimum duty cycle, the system power consumption is minimized but only 100 samples per second are acquired. This is possible because of the huge speed difference between instrumentation circuitry and the required 100 Hz sampling rate for bladder pressure. The PMU state machine applies power activation to individual circuits, creating a sample pipeline in which sensed information is passed between sequential stages as charge stored on switched capacitors. Stages are switched off after settling the next sample/hold capacitor to minimize the active time for each stage and to reduce IC peak current. The PMU state machine timing diagram illustrated in Figure 11.4 corresponds to the power gating signals in Figure 11.3.

Fast-settling elements, such as the piezoresistive pressure transducer and IDAC, have lower duty factors than switched capacitor and bias circuits, which have longer warm-up, step response, and settling limitations. To reduce peak current draw and RF interference, the FSK transmitter is separately activated for 450 μs after each sample period as determined by the adaptive transmission activity detector described later.

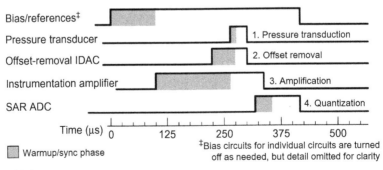

**FIGURE 11.4**

Timing diagram of PMU signals used during sample acquisition (transmission occurs in the next 450 μs). Circuits are sequentially activated to minimize active time per stage.

*This figure was modified from that presented in Ref. [19].*

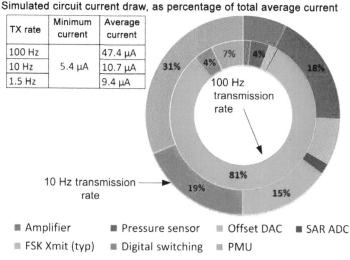

| TX rate | Minimum current | Average current |
|---------|-----------------|-----------------|
| 100 Hz  |                 | 47.4 µA         |
| 10 Hz   | 5.4 µA          | 10.7 µA         |
| 1.5 Hz  |                 | 9.4 µA          |

Simulated circuit current draw, as percentage of total average current

■ Amplifier    ■ Pressure sensor    ▨ Offset DAC    ■ SAR ADC
▨ FSK Xmit (typ)    ■ Digital switching    ■ PMU

**FIGURE 11.5**

Chart showing current used by circuit elements as percentage of total IC current.

*Reproduced with permission from Majerus and Garverick [30].*

The time-averaged current for each instrumentation circuit is limited by a duty factor of 1.4–6.5% per circuit. The average current for the pressure sensor IC is dominated by the transmission rate, as shown in Figure 11.5. The PMU and piezoresistive pressure transducer together account for at least 21% of the current usage and dominate system power when the transmission rate is less than 25 Hz.

## 11.4.2 ADAPTIVE RATE TRANSMITTER

Bladder pressures change quite slowly during filling, but contractions and motion artifacts can lead to occasional rapid pressure increases. An activity detector was designed to modify the transmission rate based on the level of bladder activity. To enable real-time, closed-loop bladder control applications, the activity detector was designed to have zero latency; important samples are transmitted immediately, while unimportant samples are not transmitted. The activity detector block diagram is shown in Figure 11.6.

The activity detector is based on the first- and second-order differences of the signal, which approximate the instantaneous first and second derivatives. These terms contain frequency information that can be used to determine the optimal sample rate [22]. The activity detector was designed to have "attack" and "release" times corresponding to typical bladder contractions. Bladder contractions begin with a sharp rise in pressure, followed by a gradual relaxation back to the resting, baseline pressure. The magnitude comparison sets the filter's attack time or the transmission delay when a sample has been determined to be significant, and this can be adjusted by choosing an appropriate comparison threshold $T$.

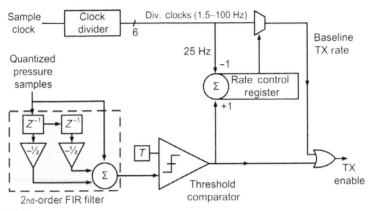

**FIGURE 11.6**

Block diagram of the activity detector.

The activity detector was designed to have a long release time so that samples are sent at a slightly increased rate for a period after a contraction starts. The long release time is controlled by a rate control register, which sets the baseline transmission rate from 1.5 to 100 Hz depending on the level of pressure activity. The rate control register is incremented when the magnitude comparator detects activity and is decremented at a constant rate of 40 ms. Thus, transmission rate remains high for a period of time after activity. Waveforms demonstrating this operation are shown in Figure 11.7 for representative, nonvoiding bladder contractions.

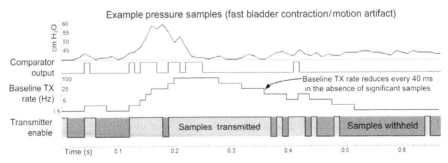

**FIGURE 11.7**

Example waveforms of the activity detector showing increased sample transmission rate during periods of high activity. The slow release of the detector provides increased transmission rates after an event.

*Modified after Majerus and Garverick [30].*

### 11.4.3 **WIRELESS BATTERY CHARGER**

A separate section of the ASIC is devoted to RF wireless battery recharge that operates at 3 MHz to prevent interference with the 27.12 MHz FSK telemetry. The battery recharge circuits capture RF energy that is provided by an external power transmitter in a resonant LC tank circuit and convert the energy to a regulated battery recharging current. The integrated battery recharge circuitry stops charging the microbattery when the capacity is reached and includes voltage-limiting circuitry to protect the system in case more RF energy is received than is needed.

The wireless RF recharge circuits use an inductive coil antenna to pick up energy from an externally applied magnetic field and convert it into a regulated charging current for the implant battery. An equivalent schematic for the RF recharge system is shown in Figure 11.8, and it consists of an external RF power transmitter that couples energy to a resonant LC circuit on the implant. A voltage doubler rectifier with output filter capacitor converts the received energy to a smooth DC voltage, $RFV_{DD}$. A current reference operating from $RFV_{DD}$ recharges the battery at a constant 100 μA rate, as long as the rectified voltage is at least as great as the battery voltage. To protect the electronics, a voltage limiter shunts excess current from $RFV_{DD}$ if the rectified voltage exceeds 4.5 V. Finally, to prevent battery overcharge, a comparator determines if the battery voltage is greater than 3.4 V; charging is prevented if the cell voltage is too great [20].

The operating frequency was chosen to maximize the voltage gain of the wireless recharging system, given by

$$\frac{V_{OUT}}{V_{IN}} = \frac{\omega^2 L_2 M}{R_1 R_2 + (\omega M)^2 + \dfrac{R_1 \omega^2 L_2^2}{R_{LOAD}}} \tag{11.1}$$

The coil parameters were chosen based on overall system size constraints and are listed in Table 11.3. Assuming quality factors greater than 20 for both coils, an operating frequency of 3 MHz was chosen. With a worst-case coupling factor, $k$, of $2 \times 10^{-4}$ (at a coil separation distance of 20 cm), the expected system power transfer

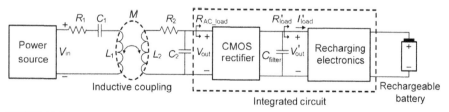

**FIGURE 11.8**

Simplified schematic of RF recharging system including coupling model between internal and external coils.

**Table 11.3** Chosen and Measured Parameters for the RF Recharging System

| Parameter | Name | Design Value |
|---|---|---|
| $\omega$ | Operating frequency | 3 MHz |
| $L_1$ | External coil inductance | 5 μH |
| $L_2$ | Internal coil inductance | 11 μH |
| $k$ | Coupling coefficient | $2 \times 10^{-4}$ |
| $R_{LOAD}$ | Rectifier load impedance | 7.5 kΩ |

efficiency is $7.5 \times 10^{-3}\%$ [20]. At this efficiency, the external coil must dissipate about 10 W of power to accomplish wireless battery recharge.

## 11.5 MICROSYSTEM ASSEMBLY AND PACKAGING

An implantable WIMM prototype was designed by combining the WIMM ASIC, a MEMS pressure sensor, a battery, and various passive components into a compact, waterproof package. A conceptual view of the component arrangement is shown in Figure 11.9. Generally, the implant was designed to be as narrow and thin as possible, so that it could fit through a standard clinical cystoscope.

Assembly of the WIMM implant prototype consisted of PCB preparation, die attachment and wire bonding, protective epoxy encapsulation, solder reflow, and manual soldering steps. First, the PCB was trimmed and sanded to its final dimensions and cleaned with DI water and acetone to remove surface contamination. The ASIC and pressure sensor were mounted and wire bonded using ultrasonic wedge bonding. Next, epoxy encapsulant (Hysol FP4650) was applied to the wire bonds and allowed to partially cure at room temperature for 12 h. Solder reflow paste (Indium 8.9HF-1) was applied to the solder pads and components were hand-placed. The PCB was next loaded into an oven and heated through the recommended solder reflow temperature profile. At this stage, the implant was mostly functional and was quickly bench-tested by attaching power leads. Once functionality was verified, the RF recharge

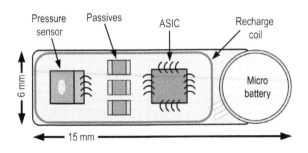

**FIGURE 11.9**

Schematic illustration of WIMM component arrangement.

**FIGURE 11.10**

Photograph of the assembled WIMM prototype prior to encapsulation steps.

coil was hand-soldered to the PCB backside along with the battery +/− terminal connections. The assembled microsystem prior to packaging is shown in Figure 11.10 with a US penny for size reference.

## 11.5.1 MICROSYSTEM PACKAGING

The microsystem packaging technique for the WIMM was evaluated on a number of early, wired prototypes, which were tested in *in vivo* animal models for acute biocompatibility. For acute implantations, polydimethylsiloxane (PDMS), a liquid silicone rubber, is an acceptable vapor barrier with excellent biocompatibility and can be used as the sole encapsulation agent. Uncured PDMS is a poor void filler and provides no mechanical rigidity for the implantable device, so Dow Corning MDX-4210 was selected only as the outermost biocompatible layer for the implantable device. Mechanical stability and additional dielectric insulation were provided by Hysol FP4650, an electronic-grade epoxy with high bulk resistivity, good void-filling characteristics, and low chemical outgassing levels. The epoxy was used to coat and underfill all electronic components, including the wire bonded ASIC and MEMS pressure sensor. A cross-sectional illustration of the encapsulated WIMM prototype is shown in Figure 11.11.

The combination of epoxy and PDMS used to protect the electronics provides a mechanically robust platform, but a "pressure port" had to be created directly above the pressure transducer. This region was designed to be equally biocompatible and waterproof, but more compliant than the surrounding encapsulation. Pressure changes cause the compliant pressure port to deflect, transmitting pressure changes inside the encapsulated device while keeping biological media safely out of contact with the silicon pressure transducer. The outer PDMS encapsulant for the prototype does not stick well to nonsilicones, so the materials used in the pressure transducer packaging

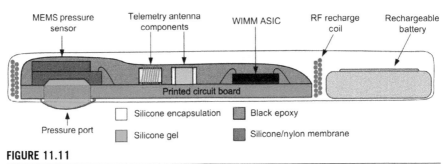

**FIGURE 11.11**

Cross section of the packaged WIMM designed for cystoscopic implantation.

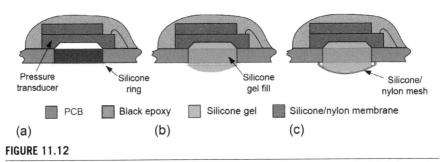

**FIGURE 11.12**

Illustration of construction process for compliant, all-PDMS pressure port in (a)–(c).

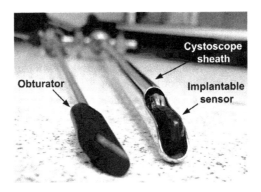

**FIGURE 11.13**

Photo of the encapsulated wireless pressure sensor within the sheath of a urological cystoscope, for delivery into the human bladder.

were largely silicone-based to prevent weak bonding at the PDMS–port interface. An illustration of the pressure port construction process is shown in Figure 11.12.

The device was designed to fit within a clinical cystoscope after packaging, as this is the tool used to access the inner bladder lumen in urological procedures. The multilayer packaging process did not add too much volume to the implant, so a small device size was maintained. The WIMM within the cystoscope sheath is shown in Figure 11.13 to demonstrate the size of the final packaged device.

## 11.6 IMPLANT *IN VIVO* ANIMAL TRIALS

All animal care and experimental procedures were performed according to NIH guidelines and were reviewed and approved by the Institutional Animal Care and Use Committee of Case Western Reserve University and the Louis Stokes Cleveland VA Medical Center. A total of three *in vivo* experiments were performed: a nonsurvival feline, a nonsurvival canine, and an ambulatory acute survival canine. The nonsurvival studies were used to investigate the accuracy of the device in its implantation location, while the ambulatory study sought to investigate how motion artifact would be superimposed on recorded lumen pressure. In these preliminary studies, a wired prototype of the device was used to limit experimental risk. Since the initial animal models were not appropriately sized for cystoscopic implantation, the use of a wired device did not significantly change the invasive implantation procedure. In all experiments, the device was implanted in an identical manner in which the bladder detrusor was cut without damaging the inner urothelial lining. The device was placed under the urothelial lining such that the power cable passed through the detrusor incision, and the organ was sutured shut, as shown in Figure 11.14.

Bladder contractions were successfully recorded in all of the animal models. In the anesthetized feline model, phasic contractions were observed during sacral nerve stimulation. After normalizing the data from the implanted device, the correlation

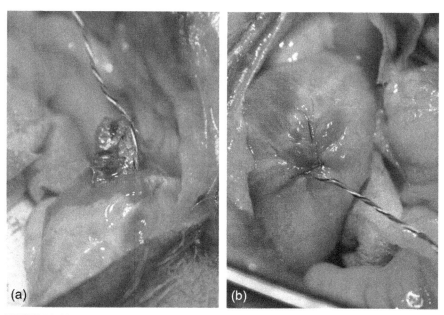

(a) (b)

**FIGURE 11.14**

(a) View of WIMM partially implanted within canine bladder and (b) the fully implanted WIMM with power/data cable leaving sutured detrusor muscle.

*Modified after Majerus et al. [31].*

coefficient between the device and the reference waveforms was at least 0.99 during sacral nerve stimulation. Rapid manual compression of the bladder produced large pressures, which correlated well to the reference catheter ($r > 0.99$). In the anesthetized canine, some variation between the reference and the implanted device was noted. The discrepancy was repeatable over several contractions, with average correlation coefficient of at least 0.87 between the reference and the implant recordings, as shown in Figure 11.15. The likely cause of this discrepancy was a trapped air bubble within the implant pocket; with the WIMM in the lumen, measured pressures correlated strongly between the device and the pressure-sensing catheter ($r > 0.95$).

Ambulatory pressure recordings in the acute implantation canine were collected every other day over 10 days of experimentation. The animal would not tolerate catheterization; therefore, the recordings were collected without a reference catheter in the bladder lumen. Early recordings were consistently within physiological range. Motion artifacts were observed, superimposed on true bladder pressure changes, as shown in Figure 11.16a and b. Voiding and nonvoiding bladder contractions were confirmed by observing animal behavior, and the corresponding device recordings during these periods generally matched expected pressure signal shapes. More variable recordings in a nonphysiological range were found in the later recordings. During necropsy, nonpositional validation revealed correlation ($r > 0.95$) with the pressure catheter, as shown in Figure 11.16b. Significant erosion of the implant through the detrusor was found.

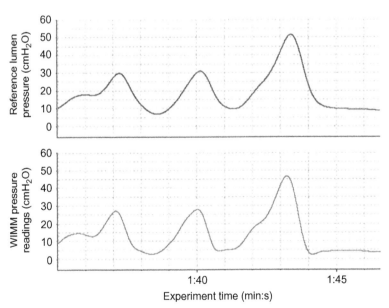

**FIGURE 11.15**

Manually induced bladder pressure changes recorded by the implanted WIMM in a canine model.

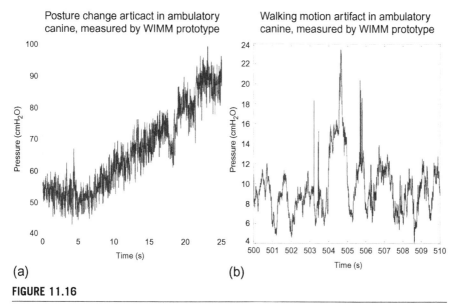

(a)  (b)

**FIGURE 11.16**

(a) Posture change motion artifacts observed in an ambulatory canine model were superimposed on bladder pressure changes, and (b) walking motion artifacts revealed high-frequency content and periodicity.

Although the device was tested in a very limited number of animals ($n=3$), the performance of the wired sensor was encouraging, and the average values of measured correlation coefficients are summarized in Table 11.4.

As evidenced in Figure 11.16, one key issue to consider with ambulatory pressure measurement is the influence of motion artifact. From a diagnostic perspective, the observed motion artifacts are not too troublesome when superimposed on bladder pressure signals, but posture change and periodic artifacts could be misinterpreted as

**Table 11.4** Summary of *In Vivo* Animal Trials Used to Validate WIMM Feasibility

| | | Average Correlation Coefficient | |
|---|---|---|---|
| Trial | Model | Submucosal | Intraluminal |
| 1 | Anesthetized feline | 0.99 | – |
| 2 | Anesthetized canine | 0.89 | 0.98 |
| 3 | Ambulatory canine | – | 0.97 |

legitimate signals. Therefore, sophisticated signal processing may be required for future ambulatory systems in order to qualify artifact from bladder event. Furthermore, the reduced longevity and complications observed with the wired device in an ambulatory model confirmed the need for a wireless implant within deep body organs. Tension caused by wires attached to an implanted sensor within an organ can lead to erosion and migration of the device, with associated serious consequences.

## 11.7 **CONCLUSION**

This case study has demonstrated the feasibility of a wireless, implantable pressure sensor suitable for chronic bladder pressure measurement in humans. WIMM dimensions were chosen to allow for cystoscopic implantation into a suburothelial location within the bladder lumen, where the sensor is shielded from the urine stream, preventing stone formation and permitting chronic use. Unlike previously described implantable pressure sensors, which were RF-powered or provided sub Hz sample rates to save energy, the WIMM is battery-powered to maximize patient mobility and provides 100 Hz sample rates to enable conditional neuromodulation or ambulatory urodynamics. The implant battery is recharged wirelessly using a coil antenna, which may be integrated into a mattress or other cushion to provide unobtrusive charging during periods of patient rest.

Traditional metal/glass encapsulation was not used to avoid RF shielding effects and to reduce the overall implant assembly cost and size. Instead, the implant was encapsulated using nonhermetic, biocompatible polymers. Wired prototype implants were tested in several animal models, and recorded bladder pressure correlated well with intraluminal catheter reference measurements $(0.88 < r \leq 0.99)$. The number of animals in the study was too low to study clinical ramifications; however, complications suffered due to the wired device indicate the need for fully wireless functionality for deep organ implants.

Future pressure-sensing applications for the WIMM are envisioned across a variety of clinical disciplines and organ implantations. Electrically, the WIMM would not require any changes to chronically sense pressure in deep organs such as the stomach, intestines, or anus. For acute applications such as a floating bladder pressure sensor or swallowable sensor for gastrointestinal diagnosis, the WIMM could be powered by a primary cell battery, permitting weeks of operation without requiring the external RF recharger. Furthermore, the WIMM implant could be used as a batteryless, RF-powered device by coupling the output of the RF rectifier to the monolithic voltage regulator. This configuration would be particularly useful in applications where very small size and weight are advantageous, such as an instrumented arterial catheter guide wire, small animal sensor platform, or pressure-sensing "smart" stent/graft. Although the WIMM was designed to measure pressure, the ASIC can interface to a wide array of voltage-output sensors (i.e., for temperature and strain), and the RF rectifier for battery recharge can receive energy from any resonant circuit, such as a piezoresistive energy scavenger. This flexibility would allow the WIMM to work in a variety of other applications where a tiny, low-power sensor would be useful.

## ACKNOWLEDGMENT

The project described was supported by Award Number 1I01RX000443-01A2 from the Rehabilitation Research and Development Service of the VA Office of Research and Development, Case Western Reserve University, and the Cleveland Clinic.

## REFERENCES

[1] Farage M, Miller K, Berardesca E, Maibach H. Incontinence in the aged: contact dermatitis and other cutaneous consequences. Contact Dermatitis 2007;57:211–7.

[2] Cakmak S, Gul U, Ozer S, Yigit Z, Gonu M. Risk factors for pressure ulcers. Adv Skin Wound Care 2009;22:412–5.

[3] Barrois B, Labalette C, Rousseau P, Corbin A, Allaert F, Saumet J. A national prevalence study of pressure ulcers in French hospital inpatients. J Wound Care 2008;17:373–9.

[4] Gelis A, Dupeyron A, Legros P, Benaim C, Pelissier J, Fattal C. Pressure ulcer risk factors in persons with spinal cord injury part 2: the chronic stage. Spinal Cord 2009;47:651–61.

[5] Kachourbous M, Creasey G. Health promotion in motion: improving quality of life for persons with neurogenic bladder and bowel using assistive technology. SCI Nurs 2000;17:125–9.

[6] Wielink G, Essinkbot M, VanKerrebroeck P, Rutten F. Sacral rhizotomies and electrical bladder stimulation in spinal cord injury 2: cost-effectiveness and quality of life analysis. Eur Urol 1997;31:441–6.

[7] Creasey G, Dahlberg J. Economic consequences of an implanted neuroprosthesis for bladder and bowel management. Arch Phys Med Rehabil 2001;82:1520–5.

[8] Creasey G, Kilgore K, Brown-Triolo D, Dahlberg J, Peckham P, Keith M. Reduction of costs of disability using neuroprostheses. Assist Technol 2000;12:67–75.

[9] Stover A, Marnejon J. Postpartum care. Am Fam Physician 1995;52:1465–72.

[10] Watanabe T, Rivas D, Smith R, Staas W, Chancellor M. The effect of urinary tract reconstruction on neurologically impaired women previously treated with an indwelling urethral catheter. J Urol 1996;156:1926–8.

[11] Shingleton W, Bodner D. The development of urologic complications in relationship to bladder pressure in spinal cord injured patients. J Am Paraplegia Soc 1993;16:14–7.

[12] Gupta A, Defreitas G, Lemack G. The reproducibility of urodynamic findings in healthy female volunteers: results of repeated studies in the same setting and after short-term follow-up. Neurourol Urodyn 2004;23:311–6.

[13] Chapple C. Primer: questionnaires versus urodynamics in the evaluation of lower urinary tract dysfunction- one, both, or none? Nat Clin Pract Urol 2005;2:555–64.

[14] Previnaire J, Soler J, Perrigot M, Boileau G, Delahaye H, Schumacker P, et al. Short-term effect of pudendal nerve electrical stimulation on detrusor hyperreflexia in spinal cord injury patients: importance of current strength. Paraplegia 1996;34:95–9.

[15] Wheeler J, Walter J, Zaszczurynski P. Bladder inhibition by penile nerve stimulation in spinal cord injury patients. J Urol 1992;147:100–3.

[16] Wenzel B, Boggs J, Gustafson K, Grill W. Closed loop electrical control of urinary incontinence. J Urol 2006;175:1559–63.

[17] Horvath E, Yoo P, Amundsen C, Webster G, Grill W. Conditional and continuous electrical stimulation increase cystometric capacity in persons with spinal cord injury. Neurourol Urodyn 2009;29(3):401–7.

[18] Cooper M, Fletter P, Zaszczurynski P, Damaser M. Comparison of air-charged and water-filled urodynamic pressure measurement catheters. Neurourol Urodyn 2011;30:329–34.

[19] Majerus S, Garverick S, Suster M, Fletter P, Damaser M. Wireless, ultra-low-power implantable sensor for chronic bladder pressure monitoring. ACM J Emerg Technol 2012;8(2):11.1–13.

[20] Suster MA, Young DJ. Wireless recharging of battery over large distance for implantable bladder pressure chronic monitoring. In: 16th international solid-state sensors, actuators and microsystems conference, Beijing; 2011.

[21] EPCOS AG. Pressure sensors—C29 series. http://www.epcos.com/inf/57/ds/c29_abs.pdf; 2009 [accessed 28.08.13].

[22] Rieger R, Taylor JT. An adaptive sampling system for sensor nodes in body area networks. IEEE Trans Neural Syst Rehabil Eng 2009;17(2):183–9.

[23] Cleven NJ, Müntjes JA, Fassbender H, Urban U, Görtz M, Vogt H, et al. A novel fully implantable wireless sensor system for monitoring hypertension patients. IEEE Trans Biomed Eng 2012;59(11):3124–30.

[24] Coosemans J, Puers R. An autonomous bladder pressure monitoring system. Sens Actuators A 2005;123(April):151–61.

[25] Wang C, Huang C, Liou J, Huang I, Li C, Lee Y, et al. A mini-invasive long-term bladder urine pressure measurement ASIC and system. IEEE Trans Biomed Circuits Syst 2008;2(1):44–9.

[26] Axisa F, Jourand P, Rymarczyk-Machal M, Smet ND, Schacht E, Vanfleteren J, et al. Design and fabrication of a low cost implantable bladder pressure monitor. In: Annual international conference of the IEEE engineering in medicine and biology society (EMBC); 2009.

[27] Valdastri P, Menciassi A, Arena A, Caccamo C, Dario P. An implantable telemetry platform system for in vivo monitoring of physiological parameters. IEEE Trans Inf Technol Biomed 2004;8(3):271–8.

[28] Chow EY, Chlebowski AL, Irazoqui PP. A miniature-impantable RF-wireless active glaucoma intraocular pressure monitor. IEEE Trans Biomed Circuits Syst 2010;4(6):340–9.

[29] Schulman J. The feasible FES system: battery powered BION stimulator. Proc IEEE 2008;96(7):1226–39.

[30] Majerus S, Garverick S. Power management circuits for a 15-μA, implantable pressure sensor. In: Custom integrated circuits conference, San Diego; 2013.

[31] Majerus S, Fletter P, Damaser M, Garverick S. Low-power wireless micromanometer system for acute and chronic bladder-pressure monitoring. IEEE Trans Biomed Eng 2011;58(3):763–8.

# Neural recording interfaces for intracortical implants*

# 12

**Manuel Delgado-Restituto, Alberto Rodríguez-Pérez**

*Institute of Microelectronics of Sevilla, Sevilla, Spain*
*University of Seville, Sevilla, Spain*

## CHAPTER CONTENTS

## 12.1 INTRODUCTION

In the last years, there has been a growing interest on the design of multichannel neural recording interfaces with wireless transmission capabilities for the untethered measurement of brain activity [1–3]. These interfaces are expected to play a significant role in both clinical (as part of therapeutic procedures in patients with neurological diseases) and neuroscience (such as brain–machine interfaces) applications. As these recording interfaces are implanted below the skull, the use of ultralow-power-consumption techniques is mandatory, not only to prevent from harmful effects in

*This work has been supported by the Spanish Ministry of Economy and Competitiveness under Grant TEC2012-33634 and the FEDER Program.

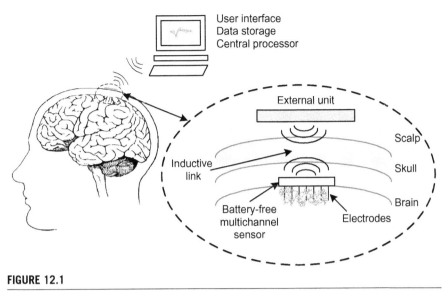

**FIGURE 12.1**

Implanted solution of the wireless neural array.

the brain but also to avoid the need for batteries. Thus, by making the power dissipation low, it becomes feasible to use energy-harvesting strategies for supplying the implant. This is illustrated in Figure 12.1, in which the intracranial device is powered via a wireless inductive link from an external unit placed on the head. The same link or a dedicated one could be also employed for data transfer to such external unit from where information could be communicated to a specific hub for compiling and processing the recorded brain activity.

This chapter aims to contribute to this scenario and presents a multichannel wireless neural sensor array designed in a standard $0.13\,\mu m$ CMOS process. It is composed of 64 channels in which neural signals are acquired, filtered, digitized, and optionally compressed. The system has two transmission modes; in one case, the information captured from a selected set of channels is transmitted as uncompressed raw data; in the other, feature vectors are extracted from the detected neural spikes at every channel and transmitted to the external unit for further processing. A single wireless inductive link, inspired on RFID technologies, is used both for powering the implant and for data transfer to/from the external unit. This link uses a $40.68\,MHz$ carrier signal and employs on–off keying (OOK) modulation for data transfer from the external unit to the implant (forward link) and load-shift keying (LSK) in the reverse direction (backward link). A $4\,MHz$ clock is used to send information through the backward link. This is enough for the implant operated in the feature extraction mode to characterize and serialize the detected spikes even in the unlikely case all the channels fire at the same instant. The total power consumption of the system operated in feature extraction mode, including the recording array and the communication protocol, is only $377\,\mu W$.

The chapter is organized as follows. First, a review of the state of the art that summarizes the evolution and the current tendencies in this topic is given in Section 12.2.

Details of the proposed neural sensor array and of the individual recording channels are presented in Sections 12.3 and 12.4, respectively. Section 12.5 describes the design of the RF front end, while the experimental results are reported in Section 12.6. Finally, Section 12.7 ends the chapter with some conclusions.

## 12.2 STATE-OF-THE-ART REVIEW

The increasing interest of the medical community in the study of brain activity through implantable devices has propelled an emerging research area among electrical engineers focused on the development of innovative solutions for implantable biosensors. In this new discipline, designers have to face aspects that are not a priority in other fields of the electronics where emphasis is placed on high-speed solutions or the reduction of the transistor sizes. In contrast, these implantable biomedical devices and sensors for neural sensing need to solve the following challenges:

- Biocompatibility. Implantable neural recording systems have not to elicit any undesirable local or systemic effects in the brain tissue. It means that all the exposed parts of the implant, including the microelectrodes, have to be made, sealed, or caged with appropriate materials [4].
- Miniaturization. Systems should exhibit small form factor to reduce the invasiveness of the implant. This calls for integrated prototypes, as discrete solutions would require an unacceptable volume occupation.
- Heating. The implanted device should not raise the temperature in the surroundings of the implant above 1 °C to not cause severe damages on the tissue. Translated to a circuit-level specification, it means that the maximum power density of the implanted circuitry is restricted to approximately $10\,mW/cm^2$ [5].
- Wireless transmission. First neurocortical recorders used wired connections through the skull to extract information from the brain and power the implant. This clearly represents a potential focus of infections and malfunctions and precludes the use of these devices for long-term implants. A wireless link for power/data telemetry is definitely a better solution.
- Power consumption and energy scavenging. The use of wireless telemetry demands a careful power management of the implant in order to guarantee a long lifetime. Given the limitations for circuit replacement, management has to be complemented with energy-harvesting techniques able to recharge internal batteries or supply the implant as a whole. This opens a new and interesting field of study that could lead to completely autonomous implanted systems.
- Data compression. Related to the optimization of the power consumption, the amount of data to be transmitted out through a wireless link has to be reduced as much as possible. This need is particularly noticeable in implants with a large number of recording channels. In these cases, data reduction techniques aiming to compress the captured information have to be integrated on-chip.

This section will review some of the existing prototypes that have faced the aforementioned challenges. Special emphasis will be paid in solutions providing wireless telemetry with data compression capabilities.

## 12.2.1 WIRELESS MULTICHANNEL NEUROCORTICAL RECORDING SYSTEMS

One of the first integrated circuits designed for the recording of the neurocortical signals captured from an array of electrodes was reported in the 1980s by Wise *et al.* [6,7]. Figure 12.2 shows the basic structure and different scanning electron microscope (SEM) views of the implemented neural probes. The probe consisted of a micromachined substrate, with conducting leads insulated by inorganic dielectrics and recording/stimulating sites distributed along the structure [8]. Each electrode was directly connected to a dedicated front-end circuit to amplify and filter the acquired signal. These conditioned signals were then multiplexed and sent out to an external processor placed out of the probe. Originally, wires were used for both powering the active circuitry and transmitting the acquired information.

It was not until the late 1990s when Najafi *et al.* [9] presented a wireless system for neural recording that used a passive silicon sieve electrode with multiple recording sites [10]. The system was used in studies of peripheral nerve regeneration. The implanted circuitry included the necessary circuitry to filter and amplify the neural information captured by the recording electrodes, an analog-to-digital converter (ADC) to digitize the data, and a controller to configure the different blocks. As a major novelty, the system included a power telemetry system able to supply the implanted system and a wireless transmitter to communicate the implanted device to an external processor. The main limitation was the large power consumption of the system (around 90 mW in 24 mm²), which compromised the maximum power density restriction for not damaging the surrounding brain tissues.

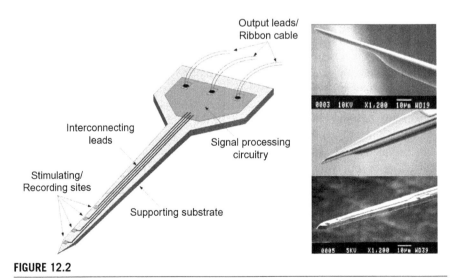

**FIGURE 12.2**

Active multielectrode probe for recording or stimulation in the central nervous system.

The Utah Intracortical Electrode Array [11], manufactured using silicon micromachining techniques, marked a new milestone in the design of multichannel neural recording solutions. Compared to the microprobe of Figure 12.2, the Utah array depicted in Figure 12.3 allows to simultaneously record the activity of neurons, which are spatially distributed across the brain cortex, and not only in just one specific spot.

Table 12.1 summarizes some of the published works that implement complete wireless solutions for multichannel neurocortical recordings. Harrison *et al.* [12] developed a 100-channel neural recording system prototype that was *flip-chip* connected to a Utah array through dedicated internal pads aligned to the electrodes. The prototype amplifies and time-multiplexes the neural signals from each electrode, digitizes selected waveforms by an integrated ADC, and has the capability to detect the occurrence of neural spikes. An embedded digital processor arranges the generated information for wireless transmission through an integrated 433 MHz frequency-shift keying (FSK) transmitter. Additionally, an inductive link is employed for powering the implant. The system consumes 13.5 mW, with a mean of 135 µW/ch.

After this work, others similar appeared in which the acquisition circuitry was arranged in arrays. In 2009, Sodagar *et al.* [13] developed an integrated 64-channel recording array, with a similar spike detection approach and with an inductively powered scheme. The main difference is in the wireless telemetry, using FSK modulation at 2 MHz for the forward transmission and OOK at a programmable frequency between 70 and 200 MHz for the reverse telemetry. This system is not integrated in single die; the signal acquisition, the signal processing, and the RF telemetry are implemented in different chips and then joined in a common platform, which occupies 1.4 cm × 1.55 cm. The total power dissipation of the microsystem is 14.4 mW, with a mean of 225 µW/ch.

**FIGURE 12.3**

Views of the Utah intracortical electrode array.

**Table 12.1** Wireless Multichannel Recording Systems State of the Art

| | [12] | [13] | [17] | [18] | [19] | [20] | [21] |
|---|---|---|---|---|---|---|---|
| CMOS process (μm) | 0.5 | 0.5 | 0.35 | 0.35 | 0.5 | 0.13 | 0.13 |
| Supply voltage (V) | 3.3 | 1.8 | 3.3 | 3 | 3 | 1.2 | 1.2 |
| Number of channels | 100 | 64 | 128 | 64 | 32 | 64 | 96 |
| Area occupation (mm²) | 4.7×5.9 | N/A | 8.8×7.2 | 3.1×2.7 | 4.93×3.3 | 4×3 | 5×5 |
| Total power (mW) | 13.5 | 14.4 | 6 | 17.2 | 5.85 | 5.03 | 6.5 |
| Power per channel (μW) | 135 | 225 | 46.87 | 269 | 182.82 | 78.6 | 67.7 |
| Amplifier gain (dB) | 39.5 | 60 | 40 | 65–83 | 67.8–78 | 54–60 | 56 |
| Low-frequency corner (Hz) | 0.025 | 10 | 0.1 | 1 | 1 | 10 | 280 |
| High-frequency corner (kHz) | 7.2 | 10 | 20 | 10 | 10 | 5 | 10 |
| Input-referred noise ($\mu V_{rms}$) | 2.2 | 8 | 4.9 | 3.05 | 4.62 | 6.5 | 2.2 |
| ADC ENOB (bits) | 9[a] | 8[a] | 9[a] | 7.2 | 8.1 | 7.8 | 10[a] |
| Sampling freq. (k-Hz) | 15 | 62 | 40 | 20 | N/A | 54 | 31.25 |
| Power source | Inductive | Inductive | Battery | Battery | Inductive | Battery | Battery |
| Fwd data telemetry carrier frequency (MHz) | ASK | OOK | N/A | OOK | | FSK | UWB |
| 2.64 | | | | | | | |
| 4–8 | | | | | | | |
| 915 | | | | | | | |
| Rvrs data telemetry | FSK | OOK | UWB | FSK | FSK | FSK | UWB |
| Carrier frequency (MHz) | 433 | 70–200 | 4000 | 400 | 915 | 915 | 30 |
| Max. output data bit rate (Mbps) | 0.35 | 2 | 90 | 1.25 | 0.71 | 1.5 | N/A |
| Data compression | Binary spike detection | Binary spike detection | On-the-fly extraction | Spike reconst. | PWM | FIR filters | |

[a]Output resolution.

Detecting the occurrence of spikes is not enough in some applications, which require information on the shape of the waveforms to perform spike sorting [14–16]. For that reason, an on-the-fly spike detection and spike-sorting solution were proposed in Ref. [17]. This prototype implements a 128-channel neural recording array with a structure similar to Harrison *et al.* [12] and Sodagar *et al.* [13]. Each neural channel includes a preamplification and filtering circuitry, which is multiplexed prior to the ADC. The digitized signal is then analyzed by a digital signal processing, which performs spike detection, feature extraction, and spike-sorting tasks. The system also offers the possibility of extracting raw data from all the channels by means of an ultra-wideband (UWB) transmitter at 4 GHz. It is remarkable to note the high achievable output data bit rate of this transmitter, 90 Mbps. The total power consumed by the system is 6 mW, that is, 46.87 µW/ch. No energy-harvesting circuit is included, so the system is supplied by batteries.

The multichannel system proposed by Bonfanti *et al.* [18] records neural signals from a 64-channel array (from which only 16 of them are available). It uses a 400 MHz FSK wireless link to transmit the information to the output. The configuration of the array is pretty similar to other works; each channel contains the necessary circuitry to adapt the acquired signals, while a shared ADC is in charge of digitizing the time-multiplexed analog outputs. In this work, the neural data are compressed by detecting the neural spikes and transmitting just the sampled information of their waveforms. The system consumes 17.2 mW and thus 269 µW/ch.

The work presented by Ghovanloo *et al.* [19] consists of an inductively powered 32-channel neural recording system with a 915 MHz FSK wireless link for data transmission and a 13.56 MHz link for powering. In this work, instead of using a standard ADC, the conditioned neural signals from the channels are used to construct a pulse width-modulated (PWM) signal from a predefined triangular waveform, in a sort of analog-to-time conversion. Each of these modulated signals is multiplexed in time and afterward transmitted to the output by means of the communication wireless link. It consumes 5.85 mW, that is, 182.82 µW/ch, on average.

A recent work reported a 64-channel neural recording interface with a fully integrated 915 MHz FSK/OOK transmitter [20]. The main innovation is the inclusion of a channel-embedded FIR filter to separate various bands in the spectrum. In this work, channels embed all the standard front-end circuitry, together with an ADC, which can also work also as FIR filter. The system can be configured either to extract raw information from the neural electrodes or to filter just the main frequency components. The system is powered by batteries and consumes 5.03 mW in total, 78.6 µW/ch.

Another multichannel system was presented by Gao *et al.* [21]. It consists of a 96-channel neural interface with a UWB transmitter (not integrated with the neural recording interface). The distribution of the channels follows the same approximation as in the latest work, as it has a dedicated ADC per channel. This system is able to transmit raw data information from all the channels of the array at a 30 Mbps output data bit rate. The total power consumption is 6.5 mW, 67.7 µW/ch.

From this review of prior art, the following conclusions can be extracted:

- The analog front-end (AFE) circuitry is, in most of the cases, organized in arrays with a separation distance similar to the pitch the underneath multielectrode array is connected to. This front-end circuitry usually consists of a band-pass filter (BPF) and a low-noise amplifier (LNA). The bandwidth of the former depends on the signal to be captured (local field potential or action potentials). In many cases, it can be configured by the user. The midband gains moved from 45 to 90 dB, while the input-referred noise must be always kept below $5\,\mu V_{rms}$ or lower, if possible.
- The conversion to the digital domain of the acquired signal is usually performed by an ADC with 8–10 bits of resolution. This ADC is, in most cases, shared by many channels. The preferred sampling conversion frequency is around 30 kS/s if spikes are to be detected and characterized.
- Data compression techniques are often embedded in the sensor interfaces to reduce the generated output data bit rate.
- Most often, inductive coupling is used as energy-harvesting strategy. Although it has some disadvantages, such as the big area requirements of the inductors and the deleterious impact of misalignment between the primary and secondary coils, it is still the only harvesting technique that has been successfully tested in an implanted device. In most of the cases, the carrier used for powering is different from the one used for communication.
- FSK transmitters at 400 or 900 MHz are the preferred solutions for the wireless transmission of the neural sensed information. However, in many cases, the achievable output data bit rate is not enough for the simultaneous transmission of the raw data generated by all the channels. Alternatively, another transmission technique, the UWB, is used to increase the maximum output data bit rate.

Although the last two points present very promising and interesting research opportunities, in this work, focus is put on the development of new and efficient solutions for the analog and mixed-signal circuits that hoard the most significant part of the total power consumption of the system, that is, the signal interfacing circuitry.

## 12.2.2 AFE INTERFACES FOR MULTICHANNEL SYSTEMS

The AFE circuitry of a neural recording channel invariably contains means for filtering and amplifying the weak signal acquired from the electrode. Most often, these means include a BPF and a LNA, followed by a programmable-gain amplifier (PGA) to maximize the output swing of the interface circuit before digitization. Depending on the position of the ADC within the recording chain, two basic channel structures can be found. In Figure 12.4a, the analog outputs from a set of channels are time-multiplexed and converted into digital words by a shared ADC placed in the periphery of the channels. Then, the generated data frames can be further compressed

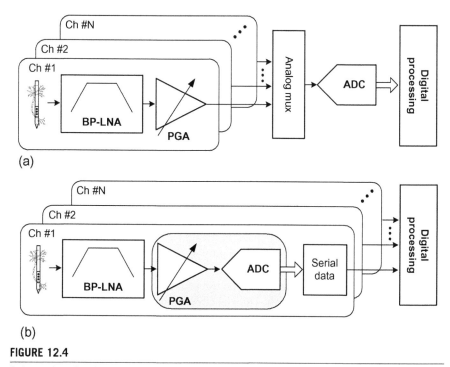

(a)

(b)

**FIGURE 12.4**

Schematics of typical neural sensor interfaces: (a) ADC sharing, (b) ADC in-channel.

or transmitted as raw data. In Figure 12.4b, instead of sharing an ADC between different channels, each recording site owns a dedicated converter. A common digital processor manages the digitized signals, classifies them, and sends them out. In terms of area occupation, this solution is obviously worse, as the area occupied by the channels is bigger compared to the ADC sharing topology. However, it has some benefits concerning power consumption [22]:

- The information sent through the array of the channels is digital, which is easily buffered and regenerated. In case of the ADC sharing topology, each channel should embed an additional analog buffer after the PGA (or redesign it accordingly) to send the analog signal across the array of neural channels to the analog multiplexer.
- In the ADC sharing topology, the sampling frequency of the A/D conversion is multiplied by the number of channels that share it ($N$). Considering the commonly employed SAR ADC topology, in which the total power consumption of the conversion is directly proportional to the sampling frequency, there is no difference, in terms of power consumption, between embedding the ADC in the channel and sharing it between $N$ channels with a sampling frequency $N$ times higher. However, a significant difference occurs in

the preceding circuitry, which must load the analog signal for the conversion
$N$ times faster in the ADC sharing scheme. Then, the ADC sharing topology
imposes much more restrictive slew-rate requirements to the analog circuitry
that loads the input of the ADC.
- The system is easily scalable by just replicating the channels in a form of array.

In the literature, as summarized in Table 12.2, examples of both strategies can be
found. Although the tendency is to reduce the area occupied per channel, the number
of shared ADCs per channel is getting lower to have a better power performance.
A meaningful example of this fact can be seen in Ref. [23], which reports a great
performance of 1.13 µW/ch. In this work, each of the outputs of the channels is connected to a dedicated S&H. Then, the outputs of these S&Hs are multiplexed to an
ADC for the digitization of the samples. Thus, although the ADC seems to be shared
between 8 channels, in fact, part of it (the S&H) is embedded in each of the channels.

Most often, SAR ADCs with capacitive-based DACs are used for signal conversion, but other approaches have been also proposed. For instance, Gosselin *et al.* [1]
used an alternative ADC circuit, based on a switched-capacitor binary search algorithm, and Muller *et al.* [24] proposed the use of a counter-based ADC. These two
works integrate the ADC inside the channel circuitry in a very low area.

Concerning the implementation of the filtering and amplification functions, one
of the preferred topologies is the one proposed by Harrison and Charles in Ref. [25].
It is based on a continuous-time feedback capacitive BPF that obtains very good
noise and power performances. In some case, derived topologies are employed (as
in Ref. [1]), or even, completely different strategies, as in Ref. [21], which uses an
SC-based filter preceded by an open-loop transconductor. However, better solutions
in terms of area, noise, and power consumption, like the one presented by Liew
*et al.* [23] or Al-Ashmouny *et al.* [26], with 1.13 and 3.3 µW/ch, respectively, use
the original feedback capacitive topology. From Table 12.2, it can be seen there is
a direct relationship between area and power consumption, on the one hand, and
input-referred noise, on the other. If this latter drops off, the first two grow up. This
is clearly illustrated by the works [21,27], which obtain the lowest noise figures, but
the highest power consumptions. In Ref. [28], an analytic study reflects the impact
of these factors in the design performance and analyzes the fundamental limits for a
given topology.

## 12.3 NEURAL SENSOR ARCHITECTURE

Figure 12.5 shows the architecture of the proposed system. It consists of an 8×8
neural recording array, each of them serially connected to an event-based processor
unit (EBPU), which stores the information generated by the channel. The data stored
in these EBPUs are read and classified by an embedded digital processor, which also
handles the timing of the implant. A communication block implements the link to/
from the external unit. Additionally, the system includes one tunable direct digital
frequency synthesizer (DDFS) per row for calibration purposes.

**Table 12.2** Analog Front End of Multichannel Neural Sensors State of the Art

| | [1] | [17] | [20] | [21] | [27] | [32] | [26] | [23] | [24] |
|---|---|---|---|---|---|---|---|---|---|
| CMOS process (µm) | 0.18 | 0.35 | 0.13 | 0.13 | 0.6 | 0.18 | 0.35 | 0.13 | 0.065 |
| Supply voltage (V) | 1.8 | 3.3 | 1.2 | 1.2 | 3.3 | 1.8 | 0.9 | 0.5 | 0.5 |
| Area/channel (mm²) | 0.1 | 0.5 | 0.09 | 0.38 | 0.16[a] | 0.035 | 0.07 | 0.073 | 0.013 |
| Low-freq. corner (Hz) | 100 | 0.1–200 | 0.1–10 | 1–180 | 250 | 350 | 0.1–1000 | 40–400 | 300 |
| High-freq. corner (kHz) | 9.2 | 2–20 | 5 | 10 | 5 | 11.7 | 1–17 | 7.5 | 10 |
| Gain (dB) | 70 | 57–60 | 54–60 | 40–56 | 60 | 49–66 | 52.4–79.8 | 48–54 | 32 |
| Input-ref. noise ($\mu V_{rms}$) | 5.8 | 4.9 | 6.5 | 2.2 | 2.2 | 5.4 | 6.76 | 5.32 | 4.9 |
| Channels/ADC | 1 | 16 | 1 | 1 | 100 | 4 | 16 | 8[b] | 1 |
| ENOB (bits) | 7[c] | 9[c] | 8[c] | 9.72 | 10[c] | 7.65 | 7[c] | 7.32 | 8[c] |
| Sampling freq. (kS/s) | 30 | 40 | 57 | 31.25 | 17.5 | 31.25 | 20.2 | 30 | 20 |
| Power/channel (µW) | 42.5 | 34 | 6.3 | 68 | 80 | 10.1 | 3.3 | 1.13 | 5.04 |
| Feature extraction bits per spike | Ana. spike det. 512 bits | Dig. spike det. 54 bits | FIR filter | N/A | Ana. spike det. 1 bit | N/A | N/A | N/A | N/A |
| Year of publication | 2009 | 2009 | 2011 | 2012 | 2009 | 2011 | 2011 | 2011 | 2011 |

[a]This area does not take into account the area of the ADC.
[b]The S&H input of each ADC is dedicated for each channel.
[c]Output resolution.

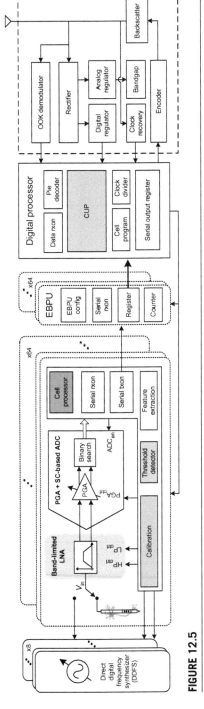

**FIGURE 12.5**

Architecture of the multichannel neural array.

Each channel embeds all the needed circuitry to acquire and digitize neural wave-forms including a LNA, a digitally tunable BPF, a PGA, an ADC, and a local digital processor to detect neural spikes and extract their features. The channel architecture is similar to that in Ref. [29], but in this version, the spike detection is accomplished in digital domain and the decision threshold is adaptively updated according to the noise floor of the captured signal. Further, in order to increase the granularity of the calibration process, three control bits are used to adjust the high-pass pole of the BPF.

### 12.3.1 MODES OF OPERATION

Together with the two already mentioned transmission modes, denoted as signal tracking and feature extraction modes, the system also offers a foreground calibration mode. They are briefly described next.

#### 12.3.1.1 Calibration

In this mode, the transfer characteristic and gain of the recording channels are individually adjusted. The objective is to tune the programming words for the high-pass (3-bit) and low-pass (2-bit) poles of the channel BPF so that its passband ranges from about 200 Hz to 7 kHz, corresponding to the spike spectral range, as well as program the output of the PGA to maximize the input swing of the ADC.

The calibration is completed for every channel by using the output signals of the DDFSs as frequency references. The schematic of the DDFS is shown in Figure 12.6. It consists of a ROM that contains the information of a sinusoidal signal, followed by a DAC and an analog attenuator to adapt the amplitude of the generated signal to the LNA's input swing. Then, by controlling the sampling frequency of the ROM, it is possible to modulate the frequency of the output sinusoidal signal. As there is one DDFS per row, passband calibration is done in a column-wise manner.

The calibration process follows the algorithm depicted in Figure 12.7. First, the DDFS is adjusted to generate a 1 kHz sinusoidal signal, in the middle of the BPF. Then, the system measures the amplitude of the LNA output and takes it as reference. After that, the DDFS switches its sampling frequency in order to generate a 200 Hz signal (desired high-pass corner of the BPF), while the corresponding control bits of the high pass pole (HPC) are configured in the highest-frequency mode. The calibra tion is completed by comparing the amplitude of the LNA output with respect to the stored reference voltage. In case that it is lower than the expected 3 dB attenuation,

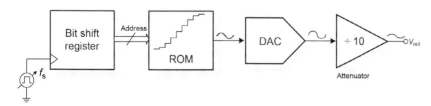

**FIGURE 12.6**

Schematic of the DDFSs.

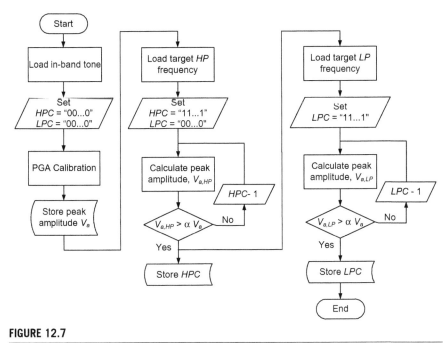

**FIGURE 12.7**

BP filter calibration process.

the HPC word is reduced by one bit, until the condition is fulfilled. This process is repeated for the LP pole using a 7 kHz sinusoidal input as reference.

Afterward, every channel starts capturing neural signals at a rate of 30 kS/s, and the gain of each PGA is adjusted so that its output fits into the input dynamic range of the corresponding ADC. Digitized signals are transmitted out column by column so that an external observer validates the completion of the calibration process. This is done because neural spiking is random by nature and channels can be silent for long periods. After validation, the observer can change to a different column or finish the calibration process by applying corresponding commands.

### 12.3.1.2 Signal tracking

In this mode, one column/row of the array is arbitrarily selected for neural signal monitoring, while the remaining channels are disabled for power saving. Neural signals are acquired at a sampling rate of 30 kS/s, 8-bit per sample, to give an overall throughput rate of 1.92 Mbps. No data compression is applied in this mode.

### 12.3.1.3 Feature extraction

In this case, the system is employed for spike detection tasks. All the 64 channels are enabled during feature extraction. Every detected spike is locally compressed at channel level by means of a piecewise linear (PWL) approximation of its waveform. This approximation involves amplitude and time interval values and results in a 47-bit representation per spike, as Figure 12.8 shows. Compared to other techniques,

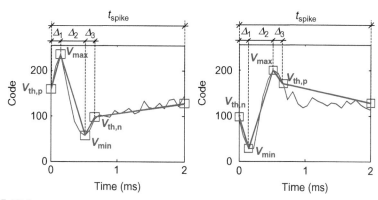

**FIGURE 12.8**

Spike feature extraction with digital spike detection.

such as principal component analysis and wavelet transform, this approach exhibits much lower computational complexity and allows a low-power low-area integrated implementation.

Figure 12.9 shows the functional flow of the feature extraction process, which is divided in two parts, the *spike detection* and the *spike processing*. During the *spike detection*, the ADC is configured to work at 30 kS/s, and the neural signal is monitored to check if it goes above a certain threshold voltage, which is calculated through a background process that averages the background noise. Once the neural signal goes above this threshold voltage, a spike is considered detected and the *spike processing* part starts. At this point, the ADC sampling mode changes to 90 kS/s to improve the resolution of the compression. The spike duration is configurable by means of a programmable variable, $t_{spike}$. The system extracts the values of the maximum and minimum voltages, as well as three temporal variables, $\Delta_1$, $\Delta_2$, and $\Delta_3$, which codify the time elapsed from the beginning of the spike (i.e., when the threshold voltage is reached) to the first peak, the second peak, and the last threshold voltage, respectively. The inclusion of a variable to codify the spike duration eases the spike-sorting flow. In case that the spike is monophasic, the second peak will be smaller than the threshold voltage. In this case, variable $\Delta_3$ is made equal to the configure spike time, as the last part of Figure 12.9 illustrates.

## 12.3.2 EVENT-BASED COMMUNICATION

EBPU units are responsible for temporarily storing the information provided by the channels. During the calibration and signal tracking modes, channels serialize and transfer data to the EBPUs, where information is retained until it is read out by the system digital processor.

In the feature extraction mode, EBPUs not only provide storing resources but also contribute to the calculation of the time intervals involved in the PWL representation of spikes. Peaking and threshold-crossing events along spikes are transmitted to corresponding EBPUs, as Figure 12.10 shows. Such units keep track of the duration

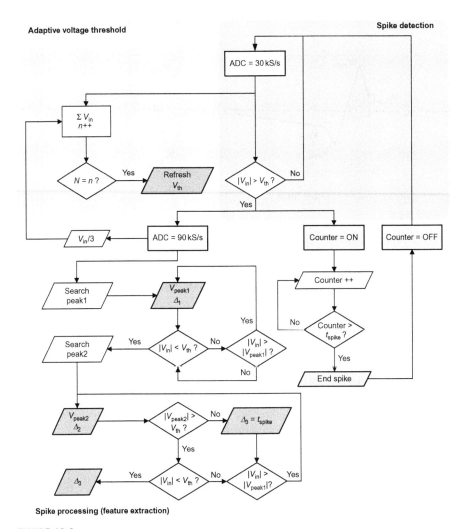

**FIGURE 12.9**

Spike detection and feature extraction process.

between the events by means of counters. When spikes end, channels send to the EBPUs the amplitude-related information to complete the associated PWL feature vectors. Once vectors are gathered, they are stored in the EBPUs ready to be read out. It is worth observing that this approach reduces the information transfer from the channels to the EBPUs by about 50%, as single events instead of complete time interval measurements (coded in 8-bit words) are transmitted.

The main digital processor cyclically reads the enabled EBPUs. If it is found that the stored information in the EBPU is complete, the digital processor retrieves data at a 4 MHz rate, builds up the transmission frame, and sends this stream to the telemetry unit for wireless transmission.

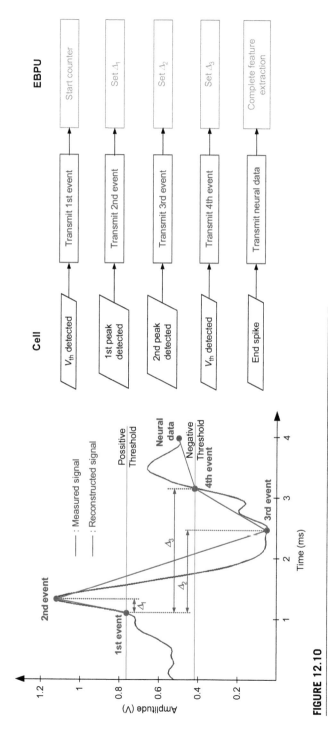

**FIGURE 12.10**

Event-based communication during the feature extraction mode.

### 12.3.3 COMMUNICATION PROTOCOL

Similar to RFID technologies, the system uses pulse interval encoding (PIE) of symbols in the forward link. Figure 12.11a shows the symbol representations for data-0 and data-1, which essentially differ on the duration of the high-level state. Figure 12.11b illustrates the structure of data frames in the forward link, that is, towards the sensor array. They are used to configure the neural recording sensor array. A forward frame consists of 24 bits, including preamble (5 bits), command (14 bits), and cyclic redundancy check (CRC) word (5 bits). As shown in Figure 12.11b, the structure and parameters included in the command word depend on the selected operation mode.

Figure 12.11c shows the structure of data frames in the backward link, that is, from the sensor array to the outside. The backward frame is 85-bit long and includes a fixed 8-bit preamble "01010101," followed by a 72-bit output data set,

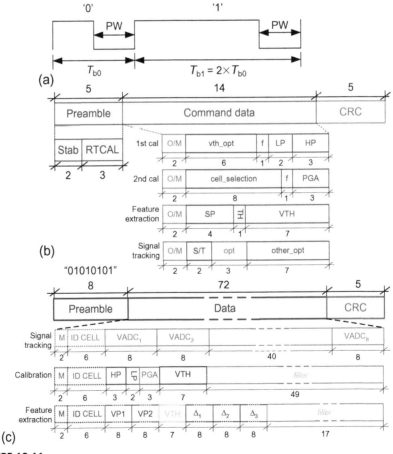

**FIGURE 12.11**

Communication protocol of the proposed system: (a) PIE format, (b) forward frame, (c) backward frame.

and completed by a 5-bit CRC word. The first 8 bits of the output data set inform about the operation mode (2 bits) and the channel identification (6 bits). In the signal tracking mode, the system collects the sampled data in groups of eight (by column or row, depending on the selected option), and only the first channel of the column/row has to be identified. In the feature extraction mode, the output data set is formed by 3 bytes of temporal information, 2 bytes of amplitude information, and 7 bits representing the applied threshold voltage. In the calibration mode, the system generates 15 bytes, which inform on the settings for the BPF, PGA, and threshold voltage.

## 12.4 CHANNEL ARCHITECTURE

Each of the channels is connected to an electrode of the multielectrode sensor array. The channels include all the circuitry to amplify, filter, digitize, and process the acquired neural signal according to the selected mode of operation. Then, they generate digital data streams, which are just collected by the central unit processor without any additional processing. The main advantages of this channel-level processing are the power consumption saving due to the suppression of long analog routing paths [3] and the scalability of the system (any array size can be easily built). The integration of all this circuitry in the shrink channel pitch ($400\,\mu m \times 400\,\mu m$) was possible through the implementation of some design methodologies and novel topologies, which are briefly described in this section.

Figure 12.12 shows the schematic of the analog front end included in the channel, which embeds the BP-LNA, the PGA, and the ADC. The former presents the classical capacitive feedback topology, proposed by Harrison in Ref. [25]. It implements and BPF function where the midband gain is given by the ratio of the input and feedback capacitances, $A_M = C_{IN}/C_F$. The band-pass transfer characteristic is defined by a low-frequency pole (the high-pass frequency corner), $f_{HP} = 1/(2\pi R_f C_f)$, while the high-frequency pole (which defines the low-pass frequency corner of the filter) is given by $f_{LP} = \alpha_m/(2\pi A_M C_L)$. In order to counteract the process variations that could modify the position of the poles, both the feedback resistance $R_f$ and the output load $C_L$ are programmable. As [29] showed, a dedicated design optimization algorithm was developed to get an optimized design in terms of power, area, and input-referred noise. Experimental results demonstrate that the nominal bandwidth of the BP filter is between 192 Hz and 6.7 kHz, with a midband gain of 47.5 dB. The total power consumption is as low as $1.92\,\mu W$.

The ADC implements an SC-based novel topology to perform a binary search algorithm that completes an 8-bit analog-to-digital conversion [30]. Additionally, it implements some PGA capabilities by means of the programming input capacitance C4. Compared to the commonly used capacitive DAC-based successive approximation ADC topology [2,3], this solution has smaller area occupation, smaller input capacitance, and large input sampling period, which compensates the excess of power consumption needed by the OTA. Furthermore, the bias current of the OTA is dynamically adapted along the conversion to the changing slew-rate requirements [29].

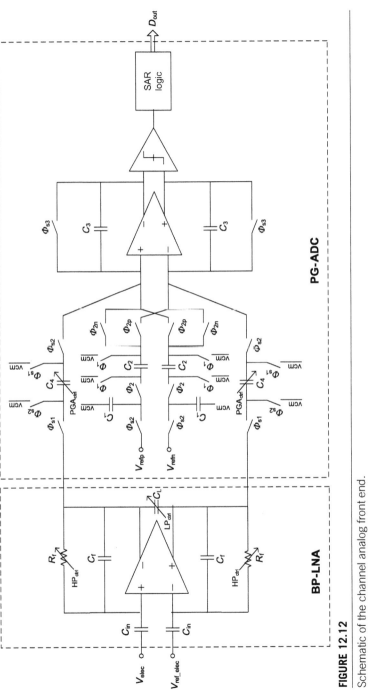

**FIGURE 12.12**

Schematic of the channel analog front end.

The ADC was optimized to the two possible sampling modes necessary to complete the defined operation modes, that is, 30 and 90 kS/s. Measurement results confirm that the ENOB of the PG-ADC is 7.6 bits in both sampling modes, while the power consumption is 515 nW and 1.52 μW, respectively.

## 12.5 TELEMETRY UNIT

Figure 12.13 shows the schematics of the power and data telemetry unit. It is based on inductive link techniques and operates in the worldwide available ISM band centered at 40.68 MHz. Data reception employs (OOK) modulation, whereas data transmission is accomplished by modulating the amplitude of the carrier by means of a switchable antenna matching network driven by the digital processor. In this latter case, the modulation depth is less than 50% and the output data are encoded using a Manchester encoder.

Not shown in the figure, the telemetry unit also includes a timing recovery circuit, which extracts the 4 MHz clock of the system from the incoming RF signal, which is also used to modulate the backward link. This is accomplished by means of divide-by-two circuits based on single transistor-clocked dynamic latches [31].

The telemetry unit also includes a power management circuitry, which harvests energy from the inductive link using a rectifier. Analog and digital supply lines of 1.2 V are obtained from corresponding regulators, while a band gap circuit generates the analog voltage references. The efficiency of the rectifier is 60% at 1 mW RF input power.

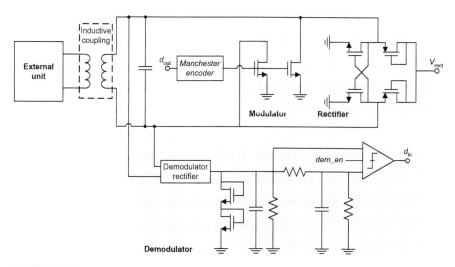

**FIGURE 12.13**

Schematic of the telemetry front end.

## 12.6 EXPERIMENTAL RESULTS

Figure 12.14a shows the microphotography of the proposed system. It has boon integrated in a 6M2P 0.13 µm standard CMOS technology. Figure 12.14b shows the layout of each channel, which includes an internal pad for flip-chip connection to a microelectrode. For the sake of testability, the channel input nodes can be also accessed from an external pad ring. Clamp cells are placed along the chip periphery to protect the microelectrode nodes from ESD damages. The system occupies 18.4 mm².

Figure 12.15a shows the frequency response of the BP-LNA under the different available configurations for both the LP and the HP pole positions. The HP pole could be adjusted between 15 and 232 Hz along eight different positions, while the LP pole could be moved between 5.2 and 10.15 kHz within four different steps. Figure 12.15b illustrates the power spectral density of the LNA's input-referred noise. Note that the $1/f$ noise contribution is attenuated at low frequencies by the HP transfer function, so it results in a flat noise-level band. The total integrated noise power is $3.8\,\mu V_{rms}$ if it is integrated between 1 Hz and 100 kHz (this gives a noise efficiency factor of 2.16), while if it is integrated in the band of interest, between 200 Hz and 7 kHz, it results to be $3.2\,\mu V_{rms}$.

Figure 12.16 illustrates the temporal evolution of the ADC output and the digital control signals during the self-calibration process. First, the PGA gain is self-calibrated at the midband of the BPF characteristic. Then, the control words for the high- and low-pass poles are adjusted. In both cases, it can be observed how the amplitude of the sinusoidal signal changes with the variation of the control words and how it is compared with respect to the reference amplitude to fix the calibrated word.

The temporal evolution of the digitized output during the controlled calibration of the PGA is shown in Figure 12.17a. The PGA starts working under the higher gain configuration. Then, the background noise that has small amplitudes don't saturate the output of the ADC, and the PGA is not readjusted until a higher peak occurs. When it happens, the PGA gain is decreased one step. It can be seen how it

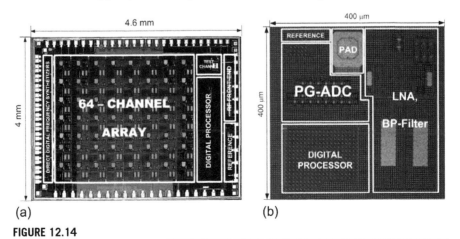

**FIGURE 12.14**

Microphotography of the: (a) neural array system, (b) channel.

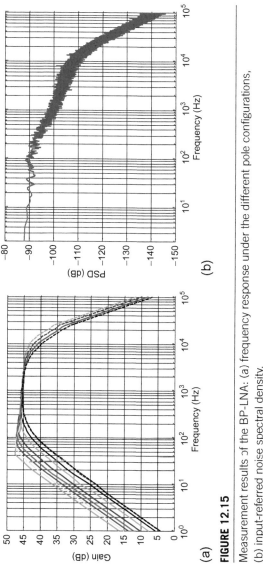

**FIGURE 12.15**

Measurement results of the BP-LNA: (a) frequency response under the different pole configurations, (b) input-referred noise spectral density.

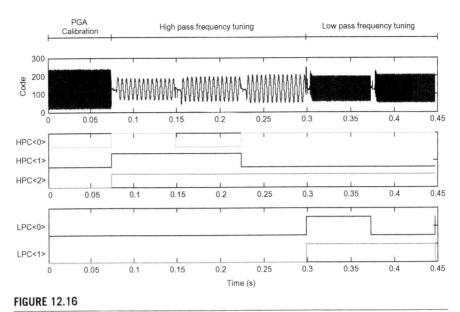

**FIGURE 12.16**

Temporal evolution of the signals during the calibration process of the BP-LNA.

is successively adapted as the ADC output saturates until it reaches a stable value. Figure 12.17b depicts the same mechanism but with a sinusoidal input signal modulated with a ramp instead of a neural input signal in order to have a more illustrative example of the adaptive adjustment.

Figure 12.18 illustrates the system operation in the feature extraction mode. Dots represent the spikes detected by the neural array in a time slot of 500 ms. Once a spike is detected in a channel and its PWL representation derived (47 bits, as Figure 12.18 illustrates), the feature vector is stored in the associated EBPU. The main digital processor cyclically reads the EBPUs every 237 μs. Considering the 85-bit length of the backward frame detailed in Section 12.3, the system requires 21.25 μs to transmit the information of one spike at 4 MHz. Therefore, we can calculate the maximum possible delay by summing up the delay of the EBPU reading and the transmission delay, which results 258.25 μs. This is much lower than a typical spike duration (around 2 ms) and, of course, much lower than the time basis for firing occurrences. It means that no information is lost not even in the unlikely case where all the channels fire at the same instant (only a small delay no larger than about 10% the duration of a spike could be observed in some of the records).

Figure 12.19 shows the power consumption distribution of the system in the feature extraction mode. The channel implementation in this chapter includes an additional digital processor, which, together with the needed buffers to communicate along the array, raises the power consumption per channel to 4.54 μW. From the simulated power consumption, it can be extracted that most of the power is consumed by the neural channels (290.56 μW). The main digital processor and EBPUs, which make extensive use of clock gating and clock frequency division techniques,

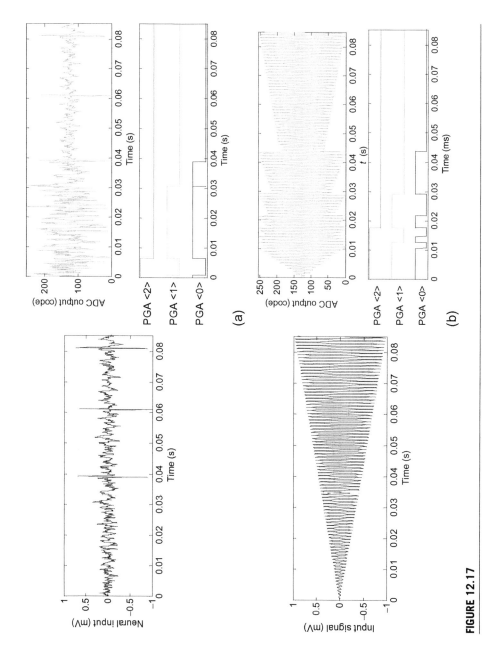

**FIGURE 12.17**

Temporal evolution during the PGA calibration: (a) with a neural input signal, (b) with a sinusoidal input signal modulated with a ramp.

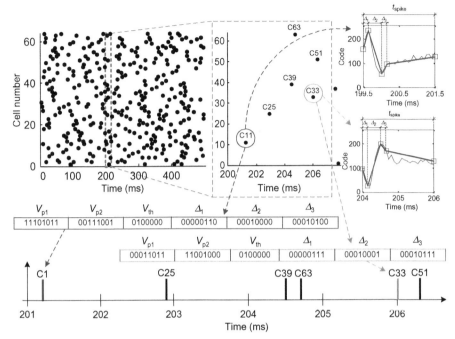

**FIGURE 12.18**

Data output stream under feature extraction mode.

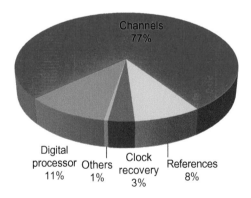

**FIGURE 12.19**

Distribution of the power consumption.

consume $40\,\mu W$ ($5\,\mu W$ of them dissipated by leakage currents). Band gap references, regulators, and current conveyors consume $32\,\mu W$. The clock recovery block, the Manchester encoder, and the demodulator require, respectively, $12.5\,\mu W$, $1.5\,\mu W$, and $400\,nW$. All in all, the total power consumption of the system sums $377\,\mu W$.

Table 12.3 summarizes the performance of the neural recording system and compares it with some state-of-the-art works. Note that the presented work presents one

**Table 12.3** Comparison of the Multichannel Neural Recording Array with the State of the Art

| | [12] | [13] | [17] | [18] | [19] | [20] | [21] | This work |
|---|---|---|---|---|---|---|---|---|
| CMOS process (μm) | 0.5 | 0.5 | 0.35 | 0.35 | 0.5 | 0.13 | 0.13 | 0.5 |
| Supply voltage (V) | 3.3 | 1.8 | 3.3 | 3 | 3 | 1.2 | 1.2 | 1.2 |
| Number of channels | 100 | 64 | 128 | 64 | 32 | 64 | 96 | 64 |
| Area occupation (mm²) | 4.7×5.9 | N/A | 8.8×7.2 | 3.1×2.7 | 4.93×3.3 | 4×3 | 5×5 | 4×4.6 |
| Total power (mW) | 13.5 | 14.4 | 6 | 17.2 | 5.85 | 5.03 | 6.5 | 0.38 |
| Power per channel (μW) | 135 | 225 | 46.87 | 269 | 182.82 | 78.6 | 67.7 | 5.9 |
| Amplifier gain (dB) | 39.5 | 60 | 40 | 65–83 | 67.8–78 | 54–60 | 56 | 47.5–65.5 |
| Low-frequency corner (Hz) | 0.025 | 10 | 0.1 | 1 | 1 | 10 | 280 | 1.5–192 |
| High-frequency corner (kHz) | 7.2 | 10 | 20 | 10 | 10 | 5 | 10 | 5.2–10.15 |
| Input-referred noise ($\mu V_{rms}$) | 2.2 | 8 | 4.9 | 3.05 | 4.62 | 6.5 | 2.2 | 3.8 |
| ADC ENOB (bits) | 9[a] | 8[a] | 9[a] | 7.2 | 8.1 | 7.8 | 10[a] | 7.65 |
| Sampling freq. (kHz) | 15 | 62 | 40 | 20 | N/A | 54 | 31.25 | 30/90 |
| Power source | Inductive | Inductive | Battery | Battery | Inductive | Battery | Battery | Inductive |
| Fwd data telemetry carrier frequency (MHz) | ASK | OOK | N/A | OOK | | | | |
| 2.64 | | | | | | | | |
| 4–8 | | | | | | | | |
| 915 | | | | | | | | |
| 40.68 | | | | | | | | |
| Rvrs data telemetry | FSK | OOK | UWB | FSK | FSK | FSK | UWB | LSK |
| Carrier frequency (MHz) | 433 | 70–200 | 4000 | 400 | 915 | 915 | | 40.68 |
| Max. output data bit rate (Mbps) | 0.35 | 2 | 90 | 1.25 | 0.71 | 1.5 | 30 | 4 |
| Data compression | Binary spike detection | Binary spike detection | On-the-fly extraction | Spike reconst. | PWM | FIR filters | N/A | Feature extraction |

[a]Output resolution.

of the lowest power dissipation per channel, including both a wireless communication circuitry and data compression mechanisms.

## 12.7 CONCLUSIONS

A 64-channel neural array with embedded data reduction techniques, fabricated in a standard CMOS 130 nm process, has been presented. Inspired on RFID systems, an inductive link is used for both powering the implant and transferring information to/from an external unit placed on the head. A distributed digital signal processing approach, with tasks at channel and array levels, has been found an efficient solution for reducing the power consumption of the SoC and simplifying communications through the array. The total power consumption of the system in the feature extraction mode is 377 μW from a nominal voltage supply of 1.2 V.

## REFERENCES

[1] Gosselin B, Ayoub AE, Roy JF, Sawan M, Lepore F, Chaudhuri A, et al. A mixed-signal multichip neural recording interface with bandwidth reduction. IEEE Trans Biomed Circuits Syst 2009;3(3):129–41.

[2] Wattanapanitch W, Fee M, Sarpeshkar R. An energy-efficient micropower neural recording amplifier. IEEE Trans Biomed Circuits Syst 2007;1(2):136–47.

[3] Wen-Sin L., Xiaodan Z, Libin Y, Yong L. A 1-V 60-μW 16-channel interface chip for implantable neural recording. In: IEEE custom integrated circuits conference; 2009. p. 507–10.

[4] Campbell P, Jones K, Huber R, Horch K, Normann R. A silicon-based, three-dimensional neural interface: manufacturing processes for an intra-cortical electrode array. IEEE Trans Biomed Eng 1991;38(8):758–68.

[5] IEEE standard C95.1-2005. IEEE standard for safety levels with respect to human exposure to radio frequency electromagnetic fields, 3 kHz to 300 GHz.

[6] Najafi K, Wise KD. An implantable multielectrode array with on-chip signal processing. IEEE J Solid-State Circuits 1986;21(6):1035–44.

[7] Ji J, Wise KD. An implantable CMOS circuit interface for multiplexed microelectrode recording arrays. IEEE J Solid-State Circuits 1992;27(3):433–43.

[8] Wise KD, Sodagar AM, Yao Y, Gulari MN, Perlin GE, Najafi K. Microelectrodes, microelectronics, and implantable neural microsystems. Proc IEEE 2008;96(7):1184–202.

[9] Akin T, Najafi K, Bradley RM. A wireless implantable multichannel digital neural recording system for a micromachined sieve electrode. IEEE J Solid-State Circuits 1998;33(1):109–18.

[10] Akin T, Najafi K. A micromachined silicon sieve electrode for nerve regeneration applications. In: International conference on solid-state sensors and actuators, 1991; 1991. p. 128–31.

[11] Nordhausen C, Maynard E, Normann R. Single unit recording capabilities of a 100 microelectrode array. Brain Res 1996;726(1):129–40.

[12] Harrison RR, Watkins PT, Kier RJ, Lovejoy RO, Black DJ, Greger B, et al. A low-power integrated circuit for a wireless 100-electrode neural recording system. IEEE J Solid-State Circuits 2007;42(1):123–33.

[13] Sodagar AM, Perlin GE, Ying Y, Najafi K, Wise KD. An implantable 64-channel wireless microsystem for single-unit neural recording. IEEE J Solid-State Circuits 2009;44(9):2591–604.

[14] Lebedev M, Nicolelis A. Brain–machine interfaces: past, present and future. Trends Neurosci 2006;29(9):536–46.

[15] Schwartz A, Cui X, Weber D, Moran D. Brain-controlled interfaces: movement restoration with neural prosthetics. Neuron 2006;52:205–20.

[16] Hochberg L, Serruya M, Friehs G, Mukand J, Saleh M, Caplan A, et al. Neuronal ensemble control of prosthetic devices by a human with tetraplegia. Nature 2006;442(13):164–71.

[17] Chae MS, Yang Z, Yuce MR, Hoang L, Liu W. A 128-channel 6 mW wireless neural recording IC with spike feature extraction and UWB transmitter. IEEE Trans Neural Syst Rehabil Eng 2009;17(4):312–21.

[18] Bonfanti A, Ceravolo M, Zambra G, Gusmeroli R, Borghi T, Spinelli AS, et al. A multi-channel low-power IC for neural spike recording with data compression and narrowband 400-MHz MC-FSK wireless transmission. In: Proceedings of the ESSCIRC 2010; 2010. p. 330–3.

[19] Lee SB, Lee H-M, Kiani M, Jow U-M, Ghovanloo M. An inductively powered scalable 32-channel wireless neural recording system-on-a-chip for neuroscience applications. IEEE Trans Biomed Circuits Syst 2010;4(6):360–71.

[20] Abdelhalim K, Genov R. 915-MHz wireless 64-channel neural recording SoC with programmable mixed-signal FIR filters. In: Proceedings of the ESSCIRC 2011; 2011. p. 223–6.

[21] Gao H, Walker RM, Nuyujukian P, Makinwa KAA, Shenoy KV, Murmann B, et al. HermesE: a 96-channel full data rate direct neural interface in 0.13 μm CMOS. IEEE J Solid-State Circuits 2012;47(4):1043–55.

[22] Chae MS, Liu W, Sivaprakasam M. Design optimization for integrated neural recording systems. IEEE J Solid-State Circuits 2008;43(9):1931–9.

[23] Liew W-S, Zou X, Lian Y. A 0.5-V 1.13 μW/channel neural recording interface with digital multiplexing scheme. In: Proceedings of the ESSCIRC 2011; 2011. p. 219–22.

[24] Muller R, Gambini S, Rabaey JM. A 0.013 mm² 5 μW DC-coupled neural signal acquisition IC with 0.5 V supply. In: IEEE international solid-state circuits conference (ISSCC 2011); 2011. p. 302–4.

[25] Harrison RR, Charles C. A low-power low-noise CMOS amplifier for neural recording applications. IEEE J Solid-State Circuits 2003;38(6):958–65.

[26] Al-Ashmouny K, Sun-Il C, Euisik Y. A 4 μW/ch analog front-end module with moderate inversion and power-scalable sampling operation for 3-D neural microsystems. In: IEEE proceedings of BioCAS 2011; 2011. p. 1–4.

[27] Harrison RR, Kier RJ, Chestek CA, Gilja V, Nuyujukian P, Ryu S, et al. Wireless neural recording with single low-power integrated circuit. IEEE Trans Neural Syst Rehabil Eng 2009;17(4):322–9.

[28] Harrison RR. The design of integrated circuits to observe brain activity. Proc IEEE 2008;96(7):1203–16.

[29] Rodriguez-Perez A, Ruiz-Amaya J, Delgado-Restituto M, Rodriguez-Vazquez A. A low-power programmable neural spike detection channel with embedded calibration and data compression. IEEE Trans Biomed Circuits Syst 2012;6(2):87–100.

[30] Rodríguez-Perez A, Delgado-Restituto M, Medeiro F. A 515 nW, 0–18 dB programmable gain analog-to-digital converter for in-channel neural recording interfaces. IEEE Trans Biomed Circuits Syst 2014;8(3):358–70.

[31] Yuan J, Svensson C. New single-clock CMOS latches and flipflops with improved speed and power savings. IEEE J Solid-State Circuits 1997;32(1):62–9.

[32] Wattanapanitch W, Sarpeshkar R. A low-power 32-channel digitally programmable neural recording integrated circuit. IEEE Trans Biomed Circuits Syst 2011;5(6):592–602.

# Implantable imaging system for automated monitoring of internal organs

**Abhishek Basak, Swarup Bhunia**

*Department of Electrical Engineering and Computer Science, Case Western Reserve University, Cleveland, Ohio, USA*

## CHAPTER CONTENTS

Bhunia et al. Implantable Biomedical Microsystems. http://dx.doi.org/10.1016/B978-0-323-26208-8.00013-3

## 13.1 INTRODUCTION

Early diagnosis of various diseases and health problems is extremely important in ensuring higher efficiency of treatment, prevention, or delay of associated complications and, even in certain scenarios, creating difference between survival and death. For example, early detection of a coronary artery blockage through screening could significantly reduce the chances of a stroke through appropriate treatment measures [1]. Besides, diagnosis of chronic liver disease at an early stage [2], through periodic tests like the liver function test, would significantly reduce the probability of progression to the advanced cirrhosis stage and the associated complications. In most of these scenarios, the diseases reach an advanced stage during the time of appearance of symptoms. Hence, the normal symptomatic detection is often too late. The other classic example in this regard is the early detection of malignant growths before they spread beyond the organs of origin to other parts of the body [3]. A significant spectrum of cancers progress to stage III/IV by the time the symptoms appear, and hence, even the state-of-the-art treatments involving chemotherapy, radiotherapy, etc. fail at that stage [4]. Figure 13.1a illustrates that the 5-year survival rates for four different cancers vary between 60% and 85% if detected in Stage I/early Stage II, depending on type and organ of origin [4]. The same index falls to below 25% in some scenarios for stage III/IV. This statistic verifies the fact that earlier the detection of cancer, the better the survival rates and in general the quality of life. On the other hand, Figure 13.1b depicts that apart from prostate cancer, the cumulative percentage of detection in stage I

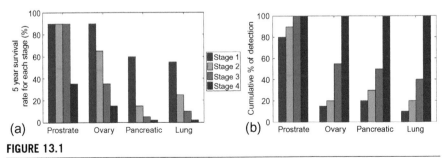

**FIGURE 13.1**

(a) Variation of 5-year survival rates with cancer staging [4]. (b) Variation of cumulative percentage of cancer detection with staging for four different cancer types [4,52].

for the other three types is less than 20% [4]. Hence, in the current diagnostic scenario, malignancies are being detected early with a very low probability.

Apart from the aforementioned, early detection of conditions such as systemic inflammatory response syndrome [5] and brain aneurysm [6] can be accomplished through periodic, automated monitoring of the concerned interstitial organs. Automated online monitoring of internal organs, such as the liver, kidneys, lungs, and gastrointestinal tracts, has extensive diagnostic applications including monitoring efficacy of treatments after detection. Apart from different blood-based screening tests [7], real-time imaging of the internal organs is well known as an effective approach to detect the onset of different abnormalities, such as an anomalous growth in internal organs. These include different imaging modalities like magnetic resonance imaging (MRI), computed tomography (CT) scan, positron-emission tomography (PET) scan, ultrasound scan, fluorescence imaging, or a combination of them.

## 13.1.1 LIMITATIONS OF EXISTING MONITORING TECHNOLOGY

Many of the prevalent health complications originate in organs, deep inside the human body. The existing diagnostic technologies including MRI, CT, and ultrasound have their limitations in imaging deep, interstitial body regions. The lack of adequate resolution is caused mainly by the specific location of an anomalous growth, deep in an internal organ, and the nonhomogeneity (chemical composition, density, attenuation, etc.) of the different intermediate tissue layers [8,9]. Besides, state-of-the-art PET, MRI, and CT are significantly expensive and can only be availed in advanced medical facilities. This reduces their accessibility for a large volume of patients, especially in developing countries. As a statistic, MRI costs over $1000 on an average [10] in the United States and Europe. Moreover, most existing diagnostic techniques involve complex operations by trained specialists, which have prevented widespread availability of these machines. They also require the patient to physically visit a medical facility for routine checkup, which may be inconvenient, in particular, for patients who lose their mobility in a disease like cancer. Hence, these techniques are mostly not suitable for periodic online monitoring of interstitial organs in an otherwise normal healthy person. As a result, most diseases/health problems are detected only through symptoms. In such scenarios, these state-of-the-art monitoring technologies play a diagnostic role only after the symptomatic detection and during the treatment phase, which is often too late. Hence, there is a critical need for an automated, localized, point-of-care technology for periodic monitoring of interstitial organs, thereby bridging this gap in medical diagnostics [9].

## 13.2 IMPLANTABLE IMAGING SYSTEM: AN OVERVIEW
### 13.2.1 PROPOSED SYSTEM

A miniaturized implantable imaging system, capable of online, periodic localized monitoring can effectively address these issues in certain scenarios. Often, people undergo surgeries in different cases such as removal of the gall bladder and kidney stones, liver transplantation, and arterial bypass. Many of these underlying health complications require periodic, automated, high-resolution monitoring of susceptible organs after surgery over the rest of the patient's life. It often so happens that different factors including expense, accessibility, and inconvenience lead to the patient not maintaining a uniform diagnostic screening schedule. So, diseases often relapse back at an advanced stage. In these scenarios, during the initial surgical procedure, a miniaturized imaging assembly could be implanted locally in the concerned deep, interstitial regions. The system, powered by an implantable battery, would periodically (depending on disease type and aggressiveness) scan the volume of interest and transmit the raw image data wirelessly after compression to a personal analysis unit, for example, a PDA and cell phone. The external unit would perform the image processing, render the volumetric scan, and compare with the previous stored scan/baseline image. If the differences, based on an automated analysis on the inferred physical and structural attributes, exceed a clinician recommended threshold, the system sets an alert to the person to visit a medical facility and take appropriate measures including state-of-the-art imaging tests. In this way, such an implantable system could potentially serve the role of this bridge in certain scenarios in present medical diagnostics.

Diagnostic ultrasound (US) imaging serves as an efficient screening tool for early detection of different anomalies. The advantages of ultrasound in terms of cost-effectiveness, safety, simplicity of operation, ability of system miniaturization, and sufficient resolution in many applications make it attractive as an efficient screening modality [11]. Ultrasound imaging has been used in practice to detect gall stones, cysts, muscle and tendon injuries and to detect malignant growths in superficial organs like the breast, ovary, prostrate, and bladder. However, it is well known that external ultrasound is often incapable of producing high-resolution images throughout the entire scan volume. The fundamental frequency versus scan depth trade-off limits its imaging capabilities in deep internal organs. Moreover, the presence of different tissue layers (skin, fat, muscles, bones, etc.) leads to greater attenuation losses and phase aberrations of the external beam. A recent, comparatively invasive ultrasound imaging approach places a miniature transducer at the tip of an endoscopic catheter to perform transesophageal, endobronchial, and transrectal imaging [12,13]. It partly mitigates the problems of attenuation and beam defocusing. However, these procedures are not suitable for continual monitoring in personalized settings. Furthermore, many inner organs, such as the right lobe of liver, lung walls, and parts of the gastrointestinal tract, remain inaccessible to a catheter used in endoscopic imaging. In our application scenario, the imager would be implanted close to the susceptible region. Hence, a miniaturized implantable system, based on diagnostic ultrasound technology and capable of localized automated, periodic scans, could alleviate these problems. Implantable medical devices (IMDs) have been researched and commercialized over the last two decades. These include cardiac devices like pacemaker, implantable defibrillator, left ventricular assist devices [14], implantable bladder pressure monitor, and drug pumps [15]. Apart from these, there have been studies on implantable sensors functioning as artificial retina [16] and arrays for brain recording and stimulation [17]. The research work, presented in Ref. [9] on the miniaturized ultrasonic imaging device, is one of the first of its kind in the field of fully implantable diagnostic imaging for automated, online high-resolution point-of-care monitoring.

A promising application of this implantable imaging device is early detection of locoregional recurrence of cancer. Often, treatment of primary tumors starts with surgical resection, followed by chemotherapy, radiation, or a combination. However, even after aggressive treatments, patients remain vulnerable to a potential recurrence, which can occur in few months to 5–10 years (depending on cancer types) during which the patients need to be rigorously monitored. At present, cancer relapses are often detected through the onset of symptoms, which is usually too late for treatments to be effective [3]. In most scenarios, symptoms appear only when the cancer has spread beyond the organ of relapse through the vascular system [4]. Hence, early detection of cancer recurrence is critical for successful treatments. The proposed imager can be surgically implanted around the site of the primary tumor after resection to periodically monitor for onset of locoregional recurrence (e.g., for soft tissue sarcomas and breast cancer). By passing an alert in case of suspected anomaly, a relapse of malignancy could be detected in stage I or early stage II, which is generally much earlier than symptomatic detection. An example scenario, depicting the relative advantages of the proposed interstitial system over external and endoscopic ultrasonic techniques for imaging the top portion of the liver right lobe, is illustrated in Figure 13.2. The localization of imaging in the volume of interest enhances the effectiveness of internal imaging. Other organs, where an implantable imager may be beneficial for early detection of relapse, are also shown in Figure 13.2. Throughout the chapter, we consider the case study of early detection of cancer recurrence. For posttreatment monitoring, highly susceptible regions for locoregional relapse include the region around the original tumor site after resection and the surrounding lymph nodes. For cancer relapse monitoring, the imager would periodically (e.g., bi-monthly) scan the 3-D volume of interest and transmit the data wirelessly to a central processing and storage system [18] for automated anomaly detection. On detection of a possible growth, the external system would generate an alert.

## 13.2.2 HOW DIAGNOSTIC ULTRASOUND IMAGING FUNCTIONS? A BRIEF RECALL

Ultrasound refers to sound waves with a frequency greater than what humans can hear, that is, beyond 20 KHz [19]. In the diagnostic domain, to produce relevant images of medical significance, the ultrasound pulse frequencies range in the order of MHz, that is, usually from 2 to 25 MHz depending on the trade-off between resolution and depth of scan in the organ of interest. In diagnostic ultrasound imaging, the transducer functions as both the transmitter and the receiver. It transmits a short ultrasound pulse and waits to receive reflections from the volume of interest, with a receive duty cycle greater than 99% compared to transmit. Inside the body, different tissue layers have different acoustic impedances due to varying speeds of sound and density of tissue composition. This leads to specular reflections from the tissue

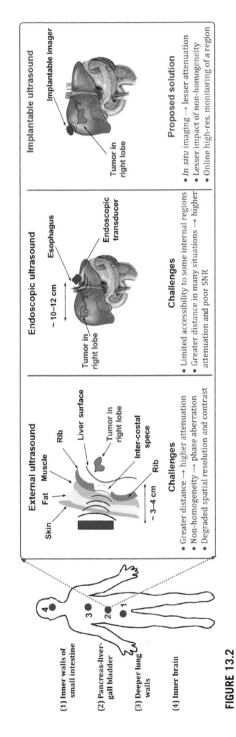

**FIGURE 13.2**

*In situ*, localized imaging can address some of the challenges associated with external and endoscopic ultrasound imaging. It is capable of providing higher-resolution online monitoring of cancer recurrence in deep internal organs [8].

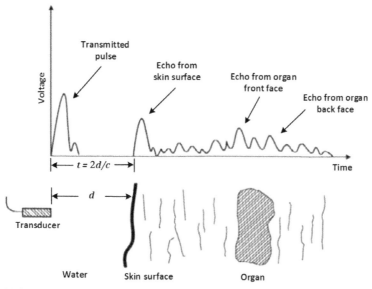

**FIGURE 13.3**

Pictorial illustration of functioning of diagnostic ultrasound imaging with reflection
signatures recorded across the receive duration [20].

interfaces. Besides, inside a particular tissue type, diffuse reflections of varying amplitudes take place in the intratissue scatterers. Different tissues have varying degrees
of ultrasonic absorption. Tissue types like the bones are highly reflective, whereas
water-filled regions are good transmitters. Normal human soft tissues vary in acoustic characteristics between these two extremes. Based on the reflections coming back
to the transducer over the receive duration, the final image is constructed after various stages of postprocessing. A rough illustration of the basis on which diagnostic
ultrasound imaging functions is given in Figure 13.3 [20].

The higher the frequency of ultrasound, the better is the resolution, that is, the
ability to distinguish two objects distinctly, but the higher is the attenuation through
the tissues as well (usually linear with frequency). Hence, there is always an inherent trade-off between resolution and scan depth of interest; that is, if the range of
interest is small like for the thyroid gland, carotid artery, and breast tissues, the diagnostic frequency used is ~20 MHz, and for obstetric organs, parts of gastrointestinal
tracts, heart, etc., a lower ultrasonic frequency of ~5–10 MHz is set to scan a greater
depth. Irrespective of frequency, ultrasound for diagnostic imaging is safe without
any adverse effects. With advances in external diagnostic ultrasound imaging over
the years, phased arrays composed of 64–256 elements are being used to utilize the
principles of beamforming (similar to principles of radar) to dynamically focus and
steer the beam, resulting in considerable improvement in lateral resolution and range
of view. 2-D arrays are also being increasingly used in high-resolution 3-D recon-
structions of organs. In general, abnormal growths in organs are detected from their

distinct reflection signatures compared to surrounding tissues, for example, high reflections from calcifications, absence of echoes from cysts (fluid-filled), and time–frequency pulse variations from tumor surfaces. The similar diagnostic concepts are used to detect fractures, tendon and muscle injuries, treatment progress, etc. Doppler ultrasound, utilizing continuous ultrasound waves and the Doppler principle of frequency shift to detect the velocity of moving objects, is used extensively to detect blood flow paths and irregularities, for example, in stenosis, clots, and angiogenesis. However, for the implantable ultrasonic imaging system described in this chapter, we restrict ourselves to the diagnostic pulsed ultrasound technology.

## 13.3 SYSTEM OVERVIEW
### 13.3.1 DESIGN SPACE EXPLORATION

For the implantable system, we consider the case study of early detection of locoregional cancer recurrence after tumor resection and treatment. The highly susceptible regions for this type of relapse include the region around the original tumor site after resection and the surrounding lymph nodes. The proposed system consists of two major components:

**(1)** *Transducer*—Due to a broad field of view, convex linear arrays are used, as illustrated in Figure 13.4a. These arrays, by shift of active apertures, can scan a plane of the region of interest. Volumetric imaging would be performed by mechanical rotation in one plane and electrical steering with natural array curvature in the other plane [9].

**(2)** *Assembly*—The assembly geometry is a hemisphere (Figure 13.4a) as the system would be mostly sutured along the organ wall and the frontal scan view is required. The array/arrays would be interlaced on the assembly surface. The transmit–receive electronics, power supply module, and micromotor for rotation would be enclosed within the assembly.

The primary objectives and design constraints of such a system are given in Table 13.1 and are as follows:

**(a)** *Spatial resolution*—The required resolution for detection of a cluster of malignant cells, in stage I/II, is usually ~5–10 mm [21], which is within high-frequency ultrasonic limits.

**(b)** *Power limit*—The peak power, dissipated for an extremely small duration during transmission (<0.01% duty cycle), is limited to 20 W, considering the specifications of an implant battery [22], charge pumps [23], and thermal limitations [24].

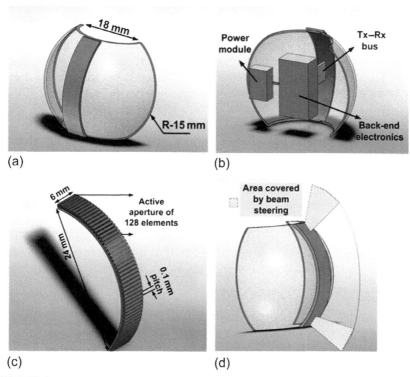

**FIGURE 13.4**

(a) Front view of proposed imaging assembly; (b) posterior view of assembly; (c) convex linear imaging array; (d) volume covered by beam steering.

**Table 13.1** System Design Objectives and Constraints

| Spatial Resolution | Peak Power | Array Pitch | Scan Range |
|---|---|---|---|
| ~5–10 mm | ~20 W | ~100 μm | ~5–10 cm |

**(c)** *Array pitch*—Present fabrication methodologies achieve a pitch limit of 100 μm with reliable yield.

**(d)** *Scan range*—A region of 5–10 cm extent around the primary tumor needs to be monitored.

The design space exploration involves the judicious selection of transducer arrays, their spatial arrangement, and assembly geometry for the best volume coverage under size and power/energy constraints. The major system components are described next.

### 13.3.1.1 Transducer

The parameters of the convex linear array are described as follows.

## Frequency

It is fixed at 15 MHz, with a bandwidth of ~70–100% based on the desired spatial resolution. With a usual dynamic range (DR) of 100 dB before depth gain compensation (DGC) and a soft tissue attenuation of 0.6 dB/MHz/cm [20], the scan range is ~6 cm. The theoretical lateral resolution, at the furthest depth (considering aperture of 1.5 cm), is within 5 mm.

## Pitch

Appearance of the grating lobes, which worsens imaging quality, depends on the array pitch and beam steering angle. The angle at which the grating lobes appear is governed by $\theta_g = \pm\sin^{-1}(\lambda/p + u_s)$ [20], where $p$ is the array pitch, $u_s$ is the maximum steering angle, and $\lambda$ is the ultrasonic wavelength. If Nyquist spatial frequency of $\lambda/2$ is maintained for the pitch, then grating lobes do not appear. However, for 15 MHz, this corresponds to a pitch of $\lambda/2 \sim 50\,\mu m$. From the fabrication constraints as given above, a pitch of $\lambda = 102\,\mu m$ is chosen. As shown in Figure 13.4d depending on the organ topology, one might require a steering of ~15–20° at the array ends. Corresponding to this, the grating lobes would appear at around 65–75°, which should not hamper the imaging quality.

## Array material

Capacitive micromachined ultrasonic transducers (CMUTs) [22] are chosen over soft PZT as our array material due to inherent advantage of easier miniaturization, lesser cost of batch fabrication, enhanced sensitivity, and integration with front-end electronics.

## Active aperture

Lateral resolution is approximated by the −6 dB full-width half-maximum beam profile, given by $LR = 0.4 \times \lambda \times F/L$, where LR is the lateral resolution, $\lambda$ is the ultrasound wavelength, $F$ is the focal depth, and $L$ is the active aperture length [20]. So, the greater the aperture, the better the LR. Besides, with increased aperture, the beam has higher acoustic energy, and hence, the image signal-to-noise ratio (SNR) is improved. An implantable battery with a supercapacitor can supply peak power in the order of ~15–20 W [23,25]. With each CMUT element requiring ~125–150 mW, the maximum number of active elements is 128. With a pitch of $\lambda$, the aperture length is 1.3 cm, which is in the higher ranges of an implantable form factor.

### 13.3.1.2 Imaging assembly

The assembly is the hemispheric structure, which houses the imaging array/arrays along with the enclosed electronics, power sources, miniature motor, etc. The major parameters of the assembly are as follows:

(1) *Arrangement of transducers*—An array would be placed along a longitudinal half circle of the hemisphere as shown in Figure 13.4a. To obtain a 3-D scan, without rotations, the entire assembly would have to be interlaced with arrays. This is not feasible due to fabrication complexities, increased interconnects, and

cost. For the convex transducer array, the circular azimuth plane (perpendicular to array surface) can be imaged due to its curvature. To obtain the volumetric image, a miniature motor [26] has to mechanically rotate the array to scan the different planes. The assembly would contain a single array, to save cost of fabrication, interconnect complexities, increased switching electronics, etc. The array would perform elevational imaging by rotations from −75° to 75° (array placed centrally as in Figure 13.4a). To allow for reliable mechanical rotations, the array extends up to 0.3 cm from the apex on both sides. The array length is ~3.76 cm, leading to ~370 elements in total. An active aperture of 128 elements is moved along the array to obtain the image for each plane. For enhanced lateral resolution, a linear shift of 1 element per aperture is set, leading to 243 scan lines in total. Due to beam steering, there would be an extra 35–40 lines, leading to a total of ~280 scan lines per azimuthal plane. The posterior view of the assembly, with the enclosed electronics is shown in Figure 13.4b. The individual array is illustrated in Figure 13.4c.

**(2)** *Embedded electronics*—The electronics include both transmit (Tx) and receive (Rx) circuitries, with a rough illustration of the components in Figure 13.5. The transmit circuit incorporates a Tx beamformer, a digital-to-analog converter (DAC), and a high-voltage (HV) amplifier to send pulses to the aperture elements. On the receive side, there are the analog front end for signal conditioning, an analog-to-digital converter (ADC), Rx beamformer,

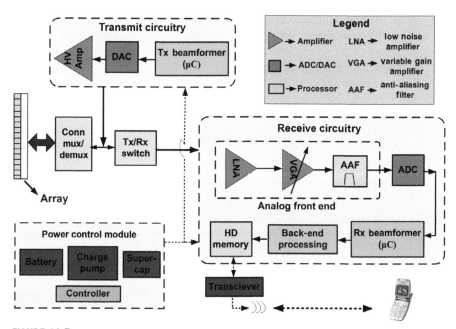

**FIGURE 13.5**

System-level block diagram, with the major electronic components.

a dedicated image processor for image rendering, a storage memory, and a transceiver to communicate with external unit. For interfacing, multiplexers and demultiplexers are used to select the array apertures, with Tx–Rx switches protecting the receive circuitry from HV transmission pulses. Detailed analysis of the electronic component design for the implantable application is out of the scope of this chapter.

## 13.3.2 IMAGING OPERATION

For uniform lateral resolution, 10 dynamic focal zones (both on transmit and receive) at 5 mm intervals are used for each scan line. The steps, followed in imaging the volume of interest, are illustrated in Figure 13.6 and described below:

**(1)** For the array in its central position, the aperture of 128 elements near one of the vertices is made active. 20 beam steering cycles with 10 focal zones per scan line are completed.

**(2)** The array works in the linear mode thereafter, with the focus along the central element of the active aperture. The active aperture is shifted by 1 element per scan iteration.

**(3)** Once the last linear-mode scan line has been imaged, another 20 steering cycles are completed. With this, the imaging of the current azimuthal plane is completed.

**(4)** Next, the array is rotated in the elevation plane (perpendicular to azimuthal) up to 75° in steps of 1°, with each plane being imaged as in steps 1–3. The miniature motor specifications [26] easily allows for a 1° resolution. The same is repeated in the opposite direction.

**(5)** With the scans of (~75×2 = 150) azimuthal planes (determined by organ topology), the 3-D rendered image of the entire susceptible region of interest is reconstructed.

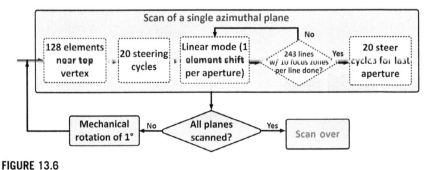

**FIGURE 13.6**

Flow diagram showing major steps in imaging a volume of interest.

### 13.3.3 **POWER ANALYSIS**

IMDs have to be designed to operate at a low power/energy budget, so that they can be powered by an internal battery, and for low thermal dissipation in the surrounding tissues. In this scenario, the rechargeable power source has to be designed, such that it can supply the peak power and the total energy required for at least one imaging cycle. Usually, for early detection of cancer recurrences, a monthly scan is sufficient, giving ample time for recharging the implantable battery through magnetic or inductive powering technologies [27,28].

#### *13.3.3.1 Power/energy requirements*

The transmit operation is the power-hungry phase of diagnostic ultrasonic imaging. The peak power is consumed during the transmit phase for a very short duration (orders of 100ns). Assuming a CMUT unipolar peak pulse voltage of 25–30V [29] with a rise time of ~10ns (15MHz frequency) and a capacitance of 1.5pF per element [29] (including the through-wafer interconnect capacitance), the peak current per array element is theoretically calculated to be ~4.5mA, at approximately our desired frequency. Hence, the transmit power per element is ~135mW. The power for an active aperture (128 elements) during transmit is ~17.5W. This is the maximum instantaneous power required for the entire imaging cycle. Besides, in diagnostic ultrasound imaging, the transmit duration is significantly less (<0.1%) as compared to the receive time. Hence, the peak power is consumed only for a short burst during the production of the acoustic pulses. The receive-side maximum power is mainly determined by the operation of all the receive channels at the same time. The ADC is the most power-hungry component of the receive electronics. With proper utilization of power down methods, maintaining a resolution of 10–12 bits, a sampling frequency of 50MSPS, and the variable gain amplifier feeding the correct input echo voltage range, the peak power per receive channel according to the specifications of a state-of-the-art receive chip is ~50–60mW [30]. For simultaneous operation of 128 channels at 50MSPS, the instantaneous power is ~7W. This is approximately 2.5× less as compared with the transmit peak power. The back-end digital beamformer and processor, implemented at a state-of-the-art technology node, would consume significantly less dynamic power in the order of tens of milliwatts. Techniques like power gating, clock gating, and near-threshold operation would minimize the leakage power at such subnanometer design modes [31]. The remaining components of the implantable system are the miniature motor and the ultralow-power transceiver. For our selected piezo motor (consumes 40% lesser power than electromagnetic motors), for a load approximating our system (~30–50g force), and with 3.3V supply, the power consumption is ~300–400mW [26]. The power consumed by a state-of-the-art, ultralow-power transceiver chip, with a fast data rate of 800Kbps, is ~15mW at 3V supply [32].

The energy required for a complete imaging iteration depends on the time required to scan the volume of interest, which is roughly calculated as follows:

**(1)**  The total number of scan lines for an azimuthal plane is 280, with 10 dynamic focusing zones per line. Every scan line would require a Tx–Rx period of ~$((2 \times 6 \times 10^{-2})/1540) = 77.9$ns. Hence, the total time for imaging an azimuthal plane is $(77.9 \times 280) = 0.23$ s. Taking into account some overhead due to memory access times (for the beamforming delay patterns), switching between azimuthal planes and other latencies, the estimated time per scan line would not exceed 0.5 s.

**(2)** The maximum number of rotations, depending on the organ topology, is ~150. Hence, the total imaging time, including the electronic latencies, is around 75 s. For a single transmit pulse or even a multicycle damped sinusoid at the desired transducer frequency, the transmit time is ~1/300 of the receive duration for a depth of 6 cm. This corresponds to ~250 ms. Hence, the total energy required during transmission is ~5 J. The rest is the time devoted for echo reception at the receive power, leading to a consumption of ~525 J.

**(3)** The back-end digital processor works at a much smaller power and hence not taken into energy calculations. Due to inertia, an approximate time for a single motor rotation is ~500 ms. For all the rotations at the motor input electrical power, the energy is ~36 J. Depending on the level of data compression, the volume of digital data transmitted by the transceiver would vary. An overestimated wireless Tx–Rx time, taking into account an ADC sampling rate of 50 MSPS, after demodulation and envelope detection (low pass filtering), would be a few minutes. The transceiver energy would be in the order of a few joules (~5–6 J), at the transceiver power.

**(4)** The total energy (the cumulative effect of all the phases, i.e., transmit, receive, beamforming, rotation, and wireless transmission) for scanning the entire volume is ~570 J, dominated by the echo reception energy. Hence, the peak power is during the acoustic transmission, whereas the maximum energy, during scanning, is consumed during the echo reception. The imaging power/energy requirements are listed in Table 13.2. The mentioned peak current is for all the array elements, and the scan time includes imaging, rotation, and wireless operation duration.

### 13.3.3.2 Power supply in implantable assembly

Regarding the power source, we consider a state-of-the-art implantable battery [22], with a capacity of ~400 mAh (volume of 2–3 cc). Under the system loading, the nominal output voltage would be ~2.8–3.2 V. To obtain the CMUT excitation voltage, one would require a series connection of three such batteries and a modified 5-stage Dickson charge pump with static charge transfer switches [33]. Functional simulation gives an output voltage of around 27 V for the chosen scheme, which is sufficient to excite the transducer. To provide the relatively high discharge current within a short burst during transmit, a capacitor of the order of 100 µF would be required at

**Table 13.2** Example Power/Energy Requirements for Implantable Imaging System

| Peak Current | Pulse Voltage | Peak Power | Imaging Time | Energy per Cycle |
|---|---|---|---|---|
| ~580 mA | ~25 V | ~17.5 W | ~5 min | ~0.6 kJ |

**Table 13.3** Specifications of the Implantable Power Source

| Capacity | # of Charge Pump Stages | Output Voltage | Total Energy | # of Cycles Before Recharge |
|---|---|---|---|---|
| 1200 mAh | 5 | ~27 V | 12.1 kJ | ~8 |

the output of the charge pump, like the operation of a capacitor at the output of the defibrillator battery during application of shock pulses [25]. The 3-battery series pack may have to be included outside the enclosure (but packaged inside the assembly). As an alternative, a hybrid cathode lithium battery incorporating a mixed silver vanadium oxide (high power densities and reliable end of service characteristics) and a carbon monofluoride (for extremely high energy densities) cathode is an extremely promising technology from Medtronic, used in their implantable defibrillators [34].

The energy available from the power source is ~12.1 KJ. The total energy calculated for a full imaging cycle is ~600 J. For a discharge till about 60–70% of the rated capacity (beyond which there would be higher battery internal resistance and hence greater loading drops) and some efficiency reduction due to charge pump and the battery leakage, the number of imaging cycles that can be performed before a recharge is ~7–8. Hence, with a monthly scan, at intervals of approximately 7–8 months, the implantable battery would need to be recharged using transcutaneous transfer using magnetic or inductive coupling. The specifications of the power source are tabulated in Table 13.3.

### 13.3.4 IMPLANTABILITY ISSUES

The ultrasonic imaging assembly is implanted locally to the susceptible volume of interest. The implant issues such as biocompatibility of packaging and heat dissipation are taken care of along the lines of existing IMDs as described below.

#### 13.3.4.1 Biocompatibility

Biocompatibility of the implant packaging material is extremely important to ensure the safety of the surrounding tissue while maintaining the reliability and sensitivity of operation. One major issue is the foreign body response induced at the tissue implant area by the host cells, leading often to a formation a fibrous collagen capsule and hence complications with the device functionality [35]. Another is the propensity towards infections around the device implant region [36]. To overcome these issues, one would use biomaterials like poly(lactic-co-glycolic acid), poly(ethylene glycol), and poly(vinyl alcohol) to package the assembly [36]. Nowadays, one of the trends is to use a biocompatible outer coating that mimics the characteristics of the body tissue to minimize the negative responses while maintaining sensor functionality. In the assembly, a surface modification would be performed in the form of a hydrogel coating, for its integrative tissue-like healing properties [35,36]. The front assembly surface would be exposed to the tissue through only the hydrogel coating to maximize the transmission efficiency of ultrasound [37]. Depending on the implant site, to prevent infection, some controlled antibiotic/anti-inflammatory drug release mechanism could be incorporated into the packaging [36]. In the absence of metals like iron and nickel, the polymer packaged implant with the hydrogel coating would not interfere with an MRI operation.

### 13.3.4.2 Attachment of device

A number of IMDs are nowadays being surgically implanted to monitor critical conditions and provide closed loop stimulation on demand. Devices are mostly sutured subcutaneously, under the skin to the organ of interest. For monitoring internal recurrence in susceptible regions like the liver right lobe, the implantable imager would be sutured to the top liver surface under the diaphragm or to the side surface along the intercostal spaces of the 7th and 8th ribs. A Dacron pouch, implanted through the device incision, would be used to hold the device in its position after suture, as in the implantable intrathecal pump [38].

### 13.3.4.3 Temperature rise in tissue

The implantable imager is a high-power implant. Usually, high-power IMDs, such as heart pumps and left ventricular assist devices, are packaged to efficiently dissipate the heat and prevent localized temperature rise beyond a few degrees. The proposed imager is in the peak power transmit mode for <0.1% of the scan operation. The system would use a heat sink coating in the form of a thermally conductive, electrically insulating material to dissipate the heat efficiently [39]. Hence, the receive power would be dissipated over the entire surface area of the assembly. The spatial peak temporal average intensity ($I_{SPTA}$) determines the thermal index of operation and hence the maximum temperature rise within the beam [40]. A strict thermal limit (in terms of $I_{SPTA}$) from FDA, for diagnostic ultrasound, is 720 mW/cm$^2$ [40]. Simulation and experimental studies have been conducted to estimate the temperature rise of the surrounding tissues during the operation of the imaging assembly. It is evident from the results [9] that either the active array surface or the focal regions have the maximum instantaneous $I_{SPTA}$ value. As the focal zones change at every iteration, the maximum $I_{SPTA}$ values would be mostly at the transducer surface over the entire imaging duration. Assuming that the entire transmit electrical power of 17.5 W is used to elevate the tissue temperature (an overestimation as losses are encountered in transducer), $I_{SPTA}$ at the active array surface with a 0.1% duty cycle is ~300 mW/cm$^2$ [9]. Hence, the thermal constraint is satisfied, thus ensuring no possibility of tissue damage during the operation. Experimental evaluation of temperature rise in tissue phantoms also verifies the same [9].

## 13.4 VERIFICATION OF ADVANTAGES OF INTERSTITIAL ULTRASONIC IMAGING

The advantages of interstitial ultrasonic imaging through the proposed implantable imaging assembly have been verified through extensive simulation and experimental trials [9]. The verification frameworks and analysis of the associated results are discussed next.

### 13.4.1 SOFTWARE SIMULATIONS

Early detection of malignant clusters entails the requirement of high-resolution imaging. Stage I/early stage II detection requires a resolution in the order of 5 mm (resolution of 1 cm suffice in some scenarios). We model the imaging array of the assembly and human tissue phantoms in Field II Ultrasound Simulation software [41] and thereby compare the imaging performances of existing external and proposed interstitial imaging of deep internal organs like the liver, pancreas, and lungs.

#### *13.4.1.1 Simulation framework*

Although the Field II Ultrasound Simulation tool [41] is based on certain approximations, for example, linear propagation of ultrasound through soft tissues and imaging of scatter points independent of each other, the pulse echo response for different apertures represents an accurate approximation of the actual imaging performance. In a real external scenario, for imaging the liver, pancreatic head, etc., the ultrasonic wave traverses primarily through the skin, fat, and muscles before the organ of interest. Field II is well suited for simulating phantom images with constant attenuation, thus enabling us to compare performances with variation in anomaly depths, imaging frequencies, and aperture lengths, in a homogenous phantom. As a case study, we focus on comparing the internal/external performances in imaging an anomaly in the liver right lobe. The thickness of the tissue layers (the skin, fat, and muscles) surrounding the liver is around 4 cm. The number of channels in the internal imager is constrained to 128 from the specifications. Aperture size for external liver imaging is limited by intercostal spaces (usually between 10 and 30 cm$^2$). For a common liver imaging frequency of 5 MHz [42], the intercostal spaces afford a maximum aperture length of around 8–10 cm or 256 elements. Usually, the high-frequency ultrasound imaging systems contain a maximum of 256 Tx–Rx channels in the transmit–receive chip [43]. Transducers with greater than 256 elements (like 32×32 2-D arrays) have to multiplex between channels. The simulation model is illustrated in Figure 13.7a.

On obtaining scan lines, back-end processing is performed to make them suitable for display, as in Figure 13.7b. It includes a flow of the steps for DGC [44]. In general, the echo DR, selected to form the final image, is dependent on the application [45,46]. The final display range is usually constrained within 20–30 dB (8 bit=255 values ~24 dB display range). To identity local tissue changes within the liver right lobe, a wide DR of echoes need to be displayed so that the weak echoes from the intra-soft tissue scattering characteristics appear relatively brighter in the image. On the other hand, for obstetric images, when the shapes of different organs are more significant, a low DR of echo amplitudes are selected to suppress the weak echoes from the subtle local scattering and amplify the amniotic fluid regions. In our application, detection of recurrent malignant growth requires monitoring local variations in tissue absorption, scattering, and elastic properties. From Ref. [46], the range of echoes that need to be displayed to get accurate diagnostic information in the liver is around 60 dB (from −20 to −80 dB with respect to tissue–air interface reflection), to include both interface reflections (liver–fat) and local variations in tissue scattering within the organ. Hence, we have chosen a DR of 60 dB, beyond which we would be mostly incorporating electronic noise, weak artifacts, and acoustic clutter.

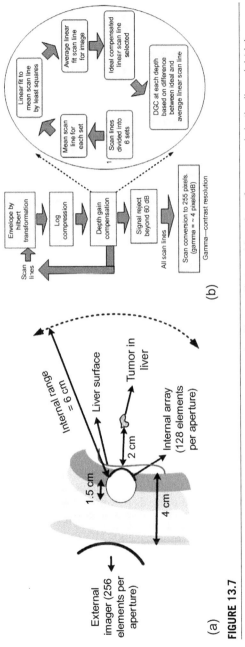

**FIGURE 13.7**

(a) Software simulation framework representing a case study of internal/external imaging of an anomaly in the liver right lobe;
(b) back-end image processing steps with a flow diagram of DGC.

### 13.4.1.2 Imaging metrics

Similar to a computational observer analysis [47] in contrast-detail phantoms under signal known exactly conditions, the objects of interest are placed at different depths within the phantom, and the imaging performances quantified. For each depth, the metrics are averaged for three lateral object positions, to take into account the randomness in the scattering amplitudes of the phantom scatterers. The objects include the following:

(a) *Lesion*—Lesions with the radius of 1.5 mm and contrasts of 6, 10, and 20 dB, above the nominal background scattering level, are considered. The detectability metric for the lesion is *lesion signal-to-noise ratio* (SNRL) [48], defined as $SNRL = |m_b - m_l|/(sd_b^2 + sd_l^2)$, where $m_b$ and $m_l$ are the mean gray levels and $sd_b$ and $sd_l$ are the gray-level variances within the surrounding background and the lesion. The background is chosen as the annular region between the lesion and a disk of twice the lesion radius. Another common metric is *contrast*, taking into account only the average gray levels, and it is defined as $contrast = |m_b - m_l|/(m_b + m_l)$, where the symbols have their usual meanings. Higher values of SNRL and *contrast* signify better lesion detectability.

(b) *Cyst*—Similarly, a cyst (anechoic object) with the radius of 1.5 mm is imaged at different depths. The detectability metric is *tissue-to-clutter ratio* (TCR) [48] defined as $TCR = 20 \times \log(m_b/m_l)$, where $m_l$ and $m_b$ are the ensemble averages of the mean cyst and background gray level, measured at the same depth and over the same circular area (a different lateral position). A higher positive value of TCR enhances the cyst detectability.

(c) *Point*—A point, with a scattering value of 20 dB higher than the nominal background scatter, is analyzed at different depths. The metrics are the axial and lateral point spread function and the point brightness with respect to the background. The −6 dB spread, on either side of the central maximum, estimates the axial and lateral resolution [48]. The brightness ratio is calculated as the ratio (in dB) between the central maximum gray level of the point blob and the average background level.

### 13.4.1.3 Simulation results

The objects of interest are placed at different depths within the calculated internal scan range of 6 cm. An equivalent scenario would be suturing the internal assembly to the liver wall and applying the external imager on the skin, as illustrated in Figure 13.7a. Objects with a diameter of 3 mm are considered (less than the required 5–7 mm, as detectability would be degraded in the real scenario due to nonhomogeneity).

(a) *Lesion*—For each depth, two lesions are placed, with centers at −3 and 3 mm, within the phantom. Another iteration is performed with the lesion center at the middle. The three values are averaged to obtain the detectability metrics at a particular depth and nominal contrast. Two internal/external lesion images of nominal contrast +6 dB, at depths of 2 and 4 cm from the liver surface, are shown in Figure 13.8a with the corresponding metrics. Qualitatively, the

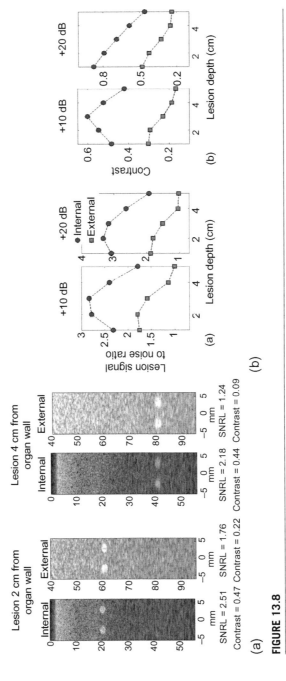

**FIGURE 13.8**

(a) Comparative internal/external images of lesions, at two different depths from the liver surface; (b) variation of SNRL and contrast with lesion depth from liver surface for two nominal lesion contrasts.

lesions corresponding to the internal images have better contrast as compared with the external ones (the background has a less noisy appearance). The two metric values also quantitatively illustrate the same [9]. The variation of SNRL and *contrast*, for both imaging modalities, is illustrated in Figure 13.8b. For the five lesion depths and three contrast values, the SNRL of internal imaging is about 1.5–2× greater than that of external scenario. Similarly, the contrast values for the interstitial case, are ~2–3.5× greater than that of external imaging. The comparatively lower values of internal metrics, corresponding to the initial lesion depth (1 cm) in a few cases, are attributed to the inaccurate Field II performance for points in the near field of the small squares, comprising the aperture for simulation.

**(b)**   *Cyst*—The cyst images at depths of 2 and 4 cm from the transducer are shown in Figure 13.9a (up) with the corresponding values of TCR. Qualitatively, the cysts are much better imaged through internal imaging, with a stark contrast at the depth of 2 cm. At both distances, the internal images depict the proper dimensions and anechoic nature of the cyst. The higher values of internal TCRs also signify the same [9]. Over all the five depths, the variation of the mean TCR is shown in Figure 13.9a (down). At lower cyst depths, the internal TCR values are ~6–8× of the external case. This contrast in values decreases significantly over depth, with the internal TCR only 2× greater than the external value at the furthest range (5 cm).

**(c)**   *Point*—For spatial resolution, a point of nominal scattering value +20 dB is considered. The corresponding internal/external images are illustrated in Figure 13.9b. The variation of the point metrics over different depths and contrasts are shown in Table 13.4. The major trends include enhanced internal axial resolution, due to operation at a higher frequency. At lower depths, the lateral resolution is significantly better for internal imaging with fast degradation with depth. This is attributed to the limited constant aperture of the internal imager. The brightness ratio is better for internal imaging (~2 dB higher) as compared to external imaging, at most depths, within the scan range. Qualitatively from the images and quantitatively from the metrics, obtained using the Field II simulation framework, we infer that the internal imager would generally perform better in terms of contrast sensitivity and spatial resolution, deep in interstitial organs, as compared with existing external techniques [9].

## 13.4.2 EXPERIMENTAL EVALUATIONS

Field II is based on the linear systems theory, applied to ultrasonic propagation [41]. Although the approach is a good approximation for ultrasonic field calculation and estimation of the pulse echo response, in actual scenarios, the ultrasound wave is nonlinearly distorted, leading to the formation of harmonics. Besides, in an organ, the density of scatterers varies spatially, leading to variations in scattering and attenuation characteristics at different positions. So, along with Field II simulation,

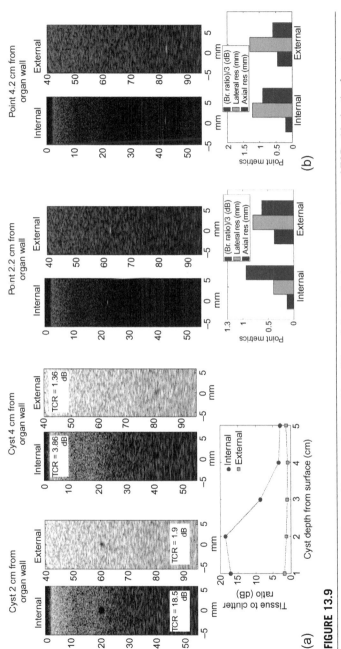

**FIGURE 13.9**

(a) Comparative internal/external images of cysts, at two different depths (up), and comparison of TCR with depth of cysts (bottom); (b) comparative internal/external images of point object, at two different depths with corresponding metrics.

**Table 13.4** Variation of Point Imaging Metrics for Internal/External Imaging

| Axial Res. (mm) | | Lateral Res. (mm) | | Br. Ratio (dB) | |
| --- | --- | --- | --- | --- | --- |
| **Internal** | **External** | **Internal** | **External** | **Internal** | **External** |
| 0.09 | 0.38 | 0.25 | 0.80 | 9.71 | 7.83 |
| 0.13 | 0.40 | 0.35 | 0.89 | 9.27 | 7.61 |
| 0.22 | 0.41 | 0.62 | 1.11 | 10.42 | 7.54 |
| 0.21 | 0.43 | 1.22 | 1.30 | 9.70 | 7.23 |
| 0.24 | 0.44 | 1.59 | 1.54 | 10.18 | 7.48 |

[9] describes an experimental framework to compare the imaging performances for an approximately homogenous medium, using a scanning system and tissue phantoms. Subsequently, the performances for the nonhomogeneous propagation case (closest emulation of actual internal/external scenarios) were also compared [9].

### 13.4.2.1 Experimental framework

The components of the experiment are enlisted in Table 13.5 and are described as follows:

**(1)** *Ultrasound system*—The system is a comparatively state-of-the-art medical assembly from Siemens, the *Sonoline Adara* series. It supports both linear scanning and phased array-mode scanning with time gain compensation controls and a maximum DR of 65 dB. It also contains a template of different preset parameters for various organ scans.

**(2)** *Transducer*—The transducer used is the *Siemens 3.5C40S*, a convex phased array of 128 elements, supporting three resonant frequencies of 2.6, 3.5, and 5 MHz. The aperture length is 4 cm.

**Table 13.5** Components of the Ultrasound Experimental Setup

| *Ultrasound system* | |
| --- | --- |
| Scanner | Siemens Sonoline Adara |
| *Ultrasound transducer* | |
| Type<br>Length and no. of elements<br>Resonant frequencies | Abdominal convex 1-D phased array<br>4 cm and 128<br>2.5, 3.6, and 5 MHz |
| *Tissue phantom* | |
| Material<br>Shape<br>Point object of interest | Silicone cure rubber (PDMS)<br>Cylindrical<br>Solder wire tip (1 mm diameter) |

(3) *Phantom*—The characteristics of the tissue phantom are as follows:

    (a) *Material*—For approximating homogenous tissue acoustic properties using simple methods, [9] used a polydimethylsiloxane (PDMS) material, obtained as soft silicone medical grade rubber [49,50], widely used as tissue phantoms in robotics-based calibration and automated surgery experiments.

    (b) *Objects of interest*—The aim was to replicate the software-based simulation framework using experimental trials. Imaging a point object of a particular nominal contrast would enable us to calculate both the spatial spreads and the contrast performance (from point brightness in image). We used a solder wire (1 mm diameter) tip as an accurate representation of a point object in our imaging characterization. Two phantoms were used, one for internal and the other for external imaging.

    (c) *Position of objects*—A side view and bottom view of the internal phantom are shown in Figure 13.10. The cylindrical internal phantom diameter was 6.5 cm and its elevation, 5.2 cm. To account for minor differences in acoustic properties and solidification across the material, we needed multiple spatially independent images of the wire tips at around iso-distances from the transducer, to characterize the performance at that depth. Hence, two wires (one L-shaped and suspended from top and the other vertical with a spiral base at the bottom) were used to obtain the results. Each wire tip was imaged from the top and a corresponding nearer phantom side surface, resulting in four azimuthal plane images (average depth of $2.3\pm0.1$ cm from the surfaces). The scan planes are shown by placement of the transducer along them in Figure 13.10. The external phantom was of diameter 11 cm, with the mean wire tip depth being ~$6.5\pm0.2$ cm.

(4) *Imaging operation*—A pair of corresponding depths were considered, namely, internal depth of 2.3 cm and external distance of 6.5 cm from the surface. From the top surface, identifying the tip was easy for the vertical wire, but with the horizontal wire, the transducer was manually moved from the opposite end and the first bright entry identified as the wire tip. Taking into account the small errors due to transducer positioning, identification of the wire tip, etc., one performed five iterations of each plane to average out the manual errors. The effect of gas bubbles in the final phantoms was reduced by desiccating the phantom during hardening. The different aperture sizes for internal imaging and external imaging were simulated by manually modulating the phantom surface by altering the pressing force accordingly; that is, to represent greater external aperture, pressure would be more; hence, initial internal/external wire distance difference is kept slightly greater than 4 cm. Holding the transducer normally on the phantom surface, without application of force, corresponded to 50 elements in contact (calibration from the image). The other elements were in contact with air. Due to significant reflections from transducer air interface, these elements were practically

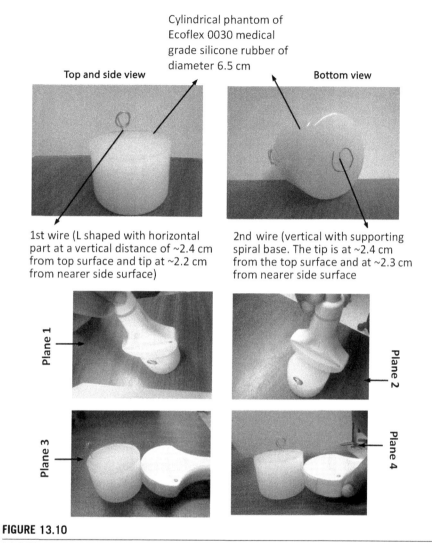

Cylindrical phantom of Ecoflex 0030 medical grade silicone rubber of diameter 6.5 cm

Top and side view

Bottom view

1st wire (L shaped with horizontal part at a vertical distance of ~2.4 cm from top surface and tip at ~2.2 cm from nearer side surface)

2nd wire (vertical with supporting spiral base. The tip is at ~2.4 cm from the top surface and at ~2.3 cm from nearer side surface

Plane 1

Plane 2

Plane 3

Plane 4

**FIGURE 13.10**

(a) Different views of the phantom with solder wires (top); four independent azimuthal planes scanned for a point object at a particular depth (bottom).

rendered ineffective for imaging. Our final internal and external apertures corresponded to ~50 and 100 elements, as shown in Figure 13.11. During imaging, the wire tip depths from the transducer were of mean 2 and 5.7 cm, respectively, for the two scenarios. The increased depth for the external imaging was 3.7 cm instead of 4, as during solidification, the wires moved a bit in different cases. For the four planes and five iterations per plane, the average metrics (axial and lateral spread and brightness ratio) and their variances were calculated in *MATLAB*.

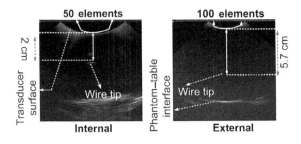

**FIGURE 13.11**

Relative dimensions in the internal/external imaging phantoms (different zooms).

**Table 13.6** Comparison of Imaging Parameters Between the Experimental Framework (External) and Internal Imaging

| Parameters | Internal | External |
|---|---|---|
| Imaging frequency | 5 MHz | 2.6 MHz |
| No. of aperture elements | 50 | 100 |
| Depth of anomaly | 2 cm | 5.7 cm |
| Dynamic range | 60 dB | 60 dB |
| Gain | 30 dB | 30 and 45 dB |
| Scan conversion | 255 pixels | 255 pixels |

The parameters for imaging in the experimental framework are given in Table 13.6. Since the position of the point object is known a priori, the focal zone is set around the corresponding depth to simulate dynamic focusing. To account for the unknown attenuation value of the medical grade rubber, one manually sets the depth gain controls separately for 2.6 and 5 MHz in Ref. [9], using a large phantom, imaging it from the top surface, such that the final image appears homogenous at all depths. The contrast resolution (gamma) after display is around 4(255/60), in all imaging scenarios.

### 13.4.2.2 Experimental results

Experimental images of the internal and external scenarios, along with the comparative metrics, are shown in Figure 13.12a. The internal image has a depth of 7.58 cm as compared with external depth of 13.65 cm (equal no. of pixels with different zooms). For each image, the top curved line represents the transducer surface, with the active imaging elements along the dark areas of the curve. The bottom bright lines represent the phantom to table interface. By calibration, it was found that the sound speed in the PDMS matrix was almost the same as that of scanner value (soft tissue speed) of 1540 m/s.

From Figure 13.12a, it is observed for all images with a back-end nominal gain of 30 dB that the background scatter is minimal. For external imaging with high gain (45 dB), background speckle is more enhanced. Qualitatively, the spatial spreads are much greater in the external 5 MHz scenario than internal imaging (~2 and ~2.3 times

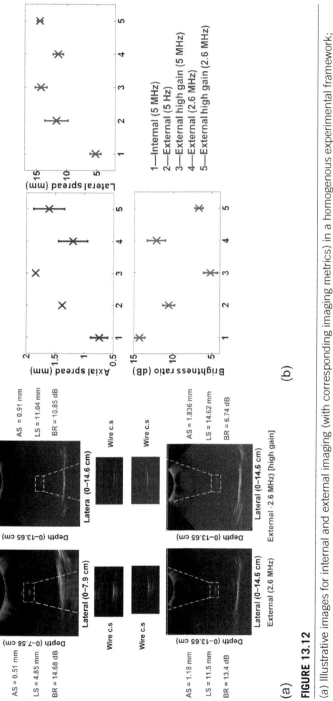

(a)

(b)

**FIGURE 13.12**

(a) Illustrative images for internal and external imaging (with corresponding imaging metrics) in a homogenous experimental framework;
(b) comparison of major performance metrics between internal and different external scenarios in case of homogenous experimental framework.

greater than internal values). One of the major issues with the external imaging at 5 MHz is the reduced brightness (poor contrast) of the wire tip, due to the greater effect of attenuation at the higher frequency. To enhance detectability, we revert to the lower frequency of 2.6 MHz for the external scenario. The spread is similar to that in the 5 MHz scenario, but the brightness is significantly improved (BR is only around 1.28 dB lower than the internal value). Further, the back-end gain was increased to the maximum value of 45 dB. Here, it is observed that the background becomes significantly noisier from mainly the acoustic clutter, as evident from the degraded brightness ratio (~8 dB lower than internal scenario). The external imaging at 5 MHz, at gain of 45 dB, is worse and hence not illustrated. From the results, it has been concluded in Ref. [9] that for a point at a depth of 2 cm and an additional external distance of 3.7 cm in a homogenous tissue-simulating medium and maintaining the actual comparative framework, the internal imaging performs better than the best case external scenario from both spatial resolution and the contrast sensitivity perspective.

For the 20 images (4 planes and 5 iterations per plane), the mean and standard deviation of the metrics were presented in Figure 13.12b. The internal metrics were mostly superior than the extracorporeal ones. On an average, the axial and lateral spreads for the best case external imaging (2.6 MHz and nominal gain) are approximately 1.7 and 2.4 times that of the internal values, signifying an enhanced spatial resolution for the internal scenario. Moreover, the brightness ratio is the highest for the internal case with an average of about 14.5 dB. It is around 2.5 dB higher than the best external imaging scenario. Hence, along with the Field II modeling and simulation, actual experimental results [9] verify the enhanced performance of interstitial imaging, compared with existing external methods, in a homogenous tissue-simulating medium.

The imaging frequencies and the number of transducer elements are different from the actual calculated design values. However, one can still perform a successful comparison of imaging performances, as long as the experimental interstitial doesn't overestimate the actual comparative scenario. Within the transducer bandwidth, the internal performance is estimated at 5 MHz, which is 1/3 of the actual design frequency. The number of active elements is 50. The best external performance, in the framework, is evaluated at 2.6 MHz, about ½ of the design frequency. The number of active external elements is 100. The active aperture determines the lateral resolution and the SNR at different depths in the image. Hence, similar to design parameters, the ratio of the internal/external apertures is ~2 to allow nearly equal degradation of the imaging performances for the two cases. The distances to the point object of interest (2 and 5.7 cm, respectively) are within the scan depths of interest. Due to greater downscaling of the internal frequency as compared with external frequency (with respect to design values), internal performance in the experimental framework is actually an underestimation of the actual imaging advantages with respect to external performance, in a real scenario. Non-homogeneity tests were also conducted with external imaging through a two-layered phantom, i.e. gelatin with flour scatterers representing soft tissue and PDMS representing subcutaneous tissue density and attenuation characteristics [9]. Interstitial imaging was emulated by scanning only through the soft tissue material. Degraded external imaging metrics, especially the lateral spread in external scenario [9] was observed due to the effect of non-homogeneity in real situations.

## 13.5 A FEW DISCUSSION POINTS

### 13.5.1 ECONOMIC FEASIBILITY

The proposed system would be economically beneficial in terms of comparison between the one-time comparatively high implantable ultrasonic device cost incurred and the cumulative costs of the routine, periodic diagnostic tests like MRI, CT, and PET over the duration of 5–10 years after the primary tumor resection. Imaging tests like MRI, CT, and PET are expensive both in the United States (MRI costs over $1000 [10]) and in Europe, with the problem escalating in developing countries. The critical component of the proposed IMD in terms of cost is the ultrasonic transducer. However, CMUTs, based on silicon, which would be used in our assembly, are comparatively of much lower cost as compared to PZT and piezo composites due to batch processing allowed by state-of-the-art surface micromachining techniques [29]. The cost of the electronics, including the LNA and ADC, would be less than the ultrasonic transducer. As an estimate, the IMD with the transceiver, memory, and the transcutaneous recharge system would cost around few hundred dollars ($500–1000). The device would be implanted during the surgery for resection. So, no significant overhead in surgical costs is incurred in such a situation. Hence, the medical costs incurred with the IMD as compared to existing diagnostic techniques would be much less considering the entire monitoring period.

### 13.5.2 DOWNSCALING OF SIZE

The upper limit for the assembly radius is estimated to be 1.5 cm. The lower limit for assembly size is constrained by the active aperture dimension and the size of the integrated electronics. Depending on electronic dimensions, the size can be scaled down to a radius of 1 cm or less, still incorporating more than the active aperture of 128 elements.

### 13.5.3 WIDER SCAN RANGE

The internal range can also be extended beyond the estimated value of 6 cm. The depth for uniform high-resolution imaging depends on the beam energy loss and hence the transducer frequency. Depending on the application, the frequency can be scaled down to around 10–12 MHz. For a frequency of 10 MHz and a range of 100 dB before gain compensation, the depth of imaging would be theoretically ~8.5 cm.

### 13.5.4 EXTENSION OF APPLICATION

The primary application of such an assembly is in the detection of cancer recurrence. It could be extended to early detection of primary tumors, in a few scenarios. A vulnerable set of population has to be identified, like people with extensive smoking habits, suffering from Crohn's disease, and with a history of high alcohol intake or gall stone diagnosis. In such a set, the imager could be implanted in a region of interest during a separate surgery, such as appendectomy or gall stone removal, for online monitoring.

### 13.5.5 IMAGE DENOISING

To reduce the effect of the background speckle, different denoising techniques are currently used in external imaging, including spatial and frequency compounding [51]. Similar techniques can be applied to the proposed assembly. The total energy required for one full scan is about 0.6 kJ. Reception in three frequency bands sequentially would increase the energy per scan to ~2 kJ, which is still much less than 60–70% of the total battery capacity.

### 13.5.6 BEYOND EARLY DETECTION OF CANCER

Depending on the area of implant, the imaging device may be beneficial for chronic monitoring applications beyond early detection of cancer. If implanted along liver walls, the implant could be able to detect fatty liver disease, internal bleeding, and in some cases gall stones. Besides, coronary artery diseases and cysts in ovaries and uterus could be detected in specific implant sites around the lungs and lower gastrointestinal tracts, respectively. Early detection of these anomalies through periodic automated monitoring reduces the chances of increased complications. Other site-specific online monitoring applications include the detection of swollen joints, soft tissue injuries, fractures, etc.

## 13.6 CONCLUSION

This chapter has presented a feasibility study of an implantable imaging assembly for automated long-term monitoring of internal organs. The proposed system uses the safe technology of diagnostic ultrasonography for online monitoring of a target region. The system is studied for early detection of recurrence of malignancy deep inside internal organs. In addition to the point-of-care diagnostic advantage, it has been verified that the interstitial system provides enhanced imaging quality (contrast sensitivity and spatial resolution) compared with existing techniques. The design space has been explored to optimize important parameters, namely, the form factor and power/energy requirement. Extensive simulations and experimental analysis using an ultrasound system validate the effectiveness of such a monitoring device.

While the proposed system is particularly attractive for monitoring cancers, which experience local recurrence (e.g., high-grade glioma and sarcoma), it can also be effective for monitoring regional recurrences by using multiple units in strategic locations. It can also be beneficial for online monitoring of a tumor during treatment (e.g., with chemotherapy), thereby improving the therapeutic efficacy. Future work would include further optimization of power, on-board processing, and animal study using a hardware prototype.

## REFERENCES

[1] Early diagnosis of coronary artery disease, http://www.totalhealth.co.uk/; [accessed 02.09.10].
[2] Riley TR, Smith JP. Preventive care in chronic liver disease. J Gen Intern Med 1999;14(11):699–704.

[3] Wulfkuhle JD. Proteomic applications for the early detection of cancer. Nat Rev Cancer 2003;3(4):267–75.

[4] Why early detection is the best way to beat cancer?, http://www.wired.com/; [accessed 22.12.08].

[5] Koch T. Monitoring of organ dysfunction in sepsis/systemic inflammatory response syndrome—novel strategies. J Am Soc Nephrol 2001;12(1):53–9.

[6] Understanding: early detection and screening, http://www.bafound.org/early-detection-and-screening.

[7] Prostrate-specific antigen (PSA) test, http://www.cancer.gov/cancertopics/; [accessed 24.07.12].

[8] Basak A. Low power implantable ultrasound imager for online monitoring of tumor growth. In: Proceedings of 33rd IEEE EMBS annual international conference; 2011. p. 2858–61.

[9] Basak A, et al. Implantable ultrasonic imaging assembly for automated monitoring of internal organs. IEEE Trans Biomed Circuits Syst (TBioCas) 2014; vol. PP(99), May.

[10] Costly diagnostic MRI tests unnecessary for many back pain patients, http://www.hopkinsmedicine.org/news/; [accessed 12.12.11].

[11] Shung KK. Diagnostic ultrasound: past, present, and future. J Med Biol Eng 2011;31(6):371–4.

[12] Clade O, et al. 10 MHz ultrasound linear array catheter for endobronchial imaging. In: Proceedings of the IEEE ultrasonics symposium, vol. 3. IEEE; 2004. p. 1942–5.

[13] Wen X, et al. Detection of brachytherapy seeds using 3-D transrectal ultrasound. IEEE Trans Biomed Eng 2010;57:2467–77.

[14] About implantable cardioverter defibrillators, http://www.medtronic.com/patients/sudden-cardiac-arrest/device/index.htm; [accessed 08.10.13].

[15] Knight KH, et al. Implantable intrathecal pumps for chronic pain: highlights and updates. Croat Med J 2007;48(1):22–34.

[16] Mathieson K, et al. Photovoltaic retinal prosthesis for high pixel density. Nat Photonics 2012;6:391–7.

[17] Ohta J, et al. Implantable CMOS biomedical devices. J Sens 2009;9(11):2073–93.

[18] Liteplo AS, et al. Real-time video transmission of ultrasound images to an iPhone. J Crit Ultrasound 2010;1(3):105–10.

[19] Szabo TL. Diagnostic ultrasound imaging: inside out. 1st ed. Boston, MA: Elsevier; 2005.

[20] Ultrasonic imaging overview, http://umanitoba.ca/faculties/medicine/units/cacs/sam/8478.html.

[21] How is liver cancer staged?, http://www.cancer.org/cancer/livercancer/; [accessed 10.02.14].

[22] Li/CFx EOL family, http://www.eaglepicher.com/.

[23] Wu J, Chang K. MOS charge pumps for low voltage operation. IEEE J Solid-State Circuits 1998;33(4):592–7.

[24] Dissanayake TD, et al. Experimental study of a TET system for implantable biomedical devices. IEEE Trans Biomed Circuits Syst 2009;3(6):370–8.

[25] Skarstad PM. Battery and capacitor technology for uniform charge time in implantable cardioverter-defibrillators. J Power Sources 2004;136:263–7.

[26] Low power piezo motion, http://www.designnews.com/; [accessed 14.05.10].

[27] Saito O, et al. Electric power-generating system using magnetic coupling for deeply implanted medical electronic devices. IEEE Trans Magn 2002;38(5):3006–9.

[28] Constandinou T. A bio-implantable platform for inductive data and power transfer with integrated battery charging. In: Proceedings of IEEE conference on circuits and systems (ISCAS); 2011. p. 2605–8.

[29] Khuri Yakub BT. Integration of 2D CMUT arrays with front-end electronics for volumetric ultrasound imaging. IEEE Trans Ultrason Ferroelectr Freq Control 2008;55(2):327–42.

[30] AFE5801—8-channel VGA with ADC, http://www.ti.com/product/afe5801; [accessed 01.05.10].

[31] Narasimhan S, et al. Ultra low-power and robust digital signal processing hardware for implantable neural interface microsystems. IEEE Trans Biomed Circuits Syst 2011;5(2):169–78.

[32] Bradley PD. An ultra low power, high performance medical implant communication system (MICS) transceiver for implantable devices. IEEE Biomed Circuits Syst Conf 2006;158–61.

[33] Wu J, Chang K. MOS charge pumps for low voltage operation. IEEE J Solid-State Circuits 1998;33(4):592–7.

[34] Skarstad PM. Lithium/silver vanadium oxide batteries for implantable cardioverter-defibrillator. In: Proceedings of 12th annual battery conference on applications and advances; 1997. p. 151–5.

[35] Ratner BD, et al. Biomaterials: where we have been and where we are going. Ann Rev Biomed Eng 2004;6:41–75.

[36] Burgess DJ, et al. A review of the biocompatibility of implantable devices: current challenges to overcome foreign body response. J Diabetes Sci Technol 2008;2(6):1003–15.

[37] Vaezy S. et al. Solid hydrogel coupling for ultrasound imaging and therapy. United States patent application, US 2003/0233045 A1, 2003.

[38] Knight KH, et al. Implantable intrathecal pumps for chronic pain: highlights and updates. Croat Med J 2007;48(1):22–34.

[39] Elwassif MM. Temperature control at DBS electrodes using a heat sink: experimentally validated FEM model of DBS lead architecture. J Neural Eng 2012;9(4):1–9.

[40] Nelson TR. Ultrasound biosafety considerations for the practicing sonographer and sonologist. J Ultrasound Med 2009;28(2):139–50.

[41] Jensen JA, Svendsen NB. Calculation of pressure fields from arbitrarily shaped, apodized, and excited ultrasound transducers. IEEE Trans Ultrason Ferroelectr Freq Control 1992;39:262–7.

[42] Outwater EK. Imaging of the liver for hepatocellular cancer. Cancer Control 2010;17(2):72–82.

[43] Next-generation ultrasound will rely on real-time compression, http://electronicdesign.com/analog/; [accessed 10.02.11].

[44] Tang M. Automatic time gain compensation in ultrasound imaging system. In: Proceedings of 3rd international conference on bioinformatics and biomedical engineering; 2009. p. 1–4.

[45] Gibbs V, Cole D, Sassano A. Ultrasound physics and technology: how, why, and when. Amsterdam, Netherlands: Elsevier Health Sciences; 2009.

[46] Hoskins P, Thrush A. Diagnostic ultrasound: physics and equipment. Cambridge, UK: Cambridge University Press; 2003.

[47] Lopez H, et al. Objective analysis of ultrasound images by use of a computational observer. IEEE Trans Med Imaging 1992;11(4):496–596.

[48] van Wijk MC, Thijssen JM. Performance testing of medical ultrasound equipment: fundamental vs. harmonic mode. Ultrasonics 2002;40:585–91.

[49] Ecoflex supersoft silicone, http://www.smoothon.com/.

[50] Pogue BW, Patterson MS. Review of tissue simulating phantoms for optical spectroscopy, imaging and dosimetry. J Biomed Opt 2006;11(4):1–16.

[51] Cincotti G, et al. Frequency decomposition and compounding of ultrasound medical images with wavelet packets. IEEE Trans Med Imaging 2001;20(8):764–71.

[52] Basak A. A wearable ultrasonic assembly for point-of-care autonomous diagnostics of malignant growth. In: Proceedings of IEEE point of care healthcare technologies (PoCHT); 2013. p. 128–31.

# Summary and future work

## SUMMARY

The design and implementation of custom hardware for biomedical implants should trade off several nonsynergistic specifications to meet the end user's needs. The consumed and dissipated powers and the hardware size are the most critical specifications to address throughout the design process due to the tight constraints imposed on these parameters. Relatively high power consumption not only quickly depletes the stored energy, for example, in a battery, but also increases the temperature of the surrounding tissues, which could lead to irreversible tissue damage. On the other hand, the highly restrained space inside the body along with the implant safety and biocompatibility issues determines the allowable size of the implantable device. The art of designing custom circuits for implantable devices consists of concurrently considering the implant clinical effectiveness and stability over time, which indeed is very difficult even without considering any other constraint, while simultaneously ensuring a long and safe device operation.

Some general design principles introduced in this book can help in the endeavor of designing signal processing hardware considering the particularities of the envisaged application and the signal processing algorithms. The first step in the hardware design is to choose the best architecture partition for the algorithms, that is, what signal processing is more convenient to be performed by analog circuits and what others to be done by digital circuits, such that the total amount of data to process is reduced by eliminating meaningless information through signal preprocessing before digitization. Once the optimal partition point is determined, power reduction techniques can be used to achieve the target specifications such as technology-optimized algorithms, optimal circuit topologies, data reduction circuits, compressed sampling, CMOS circuits operating in weak inversion, low-frequency and low-voltage operation, parallel data processing at lower speed, and self-tuning circuits that adapt their parameter to the ongoing conditions, among others. These principles are demonstrated throughout the chapter by illustrative examples of research works performed over the past years up to present.

## FUTURE WORK

In the near future, medical implantable devices can improve their clinical effectiveness and safety significantly and may expand to other clinical applications to the extent that many of the current problems such as the stability and safety of the recording interfaces (e.g., electrodes) and the detection, identification, and decoding of the information carried by the recorded signal from inside the body are improved.

Nowadays, rapid advances in nanotechnology-based materials will certainly help in the development of new tissue–sensor interfaces that will allow signal recording in a safer and more effective manner in chronically implanted devices. To improve signal processing hardware performance, research in several fields shall be performed with a multidisciplinary approach, including but not limited to biomedical sciences, micro- and nanoelectronics, and mathematics. A better understanding of how the physiological systems in the body function in healthy and pathological conditions can provide significant knowledge for designing better signal processing algorithms, circuits, and systems.

Rapid advances in electronics and sensing/actuation technology are expected to enable novel exciting applications of implantable systems. Note that, for hardware implementation of signal conditioning, telemetry, and signal processing algorithm, nanoelectronics offers great potential due to its terascale integration density, low switching power, and high performance. However, nanoscale devices also bring in a number of design challenges, such as large standby power due to leakage current and extreme parameter variations, induced reliability, and yield concerns. Hence, there is a need to devise circuit–architecture-level design solutions tailored to the computational algorithms that can leverage the benefits of nanotechnology while addressing its limitations. Efficient signal processing algorithms, amenable for real-time performance and low-area, low-power hardware implementation, are also very important for diverse implantable systems.

We strongly believe that the next generation of implantable devices needs to be driven by bioinspired circuits and systems that run algorithms dedicated to emulate, as close as possible, the biological processes. However, a direct translation of these biological processes into electronic-based hardware seems to be very difficult—and often, infeasible, at the moment—due to the extremely high complexity. However, once new research programs are carried out, an improvement in the available technology and the signal processing approaches needs to be realized to deliver the implantable solutions that millions of drug-refractive patients worldwide are waiting for.

# Index

Note: Page numbers followed by *f* indicate figures and *t* indicate tables.

Edwards Brothers Malloy
Ann Arbor MI. USA
February 27, 2015